Microbes for Restoration of Degraded Ecosystems

T0250956

Microbes for Restoration of Degraded Ecosystems

D.J. Bagyaraj

Centre for Natural Biological Resources and Community Development
University of Agricultural Sciences (UAS)
Bangalore, India

Jamaluddin

Department of Biological Science
Rani Durgavati University
Jabalpur, Madhya Pradesh, India

CRC Press
Taylor & Francis Group
Boca Raton London New York

CRC Press is an imprint of the
Taylor & Francis Group, an **informa** business

NEW INDIA PUBLISHING AGENCY
New Delhi – 110 034

CRC Press
Taylor & Francis Group
6000 Broken Sound Parkway NW, Suite 300
Boca Raton, FL 33487-2742

First issued in paperback 2023

© 2019 by New India Publishing Agency

CRC Press is an imprint of the Taylor & Francis Group, an informa business

No claim to original U.S. Government works

ISBN 13: 978-1-03-265380-8 (pbk)
ISBN 13: 978-0-367-14787-7 (hbk)

Print edition not for sale in South Asia (India, Sri Lanka, Nepal, Bangladesh, Pakistan or Bhutan)

Library of Congress Cataloging in Publication Data
A catalog record has been requested

Visit the Taylor & Francis Web site at
http://www.taylorandfrancis.com

and the CRC Press Web site at
http://www.crcpress.com

Website : www.uasbangalore.edu.in
E-mail : vc@uasbangalore.edu.in
vcuasbl 964@gmail.com
    drhshivanna@gmail.com

Off: 080-23332442
080-23330153 (Extn. 265)
Res.: 080-23331341
Mob.: 9449866900
Fax.: 91-080-23330277

UNIVERSITY OF AGRICULTURAL SCIENCES BENGALURU

Dr. H. SHIVANNA
Vice-Chancellor

Gandhi Krishi Vignana Kendra
P. B. No. 2477, Bengaluru-560 065

FOREWORD

Rapid urbanization, industrialization and other anthropogenic activities result in contamination of the environment with pesticides, crude oil, heavy metals, xenobiotics, toxic pollutants, etc. Mining activities result in removal or alteration of native vegetation and top soil. These have negative impacts on the structure and function of ecosystems. These activities have led to severe pollution and degradation of natural habitats.

Ecological restoration is the practice of renewing and restoring, within a reasonable period of time, degraded, damaged or destroyed natural ecosystems and habitats that were altered by industrial human intervention. This practice is an intentional activity used to promote and accelerate the restoration of degraded ecosystems with respect to its health, integrity and sustainability.

It has long been recognized that soil microorganisms are the major driving force behind transformation in soil, thus they have a major role in degradation of soil pollutants, and maintenance of soil fertility and ecosystem function. Scientists in recent years have been investigating on use of microorganisms for restoration of degraded ecosystems which are eco-friendly and less expensive, though such studies are limited. This book brings out the recent information available on microbes for restoration/ bioremediation of degraded ecosystems.

All the chapters in the book are written in lucid style and a comprehensive manner. I congratulate the editors of the book and all the authors who have contributed the various chapters, for their sincere efforts to bring out this useful and productive volume. I sincerely hope that this publication will have tremendous impact on the development of newer methods for restoration of degraded ecosystems using microorganisms.

Bengaluru
4th August, 2016

(H. Shivanna)
Vice-Chancellor

Preface

Fast growing human civilization, industrialization, mineral mining, oil exploration, modern agricultural practices and related anthropogenic activities in the world have resulted in elevated levels of toxic heavy metals and xenobiotic pollutants such as pesticides, petroleum hydrocarbons, electronic wastes, etc. in the environment. These pollutants impose hazardous impacts on living organisms, ecosystem health and also human health. Therefore remediation of these contaminants is becoming one of the serious environmental issues in the world. Clean up of the polluted sites can be done by physic-chemical methods which have several disadvantages such as high energy and labour requirement, generation of secondary wastes and very high cost. Recent researches have been oriented to utilize soil microorganisms either alone or together with plants to remediate and improve disturbed ecosystems. This approach is eco-friendly and relatively less expensive compared to physico-chemical methods. The investigations carried out using microbes for restoration of degraded eco-systems is limited. The proposed book covers this aspect.

The book is organized into 15 chapters. All the chapters have been contributed by renowned scientists; 13 chapters by Indian scientists, 1 chapter by Indonesian scientists, and 1 chapter by Canadian scientists. *Chapter 1* deals with reclamation of saline soils of Tamil Nadu by microorganisms. *Chapter 2* helps to understand the significance of microbes for synthesis of humification processes. *Chapter 3* deals with biosorption of heavy metals by microbes and agrowastes. *Chapter 4* is devoted to the role of mycorrhizal fungi influencing plant diversity and production in an ecosystem. *Chapters 5 to 7* provide up-to-date information on biological recovery of heavy metals by microorganisms in general and mycorrhizal fungi in particular. *Chapter 8* deals with microbial degradation of textile-dye wastes. *Chapters 9 and 10* respectively cover the role of microbes in the remediation of environments contaminated with toxic pollutants and pesticides. *Chapters 11 to 13* are devoted to bioremediation of mined-out ecosystems. *Chapter 14* discusses the role of mycorrhizal fungi in the restoration of coastal sand dune and mangrove forests. *Chapter 15* discusses the biological degradation of crude oil contaminants.

The reader may appreciate the potential of microorganisms for restoration of degraded ecosystems. The book is a comprehensive and detailed analysis of the subject. We wish to thank all the authors who have contributed their time and energy to this endeavor. Special thanks are due to Mr. R. Ashwin for his help in bringing out this book. We also wish to record our deep sense of gratitude for M/S. New India Publishing Agency, New Delhi for taking all pains and interest for timely and excellent publication.

D.J. Bagyaraj
Jamaluddin

Contents

List of Contributors

Abhishek Mundaragi, Department of Botany, Karnatak University, Dharwad-580003 Karnataka, India, *mabhishekphd@gmail.com*

Ajungla T, Department of Botany, Nagaland University, Hqrs: Lumami. 798627 P.O. Zunheboto Nagaland, India, *lkr.talijungla@gmail.com*

Anuj Kumar Singh, World Agroforestry Centre, South Asia, New Delhi, India, *ksanuj@live.com*

Asep Hidayat, Forest Plantation Fiber Technology Research Institute; Environment and Forestry Research, Development and Innovation Agency (FORDA), Ministry of Environment and Forestry, Jl. Raya Bangkinang Kuok Km. 9 Bangkinang, PO BOX 4/BKN Riau-28401, Indonesia, *ashephidayat@yahoo.com*

Asha Sahu, Division of Soil Biology, ICAR- Indian Institute of Soil Science, Nabibagh, Berasia Road, Bhopal-462038, *asha4u.bhuzone@gmail.com , ashaiiss@iiss.res.in*

Ashwin R, Centre for Natural Biological Resources and Community Development, 41 RBI Colony, Anand Nagar, Bangalore-560024, India, *ashwin.bengaluru@gmail.com*

Bagyaraj DJ, Centre for Natural Biological Resources and Community Development, 41 RBI Colony, Anand Nagar, Bangalore-560024, India

Bohre Priyanka, Rani Durgawati University, Jabalpur-482001, M.P., *pribohre@gmail.com*

Chaubey OP, State Forest Research Institute, Jabalpur-482008, M.P., *chaubey.dr@gmail.com*

Devarajan Thangadurai, Department of Botany, Karnatak University, Dharwad-580003 Karnataka, India, *drdthangadurai@gmail.com*

Dhanasekaran D, Department of Microbiology, Bharathidasan University, Tiruchirappalli-620 024, Tamil Nadu, India, *dhansdd@gmail.com*

Govintharaj Vaijayanthi, Department of Microbiology, Bharathidasan University Constituent College, Kurumbalur-621 107 Perambalur District, Tamilnadu, India *vaijayanthimicro@yahoo.com*

Harsh NSK, Forest Pathology Division, Forest Research Institute, Dehradun, 248006 *nirmalharsh57@gmail.com*

Jamaluddin, Rani Durgawati University, Jabalpur-482001. M.P.

Jeyabalan Sangeetha, Department of Environmental Science, Central University of Kerala Tejaswini Hills, Kasaragod, Kerala-671316, India, *drtsangeetha@gmail.com*

Jyoti Bisht, Department of Biotechnology, SGRR (PG) College, Dehradun-248001, India *biotech.jo@gmail.com*

Kapoor R, Department of Botany, University of Delhi, New Delhi-110007
kapoor_rupam@yahoo.com

Khanna SS, Former Chairman, Interministerial sub-group I Task Force on Integrated Nutrient Management using City Compost, Govt. of India, New Delhi-110012
khannas@gmail.com

Khasa DP, Canadian Research Chair in forest and environmental genomics, Université Laval Quebec City, Canada, G1V 0A6, G1V 0A6, *damase.Khasa@ibis.ulaval.ca*

Maman Turjaman, Forest Research and Development Centre, FORDA, Ministry of Environment and Forestry, Jalan Gunung Batu No. 5 Bogor-16680, Indonesia *turjaman@gmail.com*

MC Manna, Division of Soil Biology, ICAR- Indian Institute of Soil Science, Nabibagh, Berasia Road, Bhopal-462038, *madhabcm@yahoo.com, mcm@iiss.res.in*

Mohan V, Forest Protection Division, Institute of Forest Genetics and Tree Breeding Coimbatore-641002, *mohan@icfre.org , vmohan61@gmail.com*

Muniswamy David, Department of Zoology, Karnatak University, Dharwad-580003 Karnataka India, *mdavid.kud@gmail.com*

Nadeau MB, Centre for Forest Research, Institute for Integrative and Systems Biology. Université Laval, Quebec City, Canada, G1V 0A6, martin.*beaudoin-nadeau.1@ulaval.ca*

Narayanasamy M, Department of Microbiology, Bharathidasan University, Tiruchirappalli 620 024, Tamil Nadu, India, *samymb2013@gmail.com*

Patra AK, Division of Soil Biology, ICAR- Indian Institute of Soil Science, Nabibagh, Berasia Road, Bhopal-462038, *patraak@gmail.com*

Purushotam Prathima, Department of Botany, Karnatak University, Dharwad-580003 Karnataka, India, *prathimapacharya3@gmail.com*

Quoreshi A, Desert Agriculture and Ecosystems Program, Environment and Life Sciences Research Center, Kuwait Institute for Scientific Research, Safat, Kuwait-13109 *aquoreshi@kisr.edu.kw*

Ramasamy Vijayakumar, Department of Microbiology, Bharathidasan University Constituent College, Kurumbalur-621107 Perambalur District, Tamilnadu India *rvijayakumar1979@gmail.com*

Sangeeth Menon, Forest Protection Division, Institute of Forest Genetics and Tree Breeding Coimbatore-641002, *sangi.mini@gmail.com*

Saranya Devi K, Forest Protection Division, Institute of Forest Genetics and Tree Breeding Coimbatore-641002, *devisaranya13@yahoo.co.in*

Savitha J, Department of Microbiology and Biotechnology, Bangalore University, Jnanabharathi Campus, *drsvtj@yahoo.co.in*

Sharma GD, Bilaspur University, Bilaspur-495001, Chhattisgarh, *gduttasharma@yahoo.co.in*

Sharma S, Department of Botany, University of Delhi, New Delhi-110007
surbhi.sharma559@gmail.com

Shrinivas Jadhav, Department of Zoology, Karnatak University, Dharwad-580003, Karnataka India, *sspj123@gmail.com*

Sridhar KR, Department of Biosciences, Mangalore University, Mangalagangotri, Mangalore 574 199, Karnataka, *kandikere@gmail.com*

Sriharsha DV, Department of Microbiology and Biotechnology, Bangalore University Jnanabharathi Campus, *harshadvs.2011@gmail.com*

Srinivasan R, Forest Protection Division, Institute of Forest Genetics and Tree Breeding Coimbatore-641002, *srinivasanrbio@gmail.com*

Thajuddin N, Department of Microbiology, Bharathidasan University, Tiruchirappalli-620 024

1

Can Mycorrhizal Fungi Influence Plant Diversity and Production in an Ecosystem?

Bagyaraj, D.J. and Ashwin, R.

Abstract

The functioning and stability of terrestrial ecosystem are determined by plant biodiversity and species composition. So far, little attention has been paid to the effects of microbe-plant interactions, particularly the mycorrhizal symbiosis, on ecosystem variability, productivity and plant biodiversity. The most common type of mycorrhizal association is the arbuscular mycorrhizal (AM) type. The role of AM fungi in the uptake of diffusion limited nutrients like P and in conferring resistance to soil-borne-diseases is well documented.

One of the major components determining the success of early colonizing plants during plant seral succession is the availability of nutrients. In this context the ability of plants to associate with mycorrhizal fungi and enhance their ability to sequester nutrients from a limited resource is of benefit to the success of the plant species in the community. Tree species of the matured forest canopy and sub-canopy tend to be obligately mycotrophic or non-mycorrhizal. Mineral nutrient availability, especially of P, and the type of mycorrhizal fungi, in different habitats, are probably the primary selective factors which produced different degree of dependence on mycorrhizae.

There are two hypotheses to explain the examples of successional changes in AM fungal species in an ecosystem. One of the hypothesis suggests that the mycorrhizal fungi are the "drivers" to determine the plant species; the second suggests that the changes in mycorrhizal species which are "passengers" are dependent on the plant and environmental conditions. The role of AM fungal diversity in contributing to the maintenance of plant bio-diversity and to the ecosystem function has been studied recently.

The results showed that both plant bio-diversity and ecosystem productivity increases with the increasing number of AM fungi. Thus, it appears that the diversity of AM fungi in soil is a major factor contributing to the maintenance of plant bio-diversity and the ecosystem functioning.

Keywords: Ecosystem, Mycorrhizal Fungi, Plant Community Composition, Plant Diversity, Plant succession

1. Introduction

The functioning and stability of terrestrial ecosystem are determined by plant biodiversity and species composition (Schulze and Mooney 1993, Huang 2004). The mechanisms that control plant biodiversity are still being debated. The ability of many plant species to co-exist, and thus to determine plant biodiversity, can be explained by competitive interactions (Grace and Tilman 1990, van der Heijden *et al.* 1998b), by spatial or temporal resource partitioning (Tilman 1982, Kneitel and Chase 2004), by disturbance creating new patches for colonization (Huston 1979, Kneitel and Chase 2004)), and by interactions among different functional groups of organisms that constitute ecosystems (Brown and Gange 1989; van der Heijden *et al.* 1998b). So far, little attention has been paid to the effects of microbe-plant interactions, particularly the mycorrhizal symbiosis, on ecosystem variability, productivity and plant bio-diversity. The word 'Mycorrhiza" literally mean 'fungus root'. It is a symbiotic association between certain fungi and plant roots (Desai *et al.* 2016). Approximately 95% of all vascular plants have a mycorrhizal association (Brundrett 1991).

Traditionally, mycorrhizal associations have been divided into a range of categories, based on the taxonomy of the fungal associate and the physical forms of interactions between the root and the fungus in the mycorrhizal structures that are produced in the symbiosis. A list of mycorrhizal forms, their plant associates, and the key features of the mycorrhizae is given in Table. 1. Among the most common types of mycorrhizal association are the AM types, which are formed mainly by the fungi belonging to the phylum Glomeromycota (Sturmer 2012). These fungi are mainly associated with herbaceous vegetation, grasses, and tropical trees although a limited number of temperate woody plants may also associate with arbuscular mycorrhizae. The association is characterized by fungal penetration within the host cortical cells and the development of a variously developed, tree like branching of the hyphae between the host cell wall and plasmolemma called an arbuscule. It is here that the surface area of the interface between plant host and fungus is optimized for nutrient and carbohydrate exchange. In most of the arbuscular mycorrhizal fungi AMF, vesicles are formed in cortical cells. These consist of a swollen hyphum

Table 1: Outline of some of the features of different types of mycorrhizal associations

Mycorrhizal type	Host plant group	Characteristics	Fungal associate
Arbuscular mycorrhizae	Herbaceous plants, grasses; some trees	Formation of arbuscules within cortical cells of host root	Glomeromycetes
Ectomycorrhizae	Coniferous and deciduous trees	Formation of a sheath or mantle of fungal tissue around the root surface and a Hartig net of fungal penetration between the cortical cells to the endodermis.	Basidiomycetes Ascomycetes Deuteromycetes
Ericoid mycorrhizae	Ericaceae	Hyphal coils within the host root cortical cells	Ascomycetes
Arbutoid mycorrhizae	Arbutus	Hyphal coils within the host root cortical cells	Basidiomycetes Ascomycetes
Orchidaceous mycorrhizae	Orchids	Fungal propagule carried in the seed of the plant.	Deuteromycetes Basidiomycetes

occupying a large volume of the cell. This structure contains storage material. The AM association is formed with a large number of plant species and a relative small diversity of fungal species. Though there are 26 genera of Glomeromycetes fungi, the predominant genus forming arbuscular AM all over the world is *Glomus*.

The ectomycorrhizal habit consists of an association between, mainly, tree species and a range of fungal taxa consisting of basidiomycetes, ascomycetes and some zygomycetes. In this type, the fungus does not penetrate into the host cortical cells, but only between them, forming a Hartig net. The Hartig net exists outside the endodermis of the root. On the surface of the root, a sheath or covering of fungal material develops. This surface structures may be of varying degrees of complexity from a loose weft of hyphae to highly organized pseudoparenchymatous structures. It is the structure of the sheath, degree of branching, (induced by change in cytokinins), and nature of emanating hyphae or hyphal strands that allow morphological identification of these mycorrhizae (Agerer 1987 – 1999; Ingleby *et al.* 1990; Goodman *et al.* 1996-2000, Dighton 2016). Ectomycorrhizal associations are formed between a limited number of plant species and a huge number of fungal species.

Ericoid mycorrhizae are similar in structure to arbuscular mycorrhizae, but are associated solely with members of the Ericales (Ericaceae, Empetraceae, Epicaridaceae, Diapensiaceae and Prionotocaceae). All of these groups are sclerophyllous evergreen and reside in habitats where both nitrogen and phosphorus are sparsely available. The root system of these plants consist of very fine roots containing a single layer of cortical cells, which the mycorrhizal fungi penetrate to form hyphal coils, rather than arbuscules (Read 1996, Dighton 2016). The fungi associated with this type of symbiosis are *Hymenoscyphus* and *Oidiodendron*. Closely associated with these mycorrhizae are the arbutoid mycorrhizae.

Orchidaceous mycorrhizae are unique in terms of the obligate nature of the association. The importance of the mycorrhizal association for seed germination and the initial establishment of the plant has been reviewed by Zettler and McInnes (1992) and Rasmussen and Wigham (1994). The fungal partner is usually ascribed to the genus *Rhizoctonia*, and there has been such evolution of the obligateness of the association that the fungus is transported in the seed of the plant.

2. The Basic Functions of Mycorrhizae

Mycorrhizal association can alter morphology, wherein root hair development is suppressed and the function of the root hair is replaced by fungal hyphae. These hyphae have 2 major benefits for sequestring nutrients. They are of smaller diameter than root hairs and can penetrate more easily and to a greater distance from the root into the soil, thus exploring a greater volume of soil and presenting a greater surface area for nutrient absorption than could the root-root hair system alone (Marschner and Dell 1994, Dighton 2016). The second benefit is that it is energetically more efficient to produce a long, thin hyphum than a root hair. Further in AM type number of lateral roots produced is greater than non-mycorrhizal plant, suggesting dual benefits of mycorrhizal habit, one of immense root branching and the other of the fungal exploitation of soil for nutrients. Using radiotracer phosphate, it was shown that the phosphate depletion zone was greater (up to 110 mm) in mycorrihzal plant compared to non-mycorrhizal plant (10-20 mm), indicating that AM hyphe exploit a larger volume of soil than root hairs (Fig. 1), (Jacobson 1992).

Thus the role of AM fungi in the uptake of diffusion limited nutrients like P, Zn, Cu etc. is well documented (Bagyaraj *et al.* 2015). In addition to their role in conferring resistance to soil-borne diseases, alleviating the effects of salt and water stress, synergistic interaction with beneficial soil organisms, enhancing the abundance of soil aggregation and in improving plant growth, leading to the "big plant – little plant" syndrome is well documented (Jumpponen *et al,* 1998). Further, the derived benefit of a mycorrhizal association is predominantly nutrient acquisition if the host plant root system is poorly branched (Fig. 2) (Newsham 1995). In contrast, the benefit shifts toward pathogen prevention where the host root system is highly branched.

The ecology of mycorrhizal fungi has been reviewed by several workers (Bagyaraj 1990; 2014; Allen et al. 2003). The information on the role of mycorrhizal fungi influencing plant diversity and production in natural ecosystem remains scanty and disjointed.

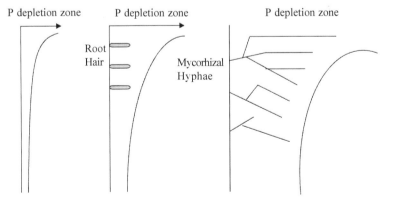

Fig. 1: Increasing P depletion zone from a root surface created by the addition of root hairs and AM hyphae as protrusions from the root surface in to the soil.

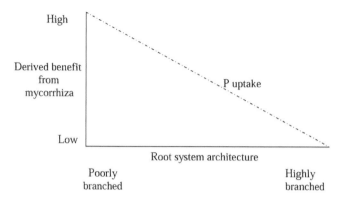

Fig. 2: Nature and magnitude of the benefit (P nutrition) derived from plant associations with arbuscular mycorrhizae, depending on the branching pattern of the host plant root system

3. Dependence on Mycorrhiza

Plants can be grouped into three categories, based on their dependence on mycorrhizae: (1) non-mycorrhizal; (2) facultative mycotrophs; and (3) obligate mycotrophs (Bagyaraj 1989).

Non-mycorrhizal plants are those which grow normally without forming mycorrhizae, even under infertile conditions. The growth of these plants is not affected by mycorrhizal inoculation (Janos 1975). Species of the Aizoceae, Amaranthaceae, Brassicaeae, Caryophyllaceae, Chenopodiaceae, Commelinaceae, Cyperaceae, Fumariaceae, Juncaceae, Nyctaginaceae, Phytolaccaceae, Polygonaceae, Portulaceae, and Urticaceae do not usually form mycorrhizae (Gerdemann 1968, Bagyaraj 1989, Sharma 2014). Non-mycorrhizal plants usually have well-developed root systems with highly branched fine roots, as well as numerous root hairs. Some of them, like *Banksia, Hakea,* Persoonia and Viminaria, produce proteoid roots (Lamont 1982; Janos 1983). Non-mycorrhizal species may secrete organic acids, thereby utilizing the bound forms of phosphorus (Graustein *et al.* 1977), or have tissues with low mineral requirements, or have slow growth rates or durable tissues which compensate for the limited nutrient supply. Non-mycorrhizal species reject mycorrhizal association either because it is incompatible with their physiology, or in order to favor their persistence in competition with mycotrophic species (Janos 1980b). They are the only species which are the effective competitors on infertile soils in the absence of mycorrhizal fungi.

Facultative mycotrophs are those plant species which are able to grow without mycorrhizae, but in which mycorrhizal colonization improves growth. In fertile soil, facultatively mycotrophic species reject mycorrhizae which are of no benefit for mineral nutrient uptake (Bowen 1980). They also produce fine roots as well as profuse root hairs (St. John 1980c). They may root more deeply, have higher root / shoot ratios or lower total mineral requirements than obligate mycotrophs. Some facultatively mycotrophic species are able to take up minerals from relatively low substrate concentrations when uninfected, and may fail to support much infection.

Obligate mycotrophs are those which can neither grow nor survive without mycorrhizae. Janos (1980a) defined them as species which would not survive to reproductive maturity at the fertility levels encountered in the natural habitats, if not colonized by mycorrhizal fungi. Obligate mycotrophs do not respond to fertilizer applications (Vozza and Hacskaylo 1971) suggesting therein obligate dependence on mycorrhizae for nutrition. Obligately mycotrophic tree species tend to have large seeds (Janos 1980b, Muthukumar and Udaiyan 2007) which favour the persistence of uninfected seedlings and the formation of large pre-infection root systems, maximizing the probability of infection. Obligate mycotrophs have coarse hairless roots. They are often light-demanding, partly because of the energy requirements of mycorrhizal association. Thus, under shade mycorrhizae become a liability; but they are still retained because of their importance to seedling growth when canopy gaps open. Tree species of the mature forest canopy and sub canopy tend to be obligately mycotrophic, while many pioneer and early-successional species are facultatively mycotrophic or non-mycorrhizal (Janos 1980b). Mineral nutrient availability, especially of phosphorus, and the type of mycorrhizal fungi in different habitats, are probably the primary selective factors which produced different degrees of dependence on mycorrhizae.

4. The Impact of Plant Community Composition on Mycorrhizal Fungi

In tropical rainforests AMF produce fewer spores (saving energy for sporulation) and exist as hyphae spreading through hyphae connecting roots. The distribution of mycorrhizal types is dependent upon the geographic distribution of the plant species and the nature of the soil. Read (1991) showed the geographical distribution of the main mycorrhizal types in the world, demonstrating that the AM habit was dominant in the temperate and tropical grasslands, tropical forests, and desert communities. Ectomycorrhizae were dominant in temperate and arctic forested ecosystems, and ericoid mycorrhizae were most common in the boreal heathland ecosystems. Mycorrhizal fungi are likely to be affected by plant-community composition (Janos 1980a, Johnson et al. 2003).

Frequent or extensive disturbance can favour non-mycorrhizal species or ectomycorrhizal species, which are effective colonizers. Non-mycorrhizal species reduce mycorrhizal fungus populations indirectly by not sustaining mycorrhizal infection. In addition, some non-mycorrhizal species can have a direct antagonistic chemical effect on the fungi (Hayman *et al.* 1975, Finlay 2008). The amounts of mycorrhizal fungus in the soil will thus be low where non-mycorrhizal plants dominate the communities, and in fertile soils where facultative mycotrophs are dominant. High populations of mycorrhizal fungi are expected in stands of facultative mycotrophs on infertile soils or where obligate mycotrophic plants are dominant. In field crops there are reports to show that in addition to the host,

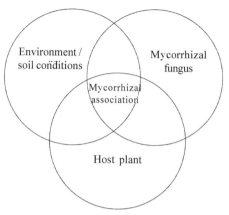

Fig. 3: Mycorrhizal associations result from three way interactions between mycorrhizal fungi, host plants and environment / soil conditions.

soil also plays an important role in determining the quality and quantity of AM population (Sreeramulu and Bagyaraj 1986). This analogy can be extended to natural ecosystem, a fungal species forming AM mycorrhiza probably dominates in a certain soil because of its ability to outgrow other species in that environment (Dayala Doss and Bagyaraj 2001). The three way interaction between mycorrhizal fungi; host plants and environment / soil conditions are illustrated by three overlapping regions in Fig. 3.

The distribution of fungal species in a mixed community of AM plant species is not homogenous. Eom *et al.* (2000) showed that the different species of plants in a tall grass prairie ecosystem have differing AMF associates (Fig. 4). Johnson *et al.* (1992) showed that the AM community differed among five plant species in garden plots in a native grassland. This information lends credence to the idea that there are feedbacks between the mycorrhizal fungal associate and the

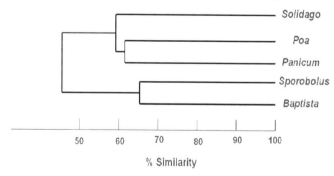

Fig. 4: Cluster Analysis of the similarity of AMF species associated with 5 host plants in a mixed species tall grass prairie ecosystem.

plant that enable the plant species to dictate the fungal species assemblage and *vice versa*.

5. Impact of Mycorrhizal Fungi on Plant Community Composition

The varied dependence of different plant species on different types of mycorrhizae can influence the composition of both seral and mature plant communities. Although there is a lack of host specificity among arbuscular mycorrhizal fungi, host preferences have been recorded (Mosse 1977; Bagyaraj 2007). Thus, the species of mycorrhizal fungus present can influence the competitive abilities of plant species. In the absence of mycorrhizal fungi, only the species which do not need mycorrhizae will be able to grow. Therefore, non-mycorrhizal species are most likely to dominate plant communities on poor soils, with no or few mycorrhizal fungi. Even if mycorrhizae form slowly, the growth lag of mycotrophic plants before infection can confer a competitive advantage on non-mycorrhizal plants. On fertile soils, the facultative mycotrophs can also occur in the absence of mycorrhizal fungi. Regardless of the availability of mycorrhizal fungi, facultatively mycotrophic species are those most likely to dominate plant communities on fertile soils.

Obligately mycotrophic species are most likely to dominate plant communities on poor soils with abundant mycorrhizal propagules. They are probably better competitors than facultative mycotrophs on infertile soils, because the adaptations of facultative mycotrophs for mineral uptake, without mycorrhiza, are redundant. For example, the hyphae can reach much farther beyond the zone of phosphorus depletion than root hairs (Janos 1980a, Smith and Read 2008). Most of the plants dependent on mycorrhizae, in tropical rain forests, do not have root hairs. Species obligately dependent on arbuscular mycorrhizae probably have similar mineral-uptake characteristics within a habitat because there are few fungal species available for this type of association compared to the large number of host species (Janos 1975). Consequently, the great species richness of tropical forests in which AM associations predominate is unlikely to reflect niche differentiation with respect to mineral uptake. In mixed plots of nine tropical tree species, including three which were non-mycorrhizal, three facultative mycotrophs, and three which were strongly mycorrhiza-dependent, the seedling survival of the last group was increased by arbuscular mycorrhizae (Janos 1983). This experiment suggested that arbuscular mycorrhizae reduce differences in competitive ability among the species, and thereby contribute to the high within-habitat species diversity characteristic of most tropical forests.

Non-mycorrhizal species and some ectomycorrhizal species may have the demographic characteristics of pioneer or fugitive species because non-mycorrhizal plants can quickly establish without awaiting infection, and ectomycorrhizal inoculum can build up very rapidly. Obligate mycotrophs probably are the best competitors under mature forest conditions.

6. Mycorrhizae and Plant Fitness

The effects of mycorrhizae on the increase in reproductive potential of plants has been noted by Koide *et al.* (1988). Stanley *et al.* (1993) and Barea et al (2002) reported increased reproductive potential leading to improvement in offspring vigor by increased seedling germination, leaf area, root: shoot ratio, and root enzyme production. Heppel *et al.* (1998) showed that offspring of AM infected *Abutiolon theophrasti* were significantly larger than offspring of non-mycorrhizal parents (Table 2). This can be a significant factor in determining plant species composition in the tropics.

Table 2: Plant fitness parameters of *Abutiolon theophrasti* offspring of mycorrhizal or non-mycorrhizal parents.

Offspring age (days)	Fitness parameters	Mycorrhizal parent	Non-mycorrhizal parent
20	Shoot height (cm)	12.5	9.4
	Shoot dry mass (g)	61.2	30.9
	Leaf number	3.6	3.0
47	Shoot height (cm)	30.6	19.8
	Shoot dry mass (g)	521	154
	Leaf number	4.4	3.4
94	Survivors per box	59.1	26.6
	Seeds per survivor	17.9	10.6

Source: Data from Heppel *et al.* (1998)

7. Mycorrhiza and Plant Succession

Pederson and Sylvia (1996) suggest that one of the major components determining the success of early colonizing plants during plant seral succession is the availability of nutrients. In this context the ability of plants to associate with mycorrhizal fungi and enhance their ability to sequester nutrients from a limited resource is of benefit to the success of the plant species in the community. Indeed, it has been shown that the dispersal of spores of hypogeous fungi by rodents and other animals like squirrels, mountain goat, mule deer is an important determinant of mycorrhizal inoculum for plants in the early stages of succession on bare ground. Small mammals defecate and are likely to deposit more spores in areas of active feeding sites than in other localities. The patchy distribution of mycorrhizal spores, and hence inoculum potential, would allow the establishment of both mycorrhizal and non-mycorrhizal plant species in the community.

Working with ectomycorrhiza Jumpponen *et al.* (2002) found that ectomycorrhizal species diversely increased to a maximum where tree canopies started to overlap. This observation at canopy closure may be related to both the relative paucity of available nutrients (P) and an increasing proportion of nutrients locked up in organic forms. Such studies with AM fungi are scanty. Indeed, Hart *et al.* (2001) propose two hypothesis to explain the examples of successional changes in AMFspecies. One of these hypotheses suggests that the mycorrhizal fungi are the driving force (drivers); the second suggests that the changes in mycorrhizal species are dependent on the plant and environmental conditions and the mycorrhizae are considered "passengers" (Fig. 5).

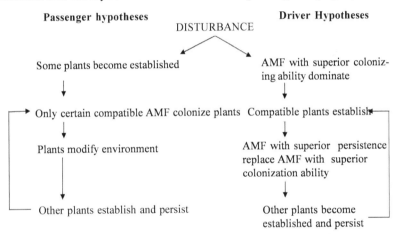

Fig. 5: A model proposing two alternate mechanisms for changes in community structure of AMF communities through time. The "passenger hypotheses" proposes that mycorrhizal communities are determined by the plant community, whereas in the "driver hypotheses" the mycorrhizae determine the plant species by interspecific differences in colonization and persistence potential of the fungi. *Source*: Hart *et al.* (2001)

8. Can Mycorrhizal Fungi Determine Plant Biodiversity in an Ecosystem?

The role of AMF diversity in contributing to the maintenance of plant biodiversity and to the ecosystem function has been studied by a few workers (van der Heijden *et al.* 1998a). They did two experiments: in the first experiment, 4 different native AMF which were all isolated from soil of a calcarious grassland, and of a combination of these 4 AMF on the species composition structure of 48 microcosms simulating European calcareous grasslands was compared. It was found that 8 *viz. Brachypodium pinnatum, Centaurium erythrea,*

Hieracium pilosella, Lotus corniculatus, Prunella grandiflora, Prunella vulgaris, Sanguisorba officinalis and *Trifolium pratense* of the 11 plant species were almost completely dependent on the presence of AMF to be successful. Different plant species benefited to different extent from different AMF suggesting host preferences by different AMF. *Carex flacca*, the only plant species that does not have a symbiotic relationship with AMF, had the highest biomass in non-mycorrhizal control treatment. Altering the AMF taxa in the soil had no significant effect on the biomass of *Bromus erectus*, the dominant plant species, in the ecosystem. From the results it was concluded that a reduction in AM biodiversity from 4 to 1 leads to a decreasing biomass of several plant species and hence it was proposed that both plant biodiversity and ecosystem productivity will increase with increasing numbers of AMF, because of added beneficial effect of each single AMF species.

In another field experiment, simulating North American old-field ecosystem in 70 macrocosms, 23 AMF species isolated from the site were inoculated containing 1,2,4, 8 or 14 AMF species (van der Heijden *et al.* 1998a). Each macrocosm [(1 x 0.75 x 0.25 m³) containing 90 kg of β-ray irradiated sand: soil (1: 1 v/v)] was showered with a seed rain consisting of 100 seeds from each of the 15 most abundant plant species of the research site. After one growing season, plants were harvested and root and shoot biomass were determined. Plant biodiversity was assessed using a Simpson's diversity index on individual species shoot – biomass data. Total shoot and root biomass, plant P, soil P and hyphal length/ g soil were also measured (Fig. 6). The results brought out lowest plant biodiversity and productivity occurred in plants without AMF or with only a few AMF species. In contrast, plant biodiversity and productivity were highest when 8 of 14 AMF species were present. Increasing AMF biodiversity lead to increased hyphal foraging capacity and more efficient exploitation of soil P, improved resource use and increased productivity and plant diversity. Increasing plant biodiversity has been shown to lead to greater ecosystem productivity (Stuwe *et al,* 1994).

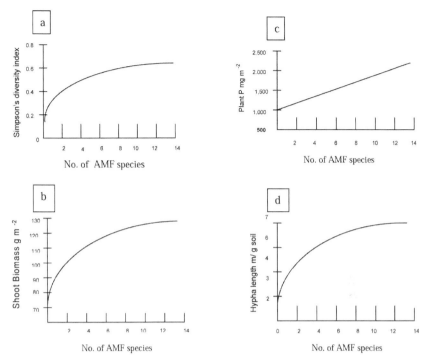

Fig. 6: The effect of AM fungal (AMF) species richness on different parameters in macrocosms simulating North American old field ecosystems. a = Simpson's diversity index, b= Shoot biomass, c= Total Plant P content, d = Length of external mycorrhizal hyphae in soil.

Conclusion

In conclusion it can be said that the functioning and stability of terrestrial ecosystems are determined by plant biodiversity and species composition (Schulze and Mooney 1993; Hooper and Vitousek 1997, Isbell *et al.* 2011). However, the ecological mechanisms by which plant biodiversity and species composition are regulated and maintained are not well understood. These mechanisms need to be identified to ensure successful management for conservation and restoration of diverse natural ecosystems. It appears that the below ground diversity of AMF is a major factor contributing to the maintenance of plant biodiversity and to ecosystem functioning. The loss of AMF biodiversity, which occurs in agricultural systems (Helgason *et al.* 1998; Johnson 1993, Jeffries and Barea 2012) could, therefore, decrease both plant biodiversity and ecosystem productivity while increasing ecosystem instability. The loss of biodiversity in soils represents an under studied field of research which requires more attention. The present reduction in biodiversity on Earth and its potential threat to ecosystem stability and sustainability (Schulze and Mooney 1993; Tilman *et al.* 1996, Berbara *et al.* 2013) can only be reversed or stopped if whole ecosystems, including ecosystems components other than plants, are protected and conserved.

References

Agerer R (1987 –1993) Colour atlas of Ectomycorrhizae 1st – 7th delivery. Einhorn – verlag, Munich.

Allen MF, Swenson W. Querejeta JI, Egerton-Warburton LM, Treseder KK (2003) Ecology of mycorrhizae: A conceptual framework for complex interactions among plants and fungi. Ann. Rev. Phytopath. 41, 271-303.

Bagyaraj DJ (1989) Mycorrhizas In: Tropical Rain Forest Ecosystems Biogeographical and Ecological Studies- A volume in Ecosystems of the World (Ed) Lieth H and Werger MJA, Elsevier, Amsterdam, pp 537–546.

Bagyaraj DJ (1990) Ecology of VA mycorrhizae In: Handbook of applied mycology (Ed) Arora DK, Bharat Rai, Mukerji KG and Knudsen GR, Vol 1 Marcel Dekker, New York , pp 3-34.

Bagyaraj DJ (2007) Arbuscular mycorrhizal fungi and their role in horticulture. In: Recent trends in horticultural biotechnology (Ed) Keshvachandran R, Nazeem PA, Girija D, John PS and Peter KV), New India Pub Agency, New Delhi, pp 53-58.

Bagyaraj DJ (2014) Ecology of arbuscular mycorrhizal fungi. In: Microbial Diversity and Biotechnology in Food Security (Ed) Kharwar RN, Upadhyay R, Dubey, N and Raghuwanshi R), Springer India, New Delhi, pp 133-148.

Bagyaraj DJ, Sharma M P and Maiti D (2015) Phosphorus nutrition of crops through arbuscular mycorrhizal fungi. Curr. Sci. 108, 1288-1293.

Barea JM, Azc´on R and Aguilar CA (2002) Mycorrhizosphere interactions to improve plant fitness and soil quality Antonie Leeuwenhoek 81: 343–351.

Berbara RLL, Goto BT, Jobim K, Covacevish F and Fonseca HM (2013) Arbuscular mycorrhizal fungi: essential belowground organisms in tropical dry environments In: Tropical Dry Forests in the Americas: Ecology, Conservation, and Management (Ed) Azofeifa AS, Powers JS, Fernandes GW and Quesada M, CRC Press, USA, pp 237-248.

Bever JD, Westover KM and Antonovics J (1997) Incorporating the soil community into plant population dynamics: the utility of the feedback approach. J. Ecol. 85, 561-571.

Bowen GD (1980) Mycorrhizal roles in tropical plants and ecosystems. In: Tropical Mycorrhiza Research (Ed) Mikola P, Clarendon, Oxford, pp 165-190.

Brown VK and Gange AC (1989) Herbivory by soil dwelling insects depresses plant species richness. Funct. Ecol. 3, 667-671.

Brundrett M (1991) Mycorrhizas in natural ecosystem. Adv. Ecol. Res. 21, 171-271.

Dayala Doss D and Bagayaraj DJ (2001) Ecosystems Dynamics of mycorrhizae In: Innovative Approaches in Microbiology (Ed) Maheshwari DK and Dubey RC, Bishen Singh Mahendra Pal Singh Pub., Dehra Dun, pp 115- 129.

Desai S, Praveen Kumar G, Amalraj LD, Bagyaraj DJ and Ashwin R (2016) Exploiting PGPR and AMF biodiversity for plant health management. In: Microbial Inoculants in Sustainable Agricultural Productivity, Vol. 1: Research Perspectives (Ed) Singh DP, Singh HB and Ratna Prabha), Springer India, pp 145-160.

Dighton J (2016) Fungi in Ecosystem Processes, Second Edition, CRC Press, USA.

Eom AH, Hartnett DC, Wilson GWT (2000) Host plant species effects on arbuscular mycorrhizal fungal communities in tall grass prairie. Oecologia 122, 435 – 444.

Finlay RD (2008) Ecological aspects of mycorrhizal symbiosis: with special emphasis on the functional diversity of interactions involving the extraradical mycelium J. Exp. Bot. 59, 1115-1126.

Gerdemann JW (1968) Vesicular-arbuscular mycorrhiza and plant growth. Annu. Rev. Phytopathol. 6, 397-418.

Goodman D, Durall D M, Trofymow J A, Berch SM (1996 – 2000). A Manual of Concise Descriptions of North American Ectomycorrhizae. Victoria, B.C.: Mycologue Publications.

Grace JD and Tilman D (1990) Perspectives on Plant Competition, Academic Press, San Diego.

Graustein WC, Cromack K and Sollins P (1977) Calcium oxalate; its occurrence in soils and effect on nutrient and geochemical cycles. Science 198, 1252-1254.

Hart MM, Reader RJ, Klironomos JN (2001) Life–history strategies of arbuscular mycorrhizal fungi in relation to their successional dynamics. Mycologia. 93, 1186-1194.

Hayman DS, Johnson AM, and Ruddlesdin I (1975) The influence of phosphate and crop species on *Endogone* spores and vesicular-arbuscular mycorrhiza under field conditions. Plant Soil 43, 489-495.

Helgason T, Daniell TJ, Husband R, Fitter AH and Young JPY (1998) Ploughing up the wood-wide web. Nature 394, 431.

Heppel KB, Shumway DL, Bateson L, Wilson, S (1998) The effect of mycorrhizal infection of *Abutilon theophrasti* on competitiveness of offspring. Funct. Ecol. 12, 171-175.

Hodge A, Helgason T and Fitter AH (2010) Nutritional ecology of arbuscular mycorrhizal fungi. Fungal Ecol. 3, 266-273.

Hooper DU and Vitousek PM (1997) The effects of plant composition and diversity on ecosystem processes. Science 277, 1302-1305.

Huang PM (2004) Soil mineral organic matter microorganism interactions: fundamentals and impacts. Adv. Agron. 82, 391–472

Huston MA (1979) General hypothesis of species diversity. Am. Nat 113, 81-101.

Ingleby K, Mason PA, Last FT, Fleming LV (1990). Identification of Ectomycorrhizas. Pub No. 5: Institute of Terrestrial Ecology Research, London.

Isbell F, Calcagno V, Hector A, Connolly J, Harpole WS, Reich PB, Lorenzen MS, Schmid B, Tilman D, Ruijven JV, Weigelt A, Wilsey BJ, Zavaleta ES and Loreau M (2011) High plant diversity is needed to maintain ecosystem services. Nature 477, 199-202

Jacobson (1992) External hyphae of vesicular-arbuscular mycorrhizal fungi associated with *Trifolium subterraneum* hyphal transport of ^{32}P over defined distances. New Phytol. 120, 509 –516.

Janos DP (1975) Effects of vesicular-arbuscular mycorrhizae on lowland tropical rain forests trees. In: Endomycorrhizas (Ed) Sanders FE, Mosse B and Tinker PB, Academic Publishers, New York, pp 437-446.

Janos DP (1980a) Mycorrhizae influence tropical succession. Biotropica 12, 56-64.

Janos DP (1980b) Vesicular-arbuscular mycorrhizae affect lowland tropical rain forest growth. Ecology 61, 151-162.

Janos DP (1983) Tropical Mycorrhizas. In: Tropical Rain Forest: Ecology and Management (Ed) Sutton SL, Whitmore TC and Chadwik AC, Blackwell, Oxford, pp 327-345.

Jeffries P and Barea JM (2012) Arbuscular mycorrhiza: a key component of sustainable plant–soil ecosystems In: Fungal Associations (Ed) Hock B, Springer Berlin Heidelberg, pp 51-75

Johnson D, Vandenkoornhuyse PJ, Leake JR, Gilbert L, Booth RE, Grime JP, Peter J, Young W and Read DJ (2003) Plant communities affect arbuscular mycorrhizal fungal diversity and community composition in grassland microcosms, New Phytol., 161: 503–515.

Johnson NC (1993) Can fertilization of soil select less mutualistic mycorrhizae? Ecol. Appl. 3, 749-757.

Johnson NC, Tilman D and Wedin D (1992) Plant and soil controls on mycorrhizal communities. Ecology 73, 2034-2042.

Jumpponen A, Trappe JM and Cazares E (2002) Occurrence of ectomycorrhizal fungi on the forefront of retreating Lyman Glacier (Washington, USA) in relation to time since deglaciation. Mycorrhiza 12, 43- 49.

Jumpponen A., Mattson KG, Trappe JM (1998) Mycorrhizal functioning of *Phialocephala fortinii* with *Pinus contorta* on glacier forefont soil: interactions with soil nitrogen and organic matter. Mycorrhiza 7, 261-265.

Kneitel JM and Chase JM (2004) Trade-offs in community ecology: linking spatial scales and species coexistence. Ecol. Lett. 7, 69–80.

Koide RT, Li M., Jewis J and Irby C (1988) Role of mycorrhizal infection in the growth and reproduction of wild versus cultivated plants. I. Wild vs. cultivated oats. Oecologia 77, 537-543.

Lamont B (1982) Mechanisms for enhancing nutrient uptake in plants, with particular reference to Mediterranean South Africa and Western Australia. Bot. Rev. 48, 597-689.

Marschner H and Dell B (1994) Nutrient uptake in mycorrhizal symbioses. Plant Soil 159, 89-102.

Mosse B (1977) The role of mycorrhiza in legume nutrition on marginal soils. In: Exploiting the legume- *Rhizobium* Symbiosis in Tropical Agriculture (Ed) Vincent JM, Whitney AS and Bose J, Univ. of Hawaii, Misc. Publ., pp 275-292.

Muthukumar T and Udaiyan K (2007) AM fungi in forest tree seedling production. In: The Mycorrhizae: Diversity, Ecology and Applications (Ed) Tiwari M and Sati SC, Daya Publishing House, Delhi, pp 200-223

Newsham KK, Filter AH, Watkinson AR (1995) Multifunctionality and biodiversity in arbuscular mycorrhizas. TREE 10, 407-441.

Pederson CT, Sylvia DM (1996) Mycorrhiza: Ecological implications of plant interactions. In: Concepts in Mycorrhizal Research (Ed) Mukerji KG, Dordecht, The Netherlands: Kluwer Academic Publishers.

Rasmussen HN and Wigham DF (1994) Seed ecology of dust seeds *in situ*: a new study technique and its application in terrestrial orchids. Am. J. Bot. 80, 1374-1378.

Read DJ (1991) Mycorrhizas in ecosystems. Experientia 47, 376-391.

Read DJ (1996) The structure and function of the ericoid mycorrhizal root. Ann Bot 77, 365- 376.

Schulze ED and Mooney HA (1993) Biodiversity and Ecosystem Function, Springer, Berlin.

Sharma MP, Sharma SK, Prasad RD, Pal KK and Dey R (2014) Application of arbuscular mycorrhizal fungi in production of annual oilseed crops. In: Mycorrhizal Fungi: Use in Sustainable Agriculture and Land Restoration (Ed) Solaiman ZM, Abbott LK and Varma A, Springer Berlin Heidelberg, pp 119-148.

Smith SE and Read DJ (2008) Mycorrhizal Symbiosis (Third Edition), Academic Press, Elsevier, NY.

Sreeremulu KR and Bagyaraj DJ (1986) Field response of chilli to VA mycorrhiza on black clayey soil. Plant Soil 93, 299-302.

St. John TV (1980) Root size, root hairs and mycorrhizal infection. A re-examination of Baylis's hypotheses with tropical trees. New Phytol 84, 483-487.

Stanley MR, Koide RT, Schumway DL (1993) Mycorrhizal symbiosis increases growth, reproduction and recruitment of *Abutilon theophrasti* Medic. in the field. Oecologia 94, 30-35.

Stew K, White L and Brown R (1994) The influence of eroding topography on steady state geotherms. Application to fission track analysis. Earth Planet Sci. Lett. 124, 63-74.

Sturmer SL (2012) A history of the taxonomy and systematics of arbuscular mycorrhizal fungi belonging to the phylum Glomeromycota. Mycorrhiza 22, 247–258.

Tilman D (1982) Resource Competition and Community Structure, Princeton Univ Press, Princeton.

Tilman D, Wedin D and Knops J (1996) Productivity and sustainability influenced by biodiversity in grassland ecosystems. Nature 379, 718-720.

van der Heijden MGA, Boller T, Wiemken A, Sanders IR (1998a) Different arbuscular mycorrhizal fungal species are potential determinants of plant community structure. Ecology 79, 2082-2091.

van der Heijden MGA, Klironomos JN, Ursic M, Moutoglis P, Streitwolf-Engel R, Boller T, Wiemken A and Sanders IR (1998b) Mycorrhizal fungal diversity determines plant biodiversity, ecosystem variability and productivity Nature 396, 69-72.

Vozzo J A and Hacskaylo E (1971) Inoculation of *Pinus caribaea* with ectomycorrhizal fungi in Puerto Rico. For. Sci. 17, 239-245.

Zettler L W and McInnes T (1992) Propogation of *Plantathera integrilabia* (Correll) Luer, an endangered terrestrial orchid, through symbiotic seed germination. Lindleyana. 7, 154-161.

2

Improvement of Growth of Tree Seedlings in Heavy Metal Polluted Soil by Ectomycorrhizal Fungi

Ajungla, T. and Sharma, G.D.

Abstract

Study was undertaken to investigate the toxicity of heavy metals on the structure and function of ectomycorrhizal fungi and their ability to colonize the pine seedlings. Ectomycorrhizal fungi were isolated, cultured and multiplied on MMN medium. Six heavy metals (Cd, Ni, Zn, Pb, Cu and Al) were selected and different concentrations viz; 0 ppm (control), 10 ppm, 50 ppm, 100 ppm, 200 ppm and 500 ppm were used for treating the seedlings. The study was continued for one year. Different growth parameters were measured. Factors such as soil moisture, pH and soil organic carbon on ectomycorrhizal colonization were also analyzed. The mycorrhizal colonization was more in those seedlings having lower concentration of heavy metal treatment. Statistical data analysis has shown that the heavy metals had negative correlation with mycorrhizal colonization. High concentration of metals were toxic to external mycelium, which ultimately resulted in reduction of ectomycorrhizal development. The accumulation of metals in the fungal mycelium and protecting root against metal toxicity was an important factor. Inspite of heavy metal contaminations, the ectomycorrhizal inoculated seedlings showed significant growth improvement ($p < 0.01$ and $p < 0.05$). There was a positive correlation between mycorrhizal colonization and the growth of pine seedlings. The percentage of inhibition of shoot height, needle length, number of needles, seedlings volume and survival of seedlings were observed in non-mycorrhizal ones. The concentration of nutrients was restricted by the presence of heavy metals. The total concentrations of NPK decreased with increased in heavy metal concentrations. A significant variation was found in the concentration of N, P and K between mycorrhizal and non-mycorrhizal one.

Keywords: Ectomycorrhiza, Heavy metals, Mycorrhizal colonization, Pine, Soil organic carbon.

1. Introduction

Mycorrhiza is a symbiotic mutualistic association between soil fungi and fine plant roots. It is neither the fungi nor the plant root, but rather the structure formed from these two partners. Since this association is mutualistic in nature, both fungi and the plant benefit from the association. The concept of forming ectomycorrhizae on tree seedlings in nurseries with specific fungi was originally developed by Moser (1968) in Austria. With modification of Moser's technique, Takacs (1967), in Argentina showed that the survival and growth of the tree seedlings with specific ectomycorrhizae exceeded performance of seedlings that lacked or had few native mycorrhizae at planting site. Within the host root, the fungi ramifies between the outer cells forming a complex network known as Hartig net with large surface area for contact between the fungus and the host, which allow efficient transfer of essential metabolites. The fungal mycelia extend and function as a link between plant root system and the soil.

The type of ectomycorrhizal formation is attributed to different fungal symbionts and may vary with changes in the physical, chemical and biotic environment. Ectomycorrhizae mainly interact with forest tree species such as Pinaceae and Fagaceae and are species of Basidiomycetes and Ascomycetes (Bonfante *et al.* 2010). Both fungus and plant rely upon the other symbiotic partner; the tree for inorganic (P) nutrient uptake and the fungus for the supply of carbohydrate. High soil organic carbon content favoured better ectomycorrhizal development thereby, promoting an enhanced growth of its host (Tzudir *et al.* 2014). Mycorrhizal mycelia in addition to its symbiotic function are capable of establishing interplant connections (Read 1992) and can act as source of inoculum for varied degree of mycorrhizal development in the non-inoculated seedlings of pine. The fact that the mycelia can form interplant connections there by initiating the mycorrhizal formation provides pathways for the interplant transfer of metabolites. Ajungla *et al.* (2010) have shown that extensive mycorrhizal roots system have positive impact on the growth performance of the host plant. Early colonization by ectomycorrhizal fungi is essential for survival and enhanced growth of the associated host plants. Application of ectomycorrhizal inoculum in the nursery helps in nutrient uptake, seedlings survival and fitness, which will ultimately lead to healthy functioning of the host plant.

Rapid urbanization, industriazation and other anthropogenic activities like, excessive use of fertilizers and pesticides along with the practice of slash and burn system of cultivation specially in the north eastern region have negative impact on the structure and function of ecosystem. These activities have led to severe degradation of natural habitats. The native vegetation and top soil have either been removed, lost or altered. Through normal successional processes, it would take long time to rehabilitate the degradated land. Successful stabilization of degraded land due to practice of shifting cultivation have been observed (as shown in Figure 1) when tree seedlings such as *Pinus insularis*, *Quercus griffithii* and *Schima wallichii* were inoculated in the nursery with different ectomycorrhizal fungi (Bendangmenla *et al.* 2005). With the advancement of inoculation techniques there are opportunities to recover these disturbed lands which can effectively be restored through plantation with mycorrhizal tree species of high adaptability and productivity. Certain ectomycorrhizal genera appear in the early succession sequence whereas others dominate in the later stages of succession (Sharma *et al.* 2004, Arenla *et al.* 2014). Forest trees and ectomycorrhizal fungi have evolved together by adopting different kinds of stresses. Mycorrhiza is now considered an important tool, which hold tremendous potential for use in agriculture and forestry.

Heavy metals such as Cd, Ni, Zn, Pb and Cu are toxic to biological system due to their persistence and destructive nature. These metals vary in their action (Gadd 1993). Recently, the use of heavy metals has increased in various forestry practices in the form of pesticides, fungicides and municipal and industrial discharges. Heavy metals like Pb and Zn are found inhibitory to microbial growth and decomposition (Joshi 1993). High concentration of dissolved Al has resulted in the screening of many forest tree species (Shaedle *et al.* 1989) and their response to the mycorrhizal fungi (Harley and Smith 1983). The reduced growth of extramycelium due to heavy metals in the soil is likely to reduce potential of the fungus to absorb the minerals and water and transfer them to the seedlings. Mycorrhizal inoculation improves the plant productivity especially in soil with low nutrient status and metal polluted condition (Ajungla *et al.* 2003). The tolerance of heavy metals depends on the accumulation of toxic ions in the root cells (Tinker 1981). The interaction between various contaminants with respect to heavy metal toxicity have been studied (Symenoidis and Karataglis 1992) However, investigation on the heavy metals toxicity to the mycorrhizal system and their role in the metal polluted soil is still scarce.

Fig. 1: Seedlings of ▲ *P. insularis*, △ *Q. griffithii* and ▲ *S. wallichii*
A) 1 year old tree seedlings. B) 8ᵗʰ month old tree seedlings (Bendangmenla, 2005)

2. Soil Moisture and pH and Ectomycorrhizal Colonization

Many timber yielding trees have ectomycorrhizal association. The ectomycorrhizal association is characteristics of woody perennials in families such as Betulaceae, Pinaceae Salicaceae, Fagaceae, Dipterocarpaceae and Myrtaceae. These families account for most of the world's wood production. Climatic factors affect the growth and development of ectomycorrhizal fungi. The studies carried out in laboratory conditions however, show differences in response to the natural environmental conditions due to complexity and interactions between edaphic, climatic and biological components (Jha, 1990). In symbiosis, the plant provides carbon and in return gets inorganic nutrient from soil. Dipterocarp trees are obligately ectomycorrhizal and extensive studies have been made on the association of ectomycorrhizal fungi with trees in mixed dipterocarp forests in Thailand, Malayan Peninsula and Indonesia (Natarajan, 2005). Some ectomycorrhizal fungi like *Suilus* is specific while *Amanita* is general mycorrhizal fungus. Ectomycorrhiza is common in temperate, alpine cold boreal ecosystem but also found in tropical, subtropical and rain forests. The root hairs of tree species having ectomycorrhiza increase in thickness after fungal infection and produce dichotomy. Nearly 2000 fungal species have been identified to form ectomycorrhiza with 10% of plants, mostly woody ones. Some of these are obligate but mostly facultative ones. Most of ectomycorrhizal fungi can be cultured but are unable to degrade complex polymers like lignin and cellulose.

In eastern Himalaya, we have conducted study with an exotic species of *Dipterocarpus retusus* inoculated with *Scleroderma citrinum* and *Russula* sp. The seedling roots were assessed for ectomycorrhizal association for one year. The soil moisture and pH were also determined during summer and spring seasons. During this, different growth parameters of the seedlings were also studied. After one year, the infected roots were identified based on morphotypic chatacters, *Russula* sp. *was* orange to gold colour and often had a distinctive layer of cystidia covering the mantle while *Scleroderma citrinum* were white morphotype with clamps (Arenla *et al.* 2014). Colonization of *D. retusus* seedling roots was higher compared to *Russula* sp. inoculated plants (Table 1).

Table 1: Ectomycorrhizal colonization on Dipterocarp roots.

Inoculum	*Dipterocarpus retusus (%)*
M1	73
M2	88
M0	30

Note: Each value is a mean of five replicates. M1= *Russula*, SP M2= *Scleroderma citrinum*, M0=Without fungal inoculation.

The summer season with high moisture harboured significantly higher level of ectomycorrhizal colonization. During spring and winter seasons, adding to low soil moisture and temperature, the new ectomycorrhizal colonization was either very less or did not form at all. There was also consistent increase in growth of the *Scleroderma citrinum* inoculated seedlings, which showed better growth (Table 2).

Table 2: Growth of the *Dipterocarpus retusus* seedlings inoculated with *S. citrinum* (M1) and *Russula* sp. (M2)

Growth parameters	04th Month			08th Month		
	M1	M2	M0	M1	M2	M0
Shoot height (cm)	23.5	25	19.1	37.4	38.8	29.8
Leaf length (cm)	10.3	8.6	9.2	22	23.2	12.8
Root length (cm)	8.2	10.2	8.1	14.4	13.2	11.6
Root color diameter (cm)	0.4	0.5	0.5	0.5	0.63	0.5

M1= *Russula* sp, M2= *Scleroderma citrinum*, M0= control. Each reading is a mean of five replicates

In eight month old tree seedlings, a significant difference between inoculated and uninoculated seedlings was observed. This shows that the ectomycorrhizal association along with enhanced ectomycorrhizal root tips positively correlate with the seedling growth. Growth response of the seedlings inoculated with *Scleroderma citrinum* was greater due to its higher colonization rate in the seedlings root. Similar report of Thompson *et al.* (1994) on ectomycorrhizal fungi was screened on the basis of their ability to colonize plant root. In subtropical forests of Nagaland, during summer when the soil pH was low, higher level of colonization was observed. (Table 3).

Table 3: Rhizospheric soil moisture and soil pH during summer and spring

	Spring	Spring	Spring	Summer	Summer	Summer
	M1	M2	M0	M1	M2	M0
Soil pH	5.1	5.8	5.2	4.2	4.5	4.7
Soil Moisture%	15	16	13	28	27	22

Note: M1= *Russula* sp, M2= *Scleroderma citrinum*, M0= uninoculated; Each value is a mean of three replicates

Here, we considered soil moisture and soil pH however, other environmental factors also affect the outcome of symbiosis. Watson *et al.* (1989) observed that, ectomycorrhiza is favoured by moderate soil moisture. In our study, the soil pH between 4 to 5 was most favourable for ectomycorrhizal colonization. Thakur (1990) observed a significant increase in mycorrhizae development when pH was lowered from 7.0 to 4.0. Better ectomycorrhizal formation with active plant root tips was higher during moderate soil moisture, low pH and high soil cover with leaf litter.

3. Influence of Soil Organic Carbon on the Growth of Ectomycorrhizal Seedlings

The type of ectomycorrhizal formation is attributed to different fungal symbionts and may also vary with changes in the physical, chemical and biotic environment. As a result of vast physiological diversity that has been existing among ectomycorrhizal fungi, community structural changes may potentially alter both carbon and nutrient allocation within the forest ecosystems. In this association, both the fungus and plant rely upon the symbiotic partner; the host plant for nutrient uptake and the fungus for a supply of carbon. Mycorrhizal fungi are often entirely dependent upon utilizing carbon directly from the roots as they do not decompose organic matter for their energy source (Malajezuk *et al.* 1994). Finlay and Soderstrom (1992) estimated that up to 20% of the photosynthetically fixed carbon may be allocated to the ectomycorrhizal fungus by the host plant. Thus, mycorrhizal fungi exhibit diverse function in carbon metabolism. The carbon storage depends on the balance between carbon inputs through photosynthesis and outputs through autotrophic (root and mycorrhiza) and heterotrophic (soil microbial) respiration (Bardgett *et al.* 1998).

A study was conducted to observe the effect of carbon on the growth of *Pinus patula* Schiede ex Schlecht and Cham. The seedlings were artificially inoculated with *Boletus edulis* and *Russula* sp. The mycorrhizal association was confirmed within 40-60 days (Imliyanger *et al.* 2014). This was based on the swollen root tips and the distinctive characteristics short roots bifurcation. The organic carbon content was determined by Wet-Digestion method (Walkey and Black's titration method). Carbon estimation was done every three months and the results were expressed on seasonal basis. The results showed that the soil organic carbon content varied in accordance with seasons. Soil organic Carbon content was highest during summer season while, it was found to be lowest during autumn of second year (Table 4).

Table 4: Seasonal variation of soil organic carbon (%) in mycorrhizal pine seedlings

Seasons	*Boletus edulis*	*Russula* sp.	Uninoculated
Summer 1	4.54	5.55	4.05
Autumn 1	4.50	5.25	3.15
Winter 1	4.35	3.65	3.85
Spring 1	4.15	3.60	3.35
Summer 2	4.80	5.45	3.85
Autumn 2	3.15	3.35	2.00
Winter 2	4.40	3.60	3.35
Spring 2	3.80	3.75	3.20

Note: Each value is a mean of five replicates. 1= 1st year, 2= 2nd year

Differences among the ectomycorrhizal species in efficiency of and utilization of allocated carbon were observed. Significant variation in organic carbon content between the ectomycorrhizal seedlings and non-mycorrhizal seedlings were observed (Table 5). Sharma *et al.* (1986) also found better development of mycorrhizae under higher content of organic matter. The fungi play a central role in biological interaction and biogeochemical cycle since the carbon they obtain for their photosynthetic plant host is allocated via the mycorrhizal mycelia to the soil ecosystem (Finlay *et al.* 1989).

Table 5: Analysis of variance (F) values of sampling periods with various independent parameters

Source of variance	Variation between sampling periods
Soil organic carbon	1120.563**
Mycorrhizal infection	8942.023**

**= significant at 0.01 level of probability

Greater selective nutrient uptake from the increased absorptive surface area provided by the mycelial network has been shown with many other ectomycorrhizal associates. The changes in the production of mycelium and the potential to colonize roots may affect competition among species, which in turn may lead to changes in ectomycorrhizal community structure. Results from this study reveal that, early colonization by ectomycorrhizal fungi is essential for survival and enhanced growth of the associated host plants. The reason could be attributed to the fact that mycelial strands can extend from plant to plant thereby, initiating the infection in the seedlings and provide functional pathways for more transfer of labelled assimilates between individuals. Ajungla *et al.* (2010) have found that, extensive mycorrhizal roots systems have positive impact on the growth performance of the host plant. The present study reveal early colonization by mycorrhizal fungi is essential for survival and enhancement in growth of the associated host plants. This suggest that, the application of mycorrhizal inoculum at the nursery stage will result in considerable contribution in essential nutrients uptake, and that the establishment and fitness depends upon the healthy functioning of these fungi and vice-versa.

4. Heavy Metals and Ectomycorrhizal *Pinus Kesiya* Seedlings

One of the effects that led to degradation of habitats including cultivated lands is due to rapid urbanization and increased human activity. However, with the advent of biotechnology, currently there are opportunities for recovery of these degraded ecosystems. It is now possible that the disturbed lands can be effectively restored by planting tree seedlings of wide adaptability and greater productivity. Successful afforestation and stabilization of degraded lands due to shifting cultivation have been observed when tree seedlings were inoculated in the nursery with different ectomycorrhizal fungi (Bendangmenla *et al.* 2005). Mycorrhizae are now considered as a potential tool available to plant scientists, which can aid overall performance in productivity of plants even in the polluted soil. The field studies on polluted and adverse sites for recovery and re-vegetation can be expedited by using tree seedlings tailored with ectomycorrhizae. It is now understood that, the liberal and over use of heavy metals due to rapid urbanization has resulted a negative impact on ecological systems of the earth. Heavy metals such as Pb, Cd, Ni, Zn and Cu are toxic to biological system due to their persistence and destructive nature. Heavy metals like Zn and Cd are found inhibitory to microbial growth and decomposition. Successful tree survival and establishment on the metal polluted sites has economic benefits. It has been suggested that ectomycorrhiza can ameliorate heavy metal toxicity (Wilkins, 1991). A field study was conducted in soil artificially treated with heavy metals on mycorrhiza formation. Another plot was also maintained, where pine seedlings were inoculated with mixed mycorrhizal fungi (*Boletus* sp., *Cenococcum graniforme* and *Suillus luteus*) as shown in Figure 2. It was observed that application of heavy metals such as Pb, Zn, Ni, Cd, Cu and Al reduce growth of *Pinus kesiya* seedlings (Table 6).

Table 6: Effect of heavy metals on the growth of pine seedlings inoculated with *Cenococcum graniforme*

Metals conc. (ppm)	Shoot height (cm/plant)	Needle length (cm)	No. of Needles	Seedlings volume (cm³/plant)	Root Collar dm (cm/plant)	Root length (cm/plant)
0.0	8.2	4.5	78	0.512	0.25	12.5
Zn 10	6.3	3.3	76	0.333	0.23	9.3
Zn 50	6.0	3.1	73	0.317	0.23	9.0
Zn 100	5.8	2.6	60	0.255	0.21	8.3
Zn 200	5.6	2.4	53	0.246	0.21	8.1
Zn 500	5.4	2.4	44	0.238	0.21	7.4
Ni 10	6.5	3.1	73	0.374	0.24	9.6
Ni 50	6.4	3.0	70	0.309	0.22	9.2
Ni 100	6.0	2.7	63	0.264	0.21	9.1
Ni 200	5.5	2.5	51	0.242	0.20	8.3
Ni 500	5.0	2.5	46	0.162	0.18	8.0
Al 10	6.3	3.7	76	0.304	0.22	8.9
Al 50	6.2	3.5	73	0.300	0.22	8.3
Al 100	6.0	3.4	59	0.290	0.22	7.1
Al 200	5.5	3.1	47	0.140	0.16	7.0
Al 500	5.4	3.1	47	0.138	0.16	6.7
Cd 10	6.6	3.5	72	0.238	0.19	9.8
Cd 50	6.3	3.2	71	0.227	0.10	8.8
Cd 100	6.0	3.0	68	0.216	0.18	8.7
Cd 200	5.9	2.8	51	0.236	0.20	8.0
Cd 500	5.5	2.5	50	0.140	0.16	7.1
Cu 10	6.5	3.6	71	0.286	0.21	11.0
Cu 50	6.2	3.2	66	0.273	0.21	10.7
Cu 100	6.0	3.0	60	0.264	0.21	10.0
Cu 200	5.7	2.7	51	0.128	0.15	9.3
Cu 500	5.5	2.5	47	0.123	0.15	8.0
Pb 10	6.2	3.2	59	0.300	0.22	9.8
Pb 50	6.0	3.0	55	0.264	0.21	9.6
Pb 100	5.6	2.7	52	0.202	0.19	9.2
Pb 200	5.3	2.5	46	0.133	0.16	8.0
Pb 500	5.0	2.5	40	0.112	0.15	7.3

Inoculation of *Pinus kesiya* seedlings with ectomycorrhizal fungi consistently stimulated an increased in shoot height, number of needles, lateral root length (Table 6) throughout the growing seasons as compared to non-mycorrhizal ones. All the growth parameters in mycorrhizal seedlings without any heavy metals treatment were vigorous. The root length was drastically reduced in the seedlings treated with 100ppm and above. It is likely that the presence of dense fungal mycelium mantle due to root colonization has enhanced the tolerance of the host plant to metal toxicity. Jentschke *et al.* (1999) found increased tolerance to Cd in *Picea abies* seedlings inoculated with *Paxillus involutus*. It is necessary

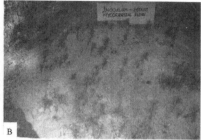

Fig 2: Pine seedlings in field with different treatment of heavy metals. A) Pine seedlings showing the treatment of Ni and Pb. B) Pine seedlings in field inoculated with mixed mycorrhizal fungi.

that the impact of heavy metals in the forest decline requires comparison of laboratory and field studies. Dixon and Bushena (1988) in their field experiments have shown that the mycorrhizal formation was strongly suppressed by Cd in oak, pine and spruce seedlings.

The results of the present investigation revealed that the *Cenococcum graniforme* inoculated pine seedlings had better survival and exhibited more vigorous growth than the non-mycorrhizal ones. This indicates the impact of mycorrhizal association on the seedlings in tolerating the toxic effect of the added metals. The retarded growth of the seedlings may have been linked to the lower level of mycorrhizal colonization. Similarly, Baar *et al.* (1994) also observed the reduction of seedlings growth due to low mycorrhizal level.

The reason for lower mycorrhizal colonization in the roots of *P. kesiya* seedlings could be due to toxic effect of Cd, Zn, Pb, Ni, Cu and Al, especially when they are present in high concentration (Table 7). It has been found that the heavy metals have suppressed the mycorrhizal colonization due to accumulation of metals in the root surface, which is also responsible for the seedlings' damage. The cytological changes observed in metal stressed pine roots were similar to those that occurred in seedlings with nutrient deficiency induced ones (Fink, 1991). At high concentration of metals, however, root extension inhibition occurred. In these circumstances, the presence of an ectomycorrhizal fungal symbiont could be of considerable importance. Extensive hyphal coils occupy most of the cortical cells of the pine roots with lower concentration of metals through which absorption occurs and this would provide greater area for the retention of metal ions. It is now very likely that the mycorrhizal fungi influenced the transport of metals in the host plants, though, not directly. After inoculation, the fungus showed rapid colonization and growth so that the toxic effects brought about by the metals were reduced quickly, thereby, allowing the root to retain its normal growth. With time, it appeared either that the plant itself or the inoculated fungus was able to bind the metals, mediate the rhizosphere pH or that the substrate or root exudates hinders the available metals so that the element was no longer toxic. However, additional research into understanding of the mechanism of protection of roots against metal toxicity is required. It is evident that mycorrhizal inoculated seedlings had better survival, better shoot height and greater number of needles as compared to non-mycorrhizal seedlings. This indicates that the mycorrhizal seedlings were more tolerant to the toxic effect of heavy metals. The main objective, therefore, must be to work out for a possible strategy, which improves the growth of forest tree in spite of heavy metal pollution. The heavy metal toxicity to the mycorrhizal system still needs careful investigation.

At present, recognition, utilization and management of mycorrhizae are necessary for the skilful production of resilient planting stock. Moreover, judicious use of agrochemicals has become a necessity and unruly disposal of both agricultural and industrial wastes to agricultural land should be avoided.

Table 7: Survival (%) of pine seedlings inoculated with mycorrhizal fungi in heavy metal polluted soil.

Metals conc. (ppm)	Survival (%)				Control	Mycorrhizal colonization (%)			
	M1	M2	M3	M4		M1	M2	M3	M4
0	93	93	93	93	90	88	88	88	88
Zn 10	76	93	76	83	70	83	64	61	64
Zn 50	76	86	70	80	66	60	61	61	53
Zn 100	73	83	66	80	63	40	52	52	49
Zn 200	70	80	63	73	60	37	39	39	46
Zn 500	60	70	60	73	58	29	36	37	39
Ni 10	86	86	93	86	90	78	71	65	59
Ni 50	80	86	86	80	80	66	70	61	55
Ni 100	80	83	73	80	73	51	55	56	54
Ni 200	76	80	73	73	73	43	46	53	38
Ni 500	73	73	66	73	64	28	42	42	36
Al 10	86	80	90	86	86	81	72	88	54
Al 50	86	80	80	86	80	75	63	52	45
Al 100	80	80	80	84	76	44	59	36	41
Al 200	80	76	73	80	73	30	55	31	33
Al 500	76	73	73	76	70	29	32	28	32
Cd 10	93	86	86	86	90	69	59	61	63
Cd 50	93	80	80	80	80	63	55	62	61
Cd 100	86	80	80	80	76	42	42	54	53
Cd 200	80	76	80	80	73	33	32	46	42
Cd 500	80	73	73	76	70	31	31	41	35
Cu 10	86	93	86	86	86	66	63	63	71
Cu 50	86	86	80	80	80	64	54	60	66

(Contd.)

Metals conc. (ppm)	Survival (%)				Control	Mycorrhizal colonization (%)			
	M1	M2	M3	M4		M1	M2	M3	M4
Cu 100	80	80	76	80	73	50	51	56	49
Cu 200	80	80	73	80	73	39	37	46	43
Cu 500	73	80	73	73	70	33	31	35	41
Pb 10	93	86	86	86	80	82	58	59	61
Pb 50	86	80	73	80	80	80	55	67	56
Pb 100	86	80	73	80	73	68	51	49	49
Pb 200	73	73	70	73	70	60	41	47	40
Pb 500	72	73	66	73	60	50	33	31	33

M1= *Suillus luteus,* M2= *Boletus* sp, M3= *Scleroderma aurantium,* M4= *Cenococcum graniforme*

Inspite of high concentration of heavy metals, *Cenococcum graniforme* significantly improved the growth of seedlings than the non-mycorrhizal ones. There was a positive correlation between mycorrhizal infection and the growth of pine seedlings. A signification variation was also found between mycorrhizal and non-mycorrhizal seedlings (Table: 8).

Table 8: Correlation coefficient (r) of mycorrhizal infection (%) with growth parameters of mycorrhizal and non-mycorrhizal seedlings.

	Mycorrhizal colonization				
Parameters	M1	M2	M3	M4	NM
Shoot height	0.98*	0.99*	0.95*	0.95*	NS
Needle length	0.91*	0.84	0.92*	0.96*	NS
Number of needles	0.93*	0.96*	0.97*	0.96*	NS
Seedlings volume	0.81**	0.79**	0.91*	0.85**	NS
Root length	0.95*	0.90*	0.89*	0.99	NS
Root collar diameter	0.81**	0.79**	0.86**	0.90*	NS
Survival	0.93*	0.94*	0.88*	0.97*	NS

M1= *C. graniforme*, M2= *S. luteus*, M3= *Boletus* sp, M4= *S. aurantium*, NM= Non-mycorrhizal, NS= Non-significant (all non-mycorrhizal seedlings were found insignificant)
*= significant at p<0.01 level **= significant at p<0.05 level

Different ectomycorrhizal fungi used for different tree species under field condition have shown a great promise for adopting inoculation techniques in artificial rehabilitation of polluted soil. Field studies on degraded sites have shown that restoration and reforestation on such soil can be expedited using seedlings tailored with ectomycorrhizae. In this study, the ability of ectomycorrhizal fungi to grow well in the presence of different concentrations of heavy metals, though to a lesser extent, is to be well noted. When plants are exposed to high concentration of heavy metals, most of them eventually fail to maintain metals homoeostasis and develop stress symptoms (Dietz *et al.* 1999). Studies have been conducted to evaluate the accumulation of Cd and chelation of metal ions occurring in the vicinity of plant roots (Naidu and Harder 1998). Accumulation of heavy metal pollutants on the plant surface is a contributing factor to the forest decline.

5. Nutrients Uptake by Ectomycorrhizal Seedlings in Heavy Metal Polluted Soil

Variation in the efficiency of mycorrhizae for metabolic activities and nutrients uptake may depend on the specific mycorrhizal fungi. The benefits of mycorrhizae for improving the growth of trees are largely related to hyphal network of extramatrical mycelia in the soil providing large surface for the absorption and uptake of nutrients (Rousseau *et al.* 1994). Mycorrhizal association is specially adapted to nutrient stress condition (Heinrich *et al.* 1989), it may therefore, be possible that under these circumstances, ectomycorrhizal roots and their mycelial network function as mineral source sink than non-mycorrhizal ones. It may also be noted that fewer extramatrical mycelium resulting from the presence of high concentration of heavy metals is likely to influence the potential of the fungus to take up essential minerals and water and transfer them to the host.

In this study, it was observed that contents of nutrients as a whole were restricted by the presence of heavy metals. The treatment of Cd, Ni, Cu, Zn, Pb and Al adversely affected the total contents of NPK in the pine seedlings. Heavy metals treatment of 50 ppm and above has shown suppressive effects, especially on non-mycorrhizal seedlings. The of NPK uptake as a whole was restricted by the high concentration of heavy metals (Table 9).

Table 9: Uptake of NPK in *Suillus luteus* inoculated pine seedlings growm in heavy metal polluted soil

Metals conc. (ppm)	Nitrogen (%)	Potassium (%)	Phosphorus (%)
	MS	MS	MS
0	2.00	0.90	0.88
Zn 10	1.83	0.84	0.86
Zn 50	1.72	0.70	0.70
Zn 100	1.56	0.78	0.69
Zn 200	1.51	0.61	0.45
Zn 500	1.05	0.45	0.42
Ni 10	1.87	0.81	0.89
Ni 50	1.76	0.59	0.80
Ni 100	1.66	0.50	0.67
Ni 200	1.63	0.42	0.65
Ni 500	1.52	0.30	0.59
Al 10	2.04	0.97	0.79
Al 50	1.94	0.77	0.74
Al 100	1.91	0.60	0.72
Al 200	1.80	0.57	0.59
Al 500	1.78	0.47	0.55
Cd 10	2.12	0.68	0.90
Cd 50	1.90	0.65	0.80
Cd 100	1.70	0.65	0.74
Cd 200	1.70	0.50	0.66
Cd 500	1.53	0.45	0.55
Cu 10	2.11	0.83	0.84
Cu 50	1.87	0.76	0.73
Cu 100	1.80	0.70	0.69
Cu 200	1.57	0.52	0.60
Cu 500	1.50	0.47	0.54
Pb 10	1.90	0.72	0.90
Pb 50	1.90	0.61	0.84
Pb 100	1.88	0.54	0.70
Pb 200	1.81	0.53	0.65
Pb 500	1.52	0.51	0.55

MS= *Suillus luteus* inoculated mycorrhizal seedlings

It has been observed that the inoculation of pine seedlings with *Suillus luteus* increased uptake of NPK at different levels of metal application. This advantage could be due to their mycorrhizal association which permits to explore more mineral resources effectively. It is also noteworthy that the lower dose of heavy metals specially 50ppm and below had the greater root development in the seedlings, which might have accounted for higher NPK accumulation.

The data processed by analysis of variance (ANOVA) showed a positive correlation between mycorrhizal colonization and NPK uptake (Table 10).

Table 10: Analysis of variance (F) values of mycorrhizal and non-mycorrhizal pine seedlings

Source of variance	Variation between mycorrhizal and non-mycorrhizal seedlings
Nitrogen	2.97*
Phosphorus	2.76**
Potassium	2.48**

*= significant at $p<0.01$ level **= significant at 0.05 level

The nitrogen concentration in ectomycorrhizal seedling roots were more than in non-mycorrhizal roots at each level of metal treatment. The main cause of reduction in growth in non-mycorrhizal seedlings may be due to their N status. Since, N is the growth limiting factor, any change in N uptake and transport would alter the growth of plant. Ingestad *et al.* (1986) concluded that nitrogen content in the shoot is directly related to the relative growth rate of plants. Our study has demonstrated that the inoculation of mycorrhizal fungi may be more advantageous to the host, which could permit the host to exploit more effectively the mineral resources. The greater fine roots development in the pine seedlings treated with 50ppm and 10ppm of heavy metals might have accounted for higher concentration of N, P and K accumulation. Tan *et al.* (1990a) concluded that the enhanced metal accumulation in non-mycorrhizal roots may be partially responsible for poor root development, which will lead to the less exploitation of mineral nutrients due to their reduced root system. Under these conditions, mycorrhizal development might be a more important factor for determining nutrient levels especially for P due to its low mobility in the soil. The main cause of the reduced nutrient concentration in the seedlings was due to the high concentration of Cu, Ni, Zn, Cd, Pb and Al in soil which restricted the mycorrhizae in the root system. The effect of heavy metal treatment in non-mycorrhizal seedlings was directly correlated to the nutrient availability (Table 11). At high levels of heavy metals, however, inhibition of nutrient contents may be due to complexing ability of mycobiont. In this instances the presence of ectomycorrhizal fungal symbionts could be of considerable importance

Table 11: Correlation coefficient (r) of mycorrhizal colonization (%) with NPK uptake by pine seedlings

Growth parameters	Mycorrhizal colonization	
	M 1	N M
Nitrogen	0.96*	0.99*
Phosphorus	0.93*	0.91*
Potassium	0.93*	0.90*

M1= *Suillus luteus* seedlings NM= Non-mycorrhizal *=significant at $p<0.01$ level

6. Heavy Metal Tolerance in Ectomycorrhizae

It has been observed that the ectomycorrhizal fungi accumulate heavy metals from highly diluted concentration in soils. The high concentration of Zn, Pb, Cd, Cu, Ni and Al limited the mycelium production and its colonization into deeper cortex. The cytological interactions in the fungal hyphae with increase in heavy metals concentration also suggested an accumulation of metals in the fungal mantle, however, this may also have occurred if mycorrhizal fungi were more sensitive to these metals than the host. Cellular and structural changes in root of pine seedlings treated with higher dose of heavy metals could be due to breakdown of cytoplasmic structure, disruption of cell walls and cortical cells and accumulation of dense compounds. Cytological changes observed in metal-stressed pine roots were similar to those that occurred in seedlings in which nutrient deficiency was induced deliberately (Fink 1991). It is clear that, the non-mycorrhizal seedlings were more sensitive to the toxic effects of heavy metals. At high concentration of added metals, however, inhibition of root extension and hence complexing ability occurred, in these, circumstances the presence of an ectomycorrhizal fungal symbiont could be of considerable importance.

Elaborate hyphal coils occupy most of the cortical cells of pine roots with lower level of metals through which absorptions occurs and these would provide a greatly increased surface for retention of metal ions. In our study, Cd and Pb were most toxic to ectomycorrhizal fungi and development of these fungi was inhibited in the moderately contaminated soil. However, a saturation point was reached in the mycelium. The uptake of metals in the plants also increased, the rapid mycelia turnover which was associated with metal exclusion and hence avoidance of toxicity. This hypothesis was supported by Colpaert and Assche (1992) for Ni and Zn tolerance in plants. Similar information about responses of various mycorrhizal fungi to toxic metals was reviewed by Hartley *et al.* (1997). The effectiveness in providing resistance to metal toxicity is dependent on different ectomycorrhizal fungal species. It is believed that the translocation and accumulation of toxic metals may be concentrated in the tissues. However, information in the heavy metals toxicity to the mycorrhizal systems is still scarce. Earlier studies (Ashida 1965) reported that fungi tend to survive heavy metals stress either by adaptation or mutation. Therefore, we can conclude that the heavy metal tolerance is a regular phenomenon which is exhibited by a number of ectomycorrhizal fungi.

References

Ajungla T, Sharma GD (2003) Accumulation of Cu and Ni in ectomycorrhizae of *Pinus kesiya* (Royle ex Gordon). Naga. Univ. Res. J. 1, 14-17.

Ajungla T, Imliyanger T (2010) Effects of ectomycorrhizal fungi on the growth performance of *Pinus patula* (Schiede ex Schlecht & cham). Environ. Biol. Conserv. 15, 29-31.

Arenla S, Ajungla T (2014) Effect of soil moisture and soil pH on ectomycorrhizal colonization *Dipterocarpus retusus* Blume seedlings. Int. J. Life Sci. Biotech. Pharmacol. 3(2), 1-6.

Ashida J, Higashi N, Kikuchi T (1965) Heavy metal tolerance in plant: Evolutionary Aspects. Protoplasma. 57, 27-32.

Baar J, Higashi N & W A, Kuyper TW (1994) Spacial distribution of *Laccaria bicolar* reflected by sporocarps after removal of litter and humus layers in a *Pinus sylvestries* forest. Mycol. Res. 98, 726-728.

Bardgett RD, Freeman C, Ostle NJ (2008) Microbial contribution to climate change through carbon cycle feedbacks. ISME J. 2, 805-814.

Bendangmenla, Ajungla T, Jamir NS (2005) Effeciency of ectomycorrhizal fungi on the growth of pine seedlings in the degraded forest. Naga. Univ. Res. J. 3, 80-83.

Bendangmenla (2007) Biology and application of mycorrhiza in reclamation of wastelands in Nagaland. Ph. D. Thesis, Nagaland University.

Bonfante P, Andrea G (2010) Mechanism underlying plant-fungus interactions in mycorrhizal symbiosis. Nat. Commun. 1 (48), 1-11.

Dietz KJ, Hartuns W (1996) The leaf epidermis: its ecophysiological significance. Prog. Bot. 57, 32-53.

Dixon J (1983) Phosphatase production by mycorrhizal fungi. Plant Soil 71, 455-462.

Colpaert JV, Van Assche JA (1992) The effect of Cd and Cd-Zn interactions on the axenic growth of ectomycorrhizal fungi. Plant Soil 145, 237-243.

Finlay RD, Ek H, Odham G, Soderstrom D (1989) Uptake, translocation and assimilation of nitrogen from 15N-labelled ammonium and nitrate sources by intact ectomycorrhizal systems of *Fogus sylvatica* infected with *Paxillus involutus.* New Phytol. 113, 47-55.

Fink S (1991) Structural changes of conifers needles due to Mg and K deficiency. Fertil. Res. 27, 23-27.

Gadd GM (1993) Interaction of fungi with toxic metals. New Phytol. 124, 25-60.

Harley JL, Smith SE (1983) Mycorrhizal Symbiosis. Acad. Press. London.

Hartley J, Cairney JWG, Meharg AA (1997) Do ectomycorrhizal fungi exhibit adaptive tolerance to potentially toxic metals in environment. Plant Soil 189, 303-319.

Heinrich PA, Mulligan DR, Patrick JW (1989) The effect of ectomycorrhizae on the phosphorus and dry weight acquisition of *Eucalyptus* seedlings. Plant Soil 109,147-149.

Imliyanger T, Ajungla T (2014) Effect of soil organic carbon on the growth of *Pinus patula inoculated* with *Boletus edulis* and *Russula* sp. Naga. Univ. Res J. 7, 313-320.

Ingestad T, Arveby AS, Kahr M (1986) The influence of ectomycorrhizae on nitrogen nutrition and growth in *Pinus sylvestris* seedlings. Physiologia Planta 68, 575-582.

Jentschke G, Winter S, Godbold DL (1999) Ectomycorrhizas and cadmium toxicity in Norway spruce seedlings. Tree Physiol. 19, 23-30.

Jha BN, Sharma GD, Mishra RR (1990) Effect of pH on the growth of ectomycorrhizal fungi in vitro. In: Current Trends in Mycorrhizal Research (Ed) Jalali BL and Chand H, Haryana Agrl Univ Press, Hissar, India, pp 66-68.

Joshi SR (1993) Ecologcal studies on the effect of air pollution on the leaf surface micro-organism and litter decompositions. Ph.D Thesis, NEHU, Shillong, Meghalaya.

Malajezuk N, Redell P, Brundrett M (1994) Role of ectomycorrhizal fungi in mine site reclamation. In: Mycorrhizae and Plant Health (Ed) Pfleger FL and Linderman RG, The American phytopathological society, St. Paul, MN.

Moser AM, Peterson CA, D'Allura AJ, Southworth D (2005) Comparasion of ectomycorrhizas of *Quercus garryana* (Fagaceae) on Serpentine and non-serpentine soils in Southwestern Oregon. Am. J. Bot. 92(2), 224-230.

Naidu R, Harder RD (1998) Effect of different organic ligands on cadmium absorption by the extractability from soils. Soil Sci. Soc. Am. J. 62(3), 644-650.

Natarajan K (2005) Biodiversity of ectomycorrhizal fungi in south Indian forests. In: The fungi: Diversity Conservation in India (Ed) Dargan JS, Atri NS and Dhingra GS, Beshin Singh Mahendra Pal Singh Press, Dehra Dun, India, pp 205-211.

Read DJ (1992) The mycorrhizae in mycelium. In: Mycorrhizal Functioning: an integrative Plant-fungal process (Ed) Allen MF, Routlege, Chapman and Hall, New York, pp 102-133.

Rousseau JVD, Sylvia DM, Fox AJ (1994) Contribution of ectomycorrhizae to the potential nutrient absorbing surface of pine. New Phytol. 128, 638-644.

Schaedle M, Thorton FC, Raynal DJ, Tepper HB (1989) Response of tree seedlings to aluminium. Tree Physiol. 5, 337-356.

Sharma SK, Sharma GD, Mishra RR (1986) Status of mycorrhizae in sub tropical forest ecosystems of Meghalaya. Acta Bot. Indica. 14, 87-92

Symeonidis L, Karataglis S (1992) Interactive effects of Cd, Pb and Zn on root growth of two metal tolerant genotypes of *Holcus lanatus* L. Biometals 5, 173-178.

Takacs EA (1967) Produccion de cultivos puros de hongos micorrizogemos en el investigaciones agropeouarias, Castelar. Idia, Suppl. For. 4, 83-87.

Tan K, Keltjens WG (1990a) Interactions between aluminium and phosphorus in sorghum plants. I. Studies with the Al- sensitive genotype- TAM428. Plant Soil 124, 25-23.

Tinker PB (1981) Levels, distribution and chemical; forms of trace elements in food plants. Physiol. Trans. Roy. Soc. 294, 41-55.

Thakur PS (1990) Studies on mycorrhizae of some conifers of Himachal Pradesh. M. Phil. Dissertation, HPU, Shimla.

Thompson BD, Grove TS, Malajczuk NS, Hardy GE (1994) The effectiveness of ectomycorrhizal fungi in increasing the growth of *Eucalyptus globulus* Labill. in relation to root colonization and hyphal development in soil. New Phytol 126, 517-524.

Watson GW, Vonder Heide-Sparvka KG, Howe VK, Von Heide-Sparvka HG (1990) Ecological significance of endo-ectomycorrhizae in oak subgenus *Erythrobalanus,* Aebori. J. 14,107-116.

Wilkens DA (1991) Der Einflub Von cadmium auf das waschstum Von Fitcher am Beipiel der mykorrhizapilze *Laccaria laccata* and *Paxillus involutus*. M. Sc Thesis. University of Gottingen, Germany.

3

Biosorption of Heavy Metals by Microbes and Agro Wastes: An Ecofriendly Inexpensive Approach for Remediation of Heavy Metal Polluted Environment

Savitha, J. and Sriharsha, D.V.

Abstract

Contamination of environment with heavy metals has become a serious global problem due to its threat to animal and human health. Removal of this non-biodegradable, hazardous heavy metals through bio-materials has received paramount interest and intense attention in the last few decades as bio-materials play crucial role in remediating heavy metal polluted environment through a process called "biosorption". The term biosorption, therefore, has become a popular theme in recent years due to its relatively simple, inexpensive, eco-friendly with rapid mechanism of removing heavy metals from polluted environment. There have been burgeoning studies around the globe for alleviating the problem of heavy metal pollution by selecting and developing an effective bio-sorbent with minimum treatment process for field application. This review aims to provide vast information on types of biosorbent available, their exploitation for better efficiency and their economic attractiveness for the restoration of heavy metal polluted environment.

Keywords: Agro wastes, Microbes, Biosorption Heavy metals, Polluted environment Restoration

1. Introduction

The environmental pollution is caused by various types of pollutants in water, air and soil. One of the major pollutants of the soil and water environment is heavy metals. The term heavy metal refers to any metallic chemical element that has a relatively high density (greater than $5.0g/cm^3$) and is toxic at low concentrations. Unlike most organic pollutants, heavy metals occur naturally in rock-forming and ore minerals and therefore a normal concentration of each metal in soil, sediments, water and living organisms were observed. With rapid industrialization, urbanization and new technological developments, anthropogenic sources of heavy metal pollution have increased, which ultimately lead to increase in concentration from the basal level to levels that pose threat to living organisms. Globally, more than 10 million sites have been reported to be polluted with different kinds of heavy metals, >50% of the sites represent contamination with heavy metals and/or metalloids (such as, arsenic) (Table 1). In addition to the known health hazardous effects, heavy metal pollution has an economic impact which has been estimated to be in excess of US $ 10 billion per year.

Table 1: Soil pollution in the world (EEA, 2007; EPAA, 2012; EPMC, 2014; USEPA, 2014, GIZ, 2012)

Country	Number of polluted sites	Per cent of heavy metal(loid)s pollution
Global	>10000000	>50
USA	>100000	>70
European Union	>80000	37
Australia	>50000	>60
China	1.0 million km²	>80
India	>350000	-

Of all the heavy metals known to pollute the environment, the primary metals considered to be highly toxic are chromium, lead, mercury, arsenic, nickel cadmium, and barium. Various industrial sources contributing to these primary heavy metals are outlined in Table 2.

Table 2: Industrial sources of heavy metals (CPCB, 2011)

Metal	Industry
Chromium (Cr)	Mining, industrial coolants, chromium salts manufacturing, leather tanning, magnetic tapes, cement, paper, rubber, pigments for paints, wood preservatives.
Lead (Pb)	Certain cosmetics, toys, lead-glazed pottery, lead acid batteries, paints, E-waste, Smelting operations, coal-based thermal.
Mercury (Hg)	Dental fillings, switches, light bulbs and batteries. Coal-burning power plants. Mercury in soil and water is converted by microorganisms to methyl mercury, a bio accumulating toxin. Power plants, ceramics, bangle industry, Chlor-alkali plants, thermal power plants, fluorescent lamps, hospital wastes (damaged thermometers, barometers, sphygmomanometers), electrical appliances etc.
Arsenic (As)	Geogenic/natural processes, smelting operations, thermal power plants, fuel burning, wood preservatives, pesticides, paints, dyes, some fertilizers.
Nickel (Ni)	Spent catalyst, sulphuric acid plant.
Cadmium (Cd)	Smelting operations, thermal power plants, battery industry. Zinc smelting, waste batteries, e-waste, paint sludge, incinerations, fuel, pigments, metal coating and plastics.
Barium (Ba)	Fluorescent lamps, diagnostics medicines, fireworks, paints, bricks, ceramics, glass and rubber.

While many of the heavy metals are needed by plants and animals at very low levels (micronutrient level), higher concentrations are known to produce toxic effects. Moreover, the heavy metals enter in to food chains from polluted soil, water and air and consequently cause food contamination, thus posing a threat to human and animal health.

1.1 Toxicity of Heavy Metals

Arsenic: Inorganic arsenic is a known carcinogen and can cause cancer of the skin, lungs, liver and bladder. Lower level exposure can cause nausea, vomiting, decreased production of red and white blood cells, abnormal heart rhythm, damage to blood vessels and a sensation of pins and needles in hands and feet. Ingestion of very high levels results in death. Long-term low level exposure cause darkening of the skin and gives an appearance of small 'corns' or 'warts' on the palms and scales (Martin and Griswold 2009).

Regulatory limit: Environmental protection Agency (EPA); 0.01ppm in drinking water.

Barium: It is not carcinogenic but short term exposure cause vomiting, abdominal cramps, diarrhea, difficulties in breathing, increased/decreased blood pressure, numbness around the face and muscle weakness. Large amounts of barium intake cause high blood pressure, changes in heart rhythm, paralysis and ultimately death.

Regulatory limit: EPA; 20ppm in drinking water.

Cadmium: It is a known carcinogen. Smokers get exposed to higher cadmium levels than non-smokers. Severe damage to the lungs may occur through breathing high levels of cadmium. Ingesting very high levels severely irritates the stomach, leading to vomiting and diarrhea. Long-term exposure to lower levels leads to a buildup in the kidneys and possible kidney disease, lung damage and fragile bones.

Regulatory limits: EPA; 5ppb or 1.005 ppm in drinking water.

FDA (Food and Drug Administration): Concentration in bottled drinking water should not exceed 0.005ppm (5 ppb).

Chromium: Chromium (VI) compounds are human carcinogenic, whereas chromium (III) is an essential nutrient. Breathing high levels can cause irritation to the lining of the nose; nose ulcers; breathing problems; cough and wheezing. Skin contact can cause skin ulcers. Allergic reactions consisting of severe redness and swelling of the skin. Long term exposure causes damage to liver, kidney and nerve tissues and skin irritation.

Regulatory limits: EPA: 0.1 ppm in drinking water.

FDA: Should not exceed 1 mg per liter (1 ppm) in bottled water.

Lead: It is a carcinogen. It affects every organ and system in the body. Long term exposure results in the decreased performance of the nervous system; weakness in fingers, wrists or ankles. Exposure to high levels severely damage the brain, kidneys and ultimately cause death. In pregnant women it causes miscarriage. In men it damages the organs responsible for sperm production.

Regulatory limits: EPA; 15ppb in drinking water.

Mercury: The nervous system is very sensitive to all forms of mercury (*i.e.*) mercury chloride and methyl mercury. Exposure to high levels permanently damage the brain, kidneys and developing fetuses. Effects on brain functioning results in irritability, shyness, tremors, changes in vision or hearing and memory loss. Short-term exposure to high levels of metallic mercury vapors cause lung damage, nausea, vomiting and increased blood pressure, heart rate, skin rashes and eye irritation.

Regulatory limits: EPA; 2 ppb in drinking water.

FDA: 1 part of methyl mercury in a million parts of seafood.

Nickel: Higher concentration of nickel causes cancer of lungs, nose and bone. Acute poisoning of Ni (II) causes headache, dizziness, nausea and vomiting, chest pain, tightness of the chest, dry cough and shortness of breath, rapid respiration, cyanosis and extreme weakness.

Regulatory limits: EPA; 20ppm in drinking water.

Unlike organic compounds, heavy metals are persistent in the environment and cannot be degraded through biological, physical or chemical means to an innocuous byproduct. Thus, metals are neither thermally decomposable nor microbiologically degradable and are thus difficult to remove from the environment. However, over the past few decades, several novel technologies have been devised for the treatment and removal of heavy metals from aquatic and terrestrial environment. Due to the mobility of heavy metals in natural water ecosystems and their toxicity to higher life forms, heavy metal ions in surface and groundwater supplies have been prioritized as major inorganic contaminants in the environment. Even if they are present in dilute, undetectable quantities in water bodies imply natural processes such as biomagnifications, accumulation and concentrations may become elevated to such an extent that they begin exhibiting toxic characteristics.

The commonly used procedures for removing metal ions from polluted environment either in soil or in aquatic environment include i) Chemical precipitation ii) Lime coagulation iii) Ion exchange (Reverse osmosis) and iv) Solvent extraction (Rich and Cherry 1987). While these methods have been known to remove heavy metal ions, they are not free from limitations or disadvantages which include high reagent requirements, generation of toxic sludge which is often difficult to dewater, caution in its disposal, high operating cost and become less effective as metal ion concentrations fall to the low parts per million ranges (Brady and Tobin 1995).

1.2 Biosorption

In view of the several life threatening diseases arising from the heavy metal pollution, it is always a challenge for the scientists to devise inexpensive methods for metal removal from the environment. Biosorption may offer part of a solution to the problem. Biosorption is a process in which solids of natural origin are employed for binding heavy metals. It is a promising alternative method to treat industrial effluents, primarily because of its low cost and high metal binding capacity. Hence, biosorption is considered to be one of the most modern approaches in environmental biotechnology (Volesky 2007). The major advantages of biosorption over conventional treatment methods include, 1) Low cost, 2) High efficiency 3) Minimization of chemical and or biological sludge 4)

No additional nutrient requirement 5) Regeneration of biosorbent and 6) Possibility of metal recovery.

Biosorption can be divided into: 1) metabolism dependent 2) non-metabolism dependent. Metal ion binding to viable cells (metabolism dependent) can occur either by surface adsorption or intracellular accumulation (bio-accumulation) whereby the metal ions are transported across the cell membrane, whereas in the case of dead biomass (non-metabolism dependent), metal uptake is by physico-chemical interaction between the metal and the functional groups present on the microbial cell surface.

A wide variety of active and inactive organisms have been employed as biosorbents to sequester heavy metals. Biosorbents such as algae, fungi, bacteria and agro waste materials are examples of biomass tested for metal biosorption with very encouraging results (Volesky and Holen, 1995; Davis *et al.* 2003; Sud *et al.* 2008). The uptake of heavy metals by biomass can in some cases reach up to 50% of the biomass dry weight. New biosorbent can be manipulated for better efficiency and multiple re-use to increase their economic attractiveness.

2. Mechanism of Biosorption

The biosorption process involves a solid phase (Sorbent or biosorbent; biological material) and a liquid phase (solvent, normally water) containing a dissolved species to be sorbed (Sorbate, metal ions). Due to higher affinity of the sorbent for the sorbate species, the latter is attracted and removed by different mechanisms. The process continues till equilibrium is established between the amount of solid-bound sorbate species and its portion remaining in the solution. The degree of sorbent affinity for the sorbate determines its distribution between the solid and liquid phases. Biosorption of metal ions onto biomass involves a combination of several metal-binding mechanisms, including physical adsorption, ion exchange, complexation and precipitation (Ahalya *et al.* 2003).

2.1 Physical Adsorption

Van der Waal's forces (electrostatic interaction) take place between metal ions in the solution and the cell wall of the biomass. Surface chemistry of the biosorbent plays a key role since adsorption is favoured by the presence of oxygen containing functional groups which can be very different according to the nature of the biosorbent. Carboxylic, phosphate, sulphate, amino, amide and hydroxyl groups are essential for the adsorption of heavy metals due to their chelating attributes. In particular, acidic groups (such as carboxyl) for pH values greater pKa are mainly in dissociated form and can exchange H^+ with metal in solution. At pH lower than pKa values, complexation phenomenon can also occur, especially for the carboxylic groups. The reaction can occur due to ion exchange reaction of the metal cation with the hydrogen ion previously attached.

2.2 Complexation

Metal ion removal from an aqueous solution takes place by complex formation of metals on the cell surface after interaction between metal ions and active groups. Metal ions can be biosorbed or complexed by carboxyl groups found in microbial polysaccharides or other polymers.

2.3 Ion Exchange

Polysaccharides that exist on cell walls of biomass consist of counter ions, such as K^+, Na^+, Ca^{2+} and Mg^{2+}. These ions, exchange with metal ions, resulting in the metal ion uptake (Kuyucak and Volesky 1988; Muraleedharan and Venkobachar 1990).

2.4 Precipitation

This mechanism is dependent on or independent of cellular metabolism.

Metal ion removal from aqueous solutions is often associated with the active defense system of biomass. This active system produces compounds that favor the precipitation process (Veglio *et al.* 1997).

3. Algae as Biosorbents

Algae inhabiting marine environment is identified as a promising biosorbent due to their high uptake capacities and their ready availability in many parts of the world's oceans. It is approximately estimated that over 3 million tons of seaweeds are harvested globally for algal products such as agar, alginate and carrageenan. Among the marine algae, mainly, the red algae and the brown algae are exploited for biosorption than the green algae. The cell wall of brown algae generally contain three components such as cellulose, alginic acid, polymers like mannuronic and glucuronic acid complexed with light metals such as sodium, potassium, magnesium and calcium and polysaccharides. Alginic acid and some sulfated polysaccharides, fucoidan, are important components of the cell walls of brown algae. Among which, aliginates and sulfate are the predominant active groups (Sheng *et al.* 2004). Green algae have cellulose and a high content of proteins bonded to the polysaccharides. These compounds contain functional groups such as amino, carboxyl, sulfate and hydroxyl, which play vital role in biosorption. In red algae, sulfated polysaccharides made of galactans play important role in biosorption. Table-3 summarizes the type of algal biomass used in the biosorption of heavy metals reported recently. In general, the ion-exchange mechanism has been found to play a dominant role for the biosorbents that originate from sea water environment. The ion-exchange occurs between heavy metals and light metals (mainly Ca^{2+} and Mg^{2+} as mono-valent Na^+ and K^+).

Table 3: List of algae used as biosorbents

Algae	Metals	Temp. (°C)	Percentage of metal removed	Reference
Brown algae	Pb Cu Cr Cd Zn Ni	37	Not clear	Jinsong *et al.* (2014)
Red and Green algae	Pb Cu Cr Cd Zn	37	Not clear	Jinsong *et al.* (2014)
Gracilaria corticata varca tecala(Red algae)	Cr Hg Pb Cd	37	92 99 99 99	Narayanaswamy *et al.*(2013)
Gracilaria lithothila (Red algae)	Cr Hg Pb Cd	37	73 98 41 82	Narayanaswamy *et al.*(2013)
Anabaena polytricha (Cyanobacteria)	Cd Pb	Not clear	11 12	Adbel *et al.*(2013)
Ulva lactuca(Green algae)	Cd	37	85	Lupea *et al.* (2012)

4. Bacteria and Fungi as Biosorbents

When compared to algae, bacteria and fungi in general, have more complex and efficient way of adsorbing or accumulating the heavy metals. Cell wall of bacteria are *principally composed of peptidoglycans, which consist of linear chains of the disaccharide N-acetyl glucosamine-β 1*, 4-N-acetylmuramic acid with peptide chains. Peptidoglycan carboxyl groups are the main metal binding sites for Gram positive bacterial cell walls while phosphate group for Gram negative microbes (Wu *et al.* 2010) serves as Metal biding site. Fungal cell walls are rich in polysaccharides and glycoproteins such as carboxyl, sulphate, phosphate and amino groups which may act individually or synergistically to bind cations (Tobin *et al.* 1990). While carboxyl, phosphate and other moieties were identified as key binding sites in early studies (Volesky and Holen 1995; Fourest *et al.* 1996) chitin and chitosan were identified to be important binding sites in recent days (Franco *et al.* 2004; Wang and Chen 2014). This type of biosorption, that is, non-metabolism dependent is relatively rapid and reversible (Kuyucak and Volesky 1988). Nonviable biomass under certain conditions has reported to adsorb metal ions in higher quantities than viable biomass (Brady and Tobin 1995).

Upon metal exposures, metal resistant microbes synthesize intercellular or extracellular metal binding molecules. Fungi and bacteria synthesize intercellular metal binding molecules called metallothioneines (MTs). These are low molecular weight, unique cysteine rich metal binding proteins. They are synthesized *do novo* upon exposure to various metal ions. Fungi synthesize two types of MTs, class II type of MT and phytochelatin (PC, class III type of MT). PC derivatives are not primary gene products, but are glutathione-related peptides (Ow 1996; Cobbett 2000). In contrast to vertebrate MTs, which bind different metal ions, fungal MTs were reported to contain exclusive copper with a cadmium ions. An Ascomycete *Neurospora crassa* accumulate copper with a concomitant synthesis of copper binding protein, consisting of single polypeptide chain of 25 aminoacid residues with 7 cysteine residues lacking aromatic amino acid and binding exclusively copper. From a mushroom, *Agaricus bisporus*, a copper-binding protein was isolated and the chemical characterization of the protein was reported. Both copper-binding proteins from *N.crassa* and *A.bisporus* were reported to belong to class II of metallothioneines (Gadd 2007). *Trichoderma harzianum* and *Fusarium phyllophilum* accumulated cysteine and/or glutathione and showed higher OASTL gene expression in the presence of high concentration of cadmium (Raspanti *et al.* 2009). Melanins, the dark pigments located in the fungal cell wall, also reduce the toxic effect of heavy metals due to the presence of various groups with high affinity to metal ions (Ledin 2000).

The most well-known extracellular metal-binding compounds in bacteria are siderophores, which are low molecular weight ligands (500-1000) possessing a high affinity for iron III (Neilands 1981). They scavenge for iron (III) and complex and solubilize it, and thus making it available to the organisms. Although primarily produced as a means of obtaining iron, siderophores are also able to bind other metals such as magnesium, manganese, chromium (III), gallium (III) and radionucleides such as plutonium (IV) (Brich and Bachofen 1990). Organic acids such as, citric acid and oxalic acids produced by fungi are reported to immobilize soluble metal ions, as insoluble one (Gadd 1999; Gadd and Sayer 2000). *Aspergillus* form oxalate crystals on agar amended with a wide range of metal compounds including insoluble phosphate of cadmium, cobalt, copper, manganese, strontium and zinc (Sayer and Gadd 1997). The formation of oxalates containing potentially toxic metal cations provides a mechanism whereby oxalate-producing fungi can tolerate environment containing high concentrations of toxic metals. Presence of copper oxalate was observed around hyphae growing on wood treated with copper as preservatives (Sutter *et al.* 1983). Alcohol, humic and fulvic acids also play role in extracellular metal sequestration.

Biosurfactants are surface-active microbial products that are produced extracellularly or as part of the cell membrane by virtue of which surface tension of the liquid is decreased so as to make metals more available to biosorbent for remediation. Production of rhamnolipid by *Pseudomonas* spp. (*P. aeruginosa* ATCC 9027), cyclic lipopeptide surfactin by *Bacillus subtilis* and biodispersan by *Bacillus sp.* (IAF 343) are example of growth-associated biosurfactant production (Asci *et al.* 2010). *Aureobasidium pullulans accumulated Pb (II) by excreting extracellular polymeric substances EPS such as pullulan and β-glucan (Suh et al.* 1999). Tables 4 and 4a give the list of bacteria and fungi explored as biosorbents for heavy metal removal in recent years.

Table 4: List of bacteria with biosorbent capacity for heavy metal removal

Bacteria	Metals	Temp (^{0}C)	Percentage of metal removed	Reference
Bacillus thuringiensis	Ni	30	94	Oves *et al.*(2013)
	Cu		91	
	Cd		87	
Bacillus thuringiensis	Pb	30	77	Babu *et al.*(2013)
	Zn		64	
	As		34	
	Cd		9	
	Cu		8	
	Ni		8	
Paenibacillus validus	Zn	35	27	Rawat *et al.*(2012)
	Ni		16	
	Cd		16	
	Cr		15	
	Co		9	
	Pb		7	
Pseudomonas sp.	Cd	30	2	Huang *et al.*(2012)
	Pb		7	

Table 4a: List of fungi used as biosorbents

Fungi	Metals	Temp (^{0}C)	Percentage of metal removed	Reference
Aspergillus flavus	Cu	30	47	Iram *et al.* (2015)
Aspergillus niger	Pb	30	88	Iram *et al.* (2015)
Penicillium chrysogenum	Cr	30	20	Jayanthi *et al.* (2014)
Aspergillus niger	Cr	30	10	Jayanthi *et al.* (2014)
Pleurotus eryngii	Hf	30	7	Amin *et al.*(2015)
Penicillium coffeae	As	30	67	Bhargavi *et al.* (2014)
Aspergillus fumigatus	Pb	30	76	Shazia *et al.* (2013)
	Cu		69	
	Cr		40	
Trichoderma harzianum	Cr	30	65	Shoaib *et al.*(2013)
Mucor racemosus	Cu	30	30	Morsy *et al.*(2013)
	Zn		28	
	Pb		11	
Trametes versicolor	Ni	30	21	Subbaiah *et al.*(2013)
Heterogenous culture	Pb	30	37	Sulaymon *et al.*
(Bacteria, yeast, protozoa)	Cr		52	(2013)
	Cd		31	
Pleurotus mutilus	U	30	63	Mezaguer *et al.* (2013)
Lactarius salmonicolor	Ni	30	11	Akar *et al.* (2013)
Mucor rouxii	Hg	30	95	Juarez *et al.* (2012)
Auricularia polytricha	Cd	30	6	Huang *et al.*(2012)
	Cu		7	
	Pb		22	
Aspergillus cristatus	Cd	30	82	Hassan *et al.* (2012)
Aspergillus sp.*Penicillium* sp.	Cu	30	46	Hemambika *et al.* (2011)
Cephalosporium sp.	Cd		95	
	Pb		70	
Aspergillus terreus	Cu	30	34	Varshney *et al.* (2011)
Pleurotus ostreatus	Cu	30	Not clear	Javaid *et al.*(2011)
	Ni			
	Zn			
	Cr			
Aspergillus niger	Cu	30	32	Javaid *et al.* (2011)
	Ni			

5. Agricultural Wastes as Biosorbents

Agro-waste is defined as waste which is produced from various agricultural activities. These agro-wastes include manures, bedding, plant stalks, hulls, leaves, and vegetable matter. A great deal of interest has been focused within the past decade in search of agricultural wastes materials as biosorbent for the removal of heavy metals from polluted environments. Selection of potential biosorbent from the large pool of agro-waste add value to the waste materials in general, and eventually reduce the agro-wastes management problems, in particular. Agro-materials are composed of lignin and cellulose as major constituents and also include other polar functional groups of lignin, which includes alcohols, aldehydes, ketones, carboxylic, phenolic and ether (other) groups (Hossain *et al.* 2012). These groups have the ability to bind heavy metals by donation of an electron pair to form complexes with the metal ions in solution (Demirbas, 2008). Due to varied laboratory conditions (*eg.* pH, temperature, adsorbent dose, particle size), it is not easy to constructively come to conclusion which specific biosorbent would be suitable for the specific metal ion. Moreover, the biosorption efficiency of the selected material (biosorbent) could be precisely determined only during large scale treatment of effluent or soils.

Table 5: Type of agro wastes as biosorbents of heavy metals

Agro waste materials	Metals	Temp (^{0}C)	Percentage of metal removed	Reference
Agave salmiana bagasse	Cd	25	8	Jimenez *et al.*(2014)
	Pb		14	
	Zn		36	
Rice husk	Ni	Not clear	43	Sharma *et al.* (2013)
Rice husk	U	Not clear	30	Kausar *et al.* (2013)
Cactus cladodes	Cd	Not clear	30	Barka *et al.* (2013)
	Pb		99	
Olive stone	Cd	Not clear	95	Alslaibi *et al.* (2013)
Sugarcane bagasse	Cr(VI)	Not clear	80	Ullah *et al.* (2013)
	Cr(III)		41	
Watermelon rind	Ni	Not clear	35	Lakshmipathy
	Co		23	*et al.* (2013)
Hemi cellulose from	Pb	23	58	Ayoub *et al.* (2013)
Pine wood	Cu		19	
Switch grass	Ni		27	
Coastal Bermuda grass				
Poultry waste	Cd	37	24	Lim *et al.* (2013)
(egg shell and chicken bones)	Pb		10	
Teff straw (Eragrostis tef)	Ni	Not clear	88	Desta (2013)
	Cd		82	
	Cu		81	
	Cr		74	
	Pb		68	
Sunflower waste carbon	Cd	Not clear	23	Jain *et al.* (2013)
Sugarcane bagasse	Hg	Not clear	97	Khoramzadeh *et al.*(2013)
Ficus carica	Cr (VI)	Not clear	19	Gupta *et al.* (2013)
Peganum hermala seeds	Pb	Not clear	95	Zamani *et al.* (2013)
	Zi		75	
	Cd		90	

(Contd.)

Agro waste materials	Metals	Temp (^{0}C)	Percentage of metal removed	Reference
Grape fruit peel	Cd	Not clear	42	Mostaedi *et al.*(2013)
	Ni		46	
Egg shell	Cd	Not clear	67	Lee *et al.*(2013)
	Pb		93	
Rapeseed residue	Cd	Not clear	34.01	Lee *et al.*(2013)
	Pb		46.1	
Birbira (Militia Ferruginea)	Mn	30	95	Mengistie *et al.*(2012)
Banana peel	Cu	30	88	Hossain *et al.* (2012)
Rice straw	Cd	Not clear	80	Ding *et al.* (2012)
Sugarcane bagasse	Ni	25	Not clear	Aloma *et al.*(2012)
Cashew nut shell	Cu	Not clear	40	Kumar *et al.*(2012)
	Cd		43	
	Zi		45	
	Ni		45	
Corn cob	Mn	25	15.74	Adeogun *et al.*(2011)
Pea nutshell biomass	Cu	Not clear	25.39	Krowaik *et al.*(2011)
	Cr		27.89	
Waste cottons buttons	Pb	30	Not clear	Anirudhan *et al.*(2011)
	Hg			
	Cu			
Orange peel	Pb	30	47	Feng *et al.*(2011)
	Cd		29	
	Ni		16	
Lemon peel	Pb	Not clear	15	Vargas *et al.*(2011)
Orange peel	Cu	Not clear	48	Vargas *et al.*(2011)
Banana peel	Cd	Not clear	57	Vargas *et al.*(2011)

6. Biosorption in Agricultural Soil

Soil is the major sink for heavy metals released into the environment by above mentioned anthropogenic activities (Table 2). Their total concentration in soils persists for a long time after their introduction. The presence of toxic metal in soil severely inhibits the biodegradation in organic contaminants also. Therefore, these concentration of soil pose risks and hazards to humans and the ecosystem through: direct ingestion or contact with contaminated soil, the food chain (soil-plant-human or soil-plant-animal-human), drinking of contaminated ground water, reduction in food quality via phytotoxicity, reduction in land usability for agricultural production causing food insecurity (Raymond *et al.* 2011). Therefore, adequate protection and restoration of soil ecosystems contaminated by heavy metals is always a major concern. Various methods of remediating metal polluted soils exists; they range from physical, chemical and biological methods. Most physical and chemical methods such as encapsulation, solidification, stabilization, electro kinetics, vitrification, vapour extraction, soil washing and flushing are not only expensive but also make the soil not suitable for plant growth. Biological approach (Bioremediation) on the other hand, encourages the establishment/ reestablishment of plants on polluted soil. The use of plants for remediation of heavy metal soil, generally called as phytoremediation is beyond the scope of this review.

Although several remediation protocols exist for successful removal of other pollutants such as hydrocarbons, pesticides etc., from soil, removal of heavy metals from contaminated land has its own limitation due to difficulty in the recovery of metals after treatment. However, the role of microbes in soil remediation is significant due to transformation of metals from their native oxidation state to another. Due to change in their oxidation state, heavy metals can be transformed to become either less toxic, easily volatilized, more water soluble (and thus can be removed through leaching), less water soluble (which allows them to precipitate and become easily removed from the environment) or less bioavailable (Garbisu and Alkorta 1997; 2003). *Bacillus subtilis, Pseudomonas putida* and *Enterobacter cloacae* have been successfully used for the reduction of hexavalent chromium (Cr (VI)) to the less toxic trivalent chromium (Wang *et al.* 1989; Garbisu *et al.* 1998). *B. subtilis* reduce the selenite to the less toxic elemental Se (Garbisu *et al.* 1997). Bacteria facilitates the extraction of metals from the soil by producing the Fe complexing molecules, the siderophores. Sulphate reducing bacteria, *Desulfovibrio desulfuricans*, converts sulphate to hydrogen sulphate which subsequently reacts with heavy metals such as Cd and Zn to form insoluble forms of these metal sulphides (White *et al.* 1998). A very significant *in situ* microbe assisted remediation is the reduction of soluble mercuric ions Hg (II) to volatile metallic mercury and Hg (O) carried out by mercury resistant bacteria (Hobman and Brown 1997). The reduced Hg (O) can easily volatilize out of the environment and subsequently gets diluted in the atmosphere (Lovely and Lloyd 2000).

Mycorrhizal fungi have been shown to employ different mechanisms for the remediation of heavy metal polluted soils (Joner and Leyval 2001) through enhanced phytoextraction through the accumulation of heavy metals in plants and phytostabilization through metal immobilization and a reduced metal concentration in plants. Other soil microbes helpful for heavy metal contaminated soils are the plant growth promoting rhizobacteria (PGPR) that are generally found in the rhizosphere. PGPR have been used in phytoremediation studies to reduce plant stress associated with heavy metal polluted soils (Kamnev and Van der Lelie 2000). Enhanced accumulation of cadmium and nickel by hyperaccumulaters (*Brassica juncea* and *Brassica napus*) has been observed when the plants were inoculated with *Bacillus* sp. (Sheng and Xia, 2006). Increased plant growth was observed due to reduction in the accumulation of Cd and Ni in the shoot and root tissues of plant inoculated with *Methylobacterium oryzae* and *Burkholderia* spp. (Vivas *et al.* 2006)

However, the combined use of mycorrhizal fungi and PGPR for alleviating heavy metals from cultivated soil depends on the species of microbe and plants involved and to some extent on the concentration of the heavy metal in soil (Chibulika and Obiora 2014).

7. Immobilization of Biomass for Better Biosorption

Due to the fragile nature of the microbial cells, problems are encountered during industrial processing of metal sorption. To overcome this problem, some form of immobilization within a strengthening matrix is devised to ensure stability and effective separation from processed solutions (Yin et al. 1999). Both natural and synthetic gelling agents have been employed in the production of biosorbent gel beads. Greater biosorbent stability is achieved by entrapment in synthetic polymer gels. Materials generally used are polyacrylamide, polysulfone, polyethylenimine (PEI) (Veglio & Beolchini 1997). Immobilization of Lentinus sajor-caju immobilized in carboxy methyl cellulose (CMC) increased biosorption of Cr (VI) ions (Arica & Bayramoglu 2005). An another basidiomycete Trametes versicolor immobilized on to CMC in live or dead state showed very effective biosorption capacities (Saglam et al. 2002). The biosorption of cadmium (II) also increased when the Lentinus sajor caju live or dead biomass was entrapped in to alginate gel via a liquid curing method in the presence of Ca (II) ions (Bayramoglu et al. 2002). Polyacrylamide (PAN-B) is considered as an effective biopolymer for immobilizing some dead fungal biomass capable of biosorbing zinc, copper and Nickel (Zouboulis et al. 2003). Cross- linking of Penicillium biomass with formaldehyde and Rhizopus biomass with bis (ethyl) sulfone increased metal uptake capacity by 10-30% compared with untreated cells as reported by Holan and Volesky (1995). Thus, among the various immobilization methods of biomass, polymetric matrix is considered as the most common method. However, these techniques have certain disadvantages: they are expensive; limit the rate of diffusion (Hu and Reeves 1997) or produce other waste products that require disposal (Wilde and Benemano 1993).

In recent days, natural, environmental friendly plant material, luffa sponge (the dried fruit of Luffa cylindrica; commonly called as sponge gourd) is used as an effective material for immobilizing microbes and treating for metal removal. It is abundant, inexpensive, rigid, non-toxic, chemically inert and highly porous agricultural waste. It is a lignocellulosic material composed mainly cellulose, hemicellulose and lignin (Rowell et al. 2000). The fibers are composed of 60% cellulose, 30% hemicellulose and 10% lignin (Mazali and Alvas 2005). The use of luffa sponge for the immobilization of algae, fungi and yeast cells for metal removal has been successfully demonstrated (Iqbal and Edyvean, 2004; Akhtar et al. 2008; li et al. 2008; Sun et al. 2010). The adsorption capacity of microbes

Table 6: List of agro wastes used for immobilizing microbes and treated as biosorbents

Natural Material used for immobilization	Metals	Concentration of metal solution (mg/L)	Percentage removal	Reference
Luffa sponge- *Phanerochaete chrysosporium*	Cr	10	92	Verma et al. (2014)
Luffa sponge- *Aspergillus cristatus*	Cd	1000	82	Hassan et al. (2014)
Luffa sponge- *Scenedesmus obliquus*	Cd	Not clear	Not clear	Chen et al. (2014)

Table 6a: List of chemicals used for immobilizing microbes and treated as biosorbents

Chemical used for immobilization	Metals	Conc. of metal solution (mg/L)	Percentage removal	Reference
Sodium alginate beads-*Bacillus megaterium*	Pb	1000	89	Sati et al. (2014)
Sodium alginate beads-*Mucor racemosus*	Zn	100	98	Morsy et al. (2013)
	Cu		80	
	Pb		96	
Calcium alginate beads-*Trametes versicolor*	Cu	1000	66	Gochev et al. (2012)
Sodium alginate beads-*Chlorella* sp.	Cu	Not clear	Not clear	Maznah et al. (2012)
	Zn			
Sodium alginate beads-*Thermophilic Bacillus* sp.	Cu	10	87	Daghistani (2012)
	Ni		72	
	Cr		86	
Sodium alginate beads-*Rhizopus* sp.	Cr	Not clear	Not clear	Liu et al. (2012)
Poly vinyl alcohol-*Rhizopus* sp.	Cr	Not clear	Not clear	Liu et al. (2012)
Poly acrylamide *Rhizopus* sp.	Cr	Not clear	Not clear	Liu et al. (2012)
Sodium alginate beads-*Aspergillus* sp.	Cu	100	64	Hemambika et al. (2011)
Sodium alginate beads-*Penicillium* sp.	Cd	100	97	Hemambika et al. (2011)
Sodium alginate beads-*Cephalosporium* sp.	Pb	100	73	Hemambika et al. (2011)

on luffa sponge is reported to be significantly higher than the free microbes. The authors studied the efficiency of fungi immobilized on luffa sponge for lead removal under laboratory condition. The adsorption capacity was significantly higher (55%) than the free mycelial form (unpublished data). Therefore, agro wastes not only serve as biosorbent but also serve as a suitable matrix for immobilization of fungi by offering mechanical strength, rigidity, increased surface area for effective metal removal. Tables 6 and 6a summarize the natural and chemical materials used for immobilizing microbes and used as biosorbents.

8. Factors Influencing Biosorption

8.1 pH of the Solution

This is the most important regulator of the biosorptive process. It influences the removal of cationic metals, but hinders the adsorption of anionic metals (Abdel-Ghani and Chaghaby 2014). At low pH, the biosorption is generally low, due to the fact that high concentration and high mobility of hydrogen ions, the H^+ ions are preferably adsorbed rather than the metal ions. At higher pH values, the lower number of H^+ and greater number of ligands with negative charges results in greater metal ions biosorption (Abdel-Ghani and Chaghaby 2014).

8.2 Temperature

Room temperature is usually desirable for the biosorption processes while higher temperature generally cause physical damage to the biosorbent (Park et al. 2010).

8.3 Initial Metal ion Concentration

Initial metal ion concentration plays a key role as a driving force to overcome the mass transfer resistance between the aqueous and solid phases. The biosorption capacity increases as the initial metal ion concentration increases, whereas the percentage of metal removal decreases by increasing the metal ion initial concentration. At low ions concentration in the solution, the ions interact with the binding sites and thus facilitates almost 100% adsorption whereas at higher concentrations, more ions are left un-adsorbed in the solution due to the saturation of the binding sites.

8.4 Biosorbent Dosage

Biomass provides binding sites for the sorption of metal ions, and therefore its concentration strongly affects the sorption of metal ions from the solution (Crini and Badot 2008). For a fixed metal initial concentration, increasing the adsorbent dose provides greater surface area and availability of more active sites, thus leading to increase in metal ion uptake. At low biomass dosage, the amount of ions adsorbed per unit adsorbent weight is high, but, however, the adsorption capacity is reduced when the biomass dosage increase as a result of lower adsorbate to binding site ratio where the ions are distributed onto larger amount of biomass binding sites. At higher dosage, the ions adsorbed are higher due to the availability of more empty binding sites as compared to lower dosage which has less binding sites to adsorb the same amount of metal ions in the adsorbate solution (Chong *et al.* 2013).

8.5 Interfering Pollutant Concentration

Increasing concentration of competing pollutants will usually reduce biosorptive removal of the target pollutant, as there will be a competition for binding sites. However, cation loading of biomass may enhance biosorption of another cation because of pH buffering effects (Fomina and Gadd 2014).

8.6 Agitation Speed

It enhances biosorptive removal rate of adsorptive pollutant by minimizing its mass transfer resistance, but care to be taken that the physical structure of biosorbent not get damaged (Park *et al.* 2010).

9. Modification of Biosorbents for Better Adsorption

The main purpose of exploring naturally available materials for biosorption is to develop cost effective biosorbent. But these biomaterials in their native form shows less sequestering capability of metals when compared to the one modified by simple prior treatment process. As the cell surface is involved mainly in sequestration of metal ions, the modification of cell wall by pretreatment greatly alter the binding of metal ions (Park *et al.* 2010). A vast number of methods are practiced in recent days to achieve this objective *i.e.* to increase the adsorption capacity of adsorbents. Figure 1. shows various techniques of modification for the preparation of better biosorbents. Generally, physical modification is very simple and cost effective, but however, reported to be less effective than chemical methods of modifications, which has been a preferable method due to its simplicity and efficiency (Vijayaragavan and Yen 2008).

Among the chemical methods, acid-washing is reported to enhance the capacity of biosorbents for cationic metals, but, same acids also can cause serious losses of the biosorbent itself due to structural damage and ultimately drop in the biosorptive capacity. Alkali treatment of fungal biomass has shown to increase significantly the metal uptake capacity. Amines, carboxyl, hydroxyl, sulfonate, phenol and phosphate groups are few binding sites which are present on the cell wall of living organisms. Modification of these functional groups is the grafting of long polymer chains onto the surface of raw biomass (O'connell *et al.* 2008) which can enhance the biosorptive capacity (Ngah and Hanafiah, 2008) of biosorbents (Fig. 1). Alteration of growth conditions of culture possibly brings about changes in composition of the cell surface, thereby influencing metal biosorption of the biomass (Mehta and Gaur 2005). Increase in biosorptive

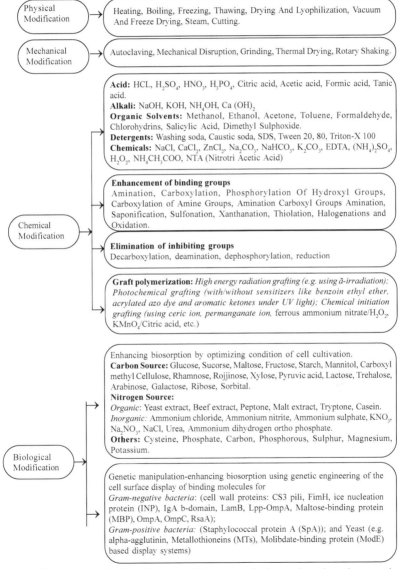

Fig. 1: Various methods of modification of biosorbents for improving adsorption capacity (Fomina and Gadd, 2014)

capacity, rigidity and pellet formation was also observed for few microbial biomass grown with clay minerals (Fomina and Gadd 2002). Genetic manipulation of microbes could also enhance biosorptive capacity (Kuroda and Ueda 2011; Li and Tao 2013), but use of sophisticated instruments and expensive reagents for this process makes it less popular. Ecological and environmental concerns and regulatory constraints are also some of the major-obstacles for testing of the genetically engineered microbes.

10. Recovery of Metals and Regeneration of Biomass by Desorption

For industrial application, the recovery of bound metal on the biosorbent and the subsequent recycling of the biosorbent should be explored. This is generally carried out with the desorption mechanism. Efficient desorption involves obtaining a concentrated eluate for metal recovery. Regeneration of the biosorbent with a minimum amount of residual metal is also desirable so that repeated adsorption-desorption cycles can be performed using the same material (Tsezos and Deutschmann 1990).

If the heavy metals are adsorbed onto the surface of a bisorbant through metabolism-independent biosorption, desorption becomes easier by simple non-destructive physical or chemical methods using chemical eluents. If biosorption is through metabolism-dependent process, *i.e.*, the bioaccumulation, the metals can be released only by destructive methods like dissolution with strong acids or alkalis (Gadd 1990; Gadd *et al.* 1998). The entire process of biosorption will be economical if a suitable eluent is used for the recovery of heavy metals from the loaded adsorbent. For a complete metal removal and recovery process, the adsorbents need to be used in a continuous sorption-desorption cycle (Mishra 2014).

Sodium carbonate (Nakajima and Sakaguchi 1993); potassium cyanide (Muzzarelli 1980); EDTA (Norris and Kelly, 1979); nitric acid (Cigdam *et al.* 2000); sulphuric and hydrochloric acid (Kuyucak and Volesky, 1988) are some of the common eluants used for the recovery of metal ions from the adsorbents after biosorption process. Although the efficiency of uranium uptake by *Penicillium* and *Actinomycetes* is reported to be the same, *Penicillium* sp. is considered as a better biosorbent as metals are eluted effectively (Sober *et al.* 1996). Therefore, selection of specific and effective eluent for specific organism is critically important.

The powdered biomass of *Rhizopus nigricans* was effective in Cr (VI) biosorption when immobilized with polymeric matrices (*viz*) calcium alginate, PVA, polyacrylamide, polyisoprene and polysulphone and desorption was successfully carried out with 0.01 N NaOH, $NaHCO_3$ and Na_2CO_3 (Bai and

Abraham 2003). The immobilized *Spirulina platensis* in calcium alginate matrix for Zn was effectively desorbed with 0.1M EDTA (Gaur and Dhankhar 2009). The agricultural waste, rice husk, efficiently adsorbed Ni from the aqueous solution at optimum pH of 6.0 with dilute HCl which seem to be effective eluent and 0.15M HCl was proved to be a suitable desorbing solution with desorption rate of about 80% (Bansal *et al.* 2009). Distilled water at pH 1, 2, 4 and 6 using dilute solution of HCl as desorbing solution was effective in removing Ni up to 81% from cinnamon plant leaf powder and 84% from Ni adsorbed jackfruit plant leaf powder. The desorption of Ni from immobilized *Escherichia coli* on bioweeds and biofilm has been studied with dilute HCl, pH of 6.0 as eluent by Saravanan *et al.* (2012). But, dilute HCl seem to be less effective for chromium (VI) desorption from *Tamarindus indica* seeds (TS) (Agarwal *et al.* 2006).

The desorption properties depend comparatively more on the adsorbent and choice of suitable eluent. It should be critically evaluated that the eluent should be metal-selective, economically feasible, with the high desorption rate.

11. Analytical Techniques to Determine Biosorption

A number of analytical techniques that have been used in recent days to study the biosorption are listed in Fig. 2. Although most of these sophisticated instruments are expensive and cost effective, it accurately determines the amount of metals adsorbed/accumulated by adsorbents.

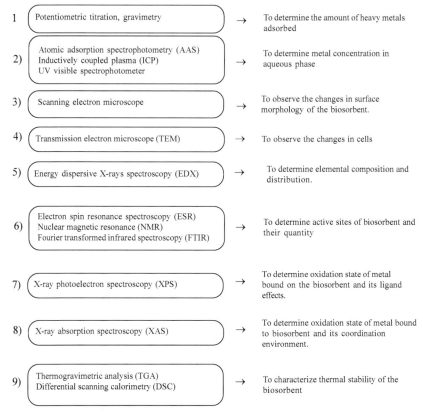

Fig. 2: Analytical techniques to determine biosorption (Park *et al.,* 2010)

12. Novel Approaches

12.1 Metal Nanoparticles

Few microbes besides having a strong biosorptive capacity also show tendency to reduce and precipitate heavy metals in their metallic form. This is one of the emerging technologies for heavy metal removal from environment, named as nanotechnology (Dixit *et al.* 2015). The transformation of the oxidation state of metals after sorption and the production of metal nanoparticles by microbes allows recovery of valued metals from waste streams in one step (Mishra and Malik 2013).

Few microbes have tremendous capability of changing the oxidation state of metals followed by deposition of metal oxides and zerovalent metals on the surface or into their cells. Manganese (Mn) and Iron (Fe) are known to be precipitated as oxides, while other precious metals such as Pd, Pt, Ag and Au gets deposited in a reduced or in the form of zerovalent nanometals. Due to this specific characteristics, such as, high specific surface areas and high catalytic reactivity, biogenic metals offer promising perspectives for the sorption.

The bacterium, *Desulfovibrio desulfuricans*, with the addition of pyruvate, formate or hydrogen gas as a hydrogen donor during its growth period, reduce Pd (II) to Pd (O) (Lloyd *et al.* 1998). Several bacteria (*Leptothrix discophora, Pseudomonas putida* MnBl, *Pseudomonas putida* GB-1, *Bacillus* sp. SG-1 *Pedomicrobium* sp. ACM) are known to oxidize soluble Mn, Mn (II), to insoluble Mn (IV) oxides (De schamphelaire *et al.* 2007). The biogenic Mn oxides produced by *Leptothrix discophora SS-1* adsorbed five times more Pb (II) than an abiotic Mn (IV) oxide and 500-5000 times more than pyrolusite oxides (Nelson *et al.* 2002). This high affinity for Pb (II) was also demonstrated for biogenic Mn oxides produced by *Pseudomonas putida* Mn B1 (Villalobes *et al.* 2005). The living cells of *Shewanella oneidensis* showed enhanced Au-removal when compared to heat-killed cells confirming the role of metabolic processes in the removal of Au (III). Moreover, it was observed, reduction to Au (O) was possible when an external electron donor was present (Mishra and Malik 2013).

A number of different genera of fungi have been investigated and proved to be extremely good candidates in the synthesis of metal and metal sulphide nanoparticles (Sastry *et al.* 2003). A species of *Verticillium* produce silver nanoparticles intracellularly while *Fusarium oxysporum* and *Aspergillus niger* have been shown to produce silver nanoparticles extracellularly (Ahmed *et al.* 2003; Duran *et al.* 2005; Sastry *et al.* 2003; Sadowski *et al.* 2008). The fungus (*Acremonium* KR 21-2), is known to oxidize soluble Mn, Mn (II), to insoluble Mn (IV) oxides (De schamphelaire *et al.* 2007)

Wide physiological diversity, small and uniform size, but with high specific surface area and controlled cultivability make microbial cells ideal producers of nanoparticles interesting for application in bioremediation. However, some of the challenges that need to be addressed are, long production time as compared to chemical production, optimization for accurate control of the particle size and authentic extraction protocols (Hennebel 2009).

12.2 Biochar

Biochar is an organic material which is currently being exploited for its potential in the management of heavy metal polluted soils. It is a carbon enriched porous material produced from pyrolysis of waste biomass residues from agricultural and forestry products (Xu *et al.* 2013). When biochar is produced from biomass, approximately 50% of the carbon that the plants absorbed as CO_2 from the atmosphere is "fixed" in the charcoal. It is similar in its appearance to charcoal and activated carbon. Biochar can stabilize heavy metals in the contaminated soils and thus improve the quality of contaminated soil (Ippolito *et al.* 2012) and thereby has a significant reduction in crop uptake of heavy metals. The "mineral" components such as phosphorous and carbonates in biochar play a significant role in stabilization of heavy metals in soils because these salts can precipitate with heavy metals and thus reduce their bioavailability (Cao *et al.* 2009). Moreover, most biochars are alkaline in nature and therefore have a liming effect which contributes to the reduction of the mobility of the heavy metals in contaminated soils (Sheng *et al.* 2005). Biochar application can also reduce the leaching of heavy metals through its effect of redox reactions of metals (Choppala *et al.* 2012). Thus, application of biochar for remediation of heavy metal contaminated environment provides a new solution to the problem of soil pollution.

12.3 Fixed Bed Column with Biosorbents

Batch experiments are the usual experimental set-up conducted at laboratory level to evaluate the ability of a material to adsorb metals. The data obtained through batch experiments are, in most cases, limited to laboratory scale and thus do not provide data and authentic methods, which can be precisely applied in industrial systems. Use of column, termed as 'Fixed Bed Column', provide data which can be applied for industrial purposes (Kumar *et al.* 2013). The design of an adsorption column depends on various important parameters such as flow rate, initial metal concentration and bed height i.e. mass of adsorbent (biomass). Understanding of adsorption characteristics, determination of break point time for adsorption operation and effective utilization of the column is possible by carrying out the mathematical modeling of fixed-bed adsorption column. Very few researchers have worked in thes area of continuous adsorption studies for heavy metal removal. Some of the important aspects such as axial dispersion, intra particle resistance and velocity variation along the bed length are taken care in this system which paves way for the industrial scale operations. Mostly the adsorbents used for packing the column are activated biosorbents. Table 8 outlines the biosorbents used in the fixed bed column in recent years.

Table 7: Types of biochar used as biosorbents for heavy metal removal

Feed Stock for Biochar production	Production Temperature (°C)	Heavy metal adsorbed	Reference
KMNO$_4$ treated Hickory wood	600	Pb, Cu, Cd	Wang et al. (2015)
Sludge	550	Cd, Pb, Zn	Liu et al. (2015)
Rice husk, organic fractions of solid wastes	Not clear	As, Cr (III), Cr (VI)	Diamadopoulos et al.(2014)
Conocarpus erectus tree wastes	400	Zn, Cd, Pb, Cu	Al-wabel et al. (2014)
Broiler Litter Manure	350	Ni, Cd	Tang et al. (2013)
Cotton seed hulls	800	Ni, Cu, Pb, Cd	Tang et al.(2013)
Chicken manure	550	Cr	Choppala et al. (2012)
Sewage sludge	500	Cu, Ni, Zn, Cd, Pb	Mendez et al. (2012)
Rice Straw	Not clear	Cu, Pb, Cd	Jiang et al. (2012)
Oak wood	400	Pb	Ahmad et al. (2012)
Anaerobically digested sugar cane bagasse	Not clear	Ni, Cd	Inyang et al. (2012)
Orchard Prune residue	500	Cd, Cr, Cu, Ni, Pb, Zn	Fellet et al. (2011)
Chicken manure and green waste	550	Cd, Cu, Pb	Park et al. (2011)
Corn Straw	600	Cu, Zn	Chen et al. (2011)
Hard wood	450	Cu, Zn	Chen et al. (2011)

Table 8: Efficiency of biosorbents in fixed bed column

Material used for column packing	Metals	Conc. of metal solution(mg/L)	Percentage removal	Reference
Rice husk	Pb	10	47	Kumar and Acharya, 2012
Coconut shell	Cu	10	Not clear	Acheampong et al. 2013
Rice straw powder	Ni	75	57	Sharma and Singh 2013
Sunflower waste carbon calcium alginate beads	Cd	10	Not clear	Jain et al. 2013
Maize tassel	Cu	20	Not clear	Sekhula et al. 2012
Oil palm fiber	Pb	100	Not clear	Nwabanne et al. 2012
Activated Tamarind seeds	Cr (VI)	1000	50	Gupta and Babu, 2010

13. Application of Biosorption with Actual Effluents

Most of the studies reported so far on metal removal by biosorption are under refined laboratory conditions with known defined concentration of metal supplemented growth media. Only a handful of studies are available to demonstrate the practical application of the biosorption process with actual industrial effluents with biomass. Most of the researchers have focused on effluents rich with chromium as a major contaminant.

Aspergillus lentulus showed 100% removal of Cr (VI) from electroplating effluents when the effluent was diluted with growth medium (1:1) (Sharma *et al.* 2011). Sepehr *et al.* (2005) demonstrated that *Aspergillus oryzae* is capable of removing chromium from tannery effluent with an efficiency of 97.6%, on the other hand, *Pseudomonas aeruginosa* showed only 42.5% chromium removal from tannery effluent as reported by Pumpel *et al.* (2001). Javaid *et al.* (2011) studied the potential use of the basidiomycetous fungus *Pleurotus ostreatus* biomass for the removal of heavy metals from electroplating industries, the removal efficiency was 46, 59, 9 and 9.5% for Cu(II), Ni(II), Zn(II) and Cr(VI) respectively. Ay *et al.* (2012) evaluated the performance of the biomass of *Punica granatum* L. peels for the removal of lead (II) ions from waste water samples of metal processing factories in Turkey. The removal percentage was estimated to be 98 under batch conditions. Gooseberry fruit (*Emblica officinalis*) biomass removes 65% copper ions at batch mode and 97.6% in column mode from the electroplating waste water (Rao and Ikram, 2011). The efficiency of the sugar cane bagasse for the removal of chromium from tannery effluent was evaluated by Ullah *et al.* (2013), the maximum removal of chromium was estimated to be 73%. The fungus *Aspergillus versicolor* effectively removed 86% of lead from effluents of battery industries located in Northern region of Kolkata, India (Bairagi *et al.* 2011*)*. The sorption capacity was effective when the effluent pH was adjusted to 5.0.

Relatively few studies are available on the treatment of actual effluents let out from industries, especially from electroplating and tannery effluents, when compared to that of batch system with synthetic solution. Unlike laboratory solutions with 'fixed' nutrients, industrial effluents contain various pollutants, interfering substances, including the metal of interest. Therefore, suitable experimental design has to be investigated for the simultaneous removal of major pollutants in the effluents. One of the possible methods would be the use of 'mixed' biosorbents, (i.e.) consisting of more than one type of biomass, for e.g. suitable agro waste with immobilized microbial biomass or consortium of microbes, although, this sort of designed biosorbents system would lead to more complicate characterization. However, in future, it may lead to the most realistic approach, considering the potential of biomaterials which are our perennial source.

Conclusion and Future Prospects

Rapid industrialization and technology developments have provided several benefits to human life. However, such developments also result in some of adverse side effects like environmental pollution/contamination. Multiple methodologies have been developed and proposed for remediation of heavy metals which range from chemical to biological, living to non-living, and microbial to phytoremediation. However, each of the methodologies has its own merits and demerits. Use of phytoremediation utilizing plants could offer cost effective and environmentally friendly remediation approach but could take longer remediation time and require ideal climatic conditions and the process of "biomagnification".

Despite the several decades of research and attempts for laboratory scale process developments to remediate heavy metal pollution, practical utility of the laboratory inventions appears to be only in its infancy and needs to define its boundaries between promise and reality. Several biotechnological approaches have been claimed as possible means of combating toxic metal pollution from industrial and other sources, although none are in widespread use. There is still much applied and field research needed. Though, there is not a single and perfect approach for mediation of heavy metal polluted environment due to the complexity associated with sites contaminated with heavy metals together with the fact that the polluted environments often contain more than one metal, a combinatorial approach will be highly beneficial. In our opinion, the bioremediation approach using native microorganisms / combination of natural waste materials with microbes grown on it may prove to be useful and would offer cost effective and environmentally safer approach. Furthermore, the existence of technologies for the genetic manipulation of microorganisms makes the tailoring of the genetic information to suit the needs. Nonetheless, success of such technologies will ultimately depend of the extensive field trials and combined efforts of biologist, soil chemists, microbiologists, geneticists, and environmental engineers to optimize the best suited parameters for a speedy bioremediation program.

Acknowledgement

Ministry of Environment, Forest and Climate Change (MOEF), Government of India, is gratefully acknowledged for providing fund for the project entitled "Biosorption by Fungi- A simple microbiological technique to remove heavy metals from e- wastes".

References

Abdel AM -Aty, Nabila S. Ammar, Hany H. Abdel Ghafar, Rizka K. Ali (2013) Biosorption of cadmium and lead from aqueous solution by fresh water alga *Anabaena sphaerica* biomass. J. Adv. Res. 4, 367–374.

Abdel-Ghani NT, El-Chaghaby GA (2014) Biosorption for metal ions removal from aqueous solutions: A review of recent studies. Int. J. Latest Sci. Technol., 3, 24-42.

Acheampong MA, Pakshirajan K, Annachhatre AP, Lens PN (2013) Removal of Cu (II) by biosorption onto coconut shell in fixed-bed column systems. J. Ind. Eng. Chem. 19, 841-848.

Adeogun AI, Ofudje AE Idowu M, Kareem SO (2011) Equilibrium, kinetics and thermodynamic studies of the biosorption of mn (ii) ions from aqueous solution by raw and acid-treated corncob biomass. BioRes. 6, 4117-4134.

Agarwal GS, Bhuptawat HK, Chaudhari, S (2006) Biosorption of aqueous chromium (VI) by *Tamarindus indica* seeds. Bioresour. Technol. 97, 949-956.

Ahalya N, Ramachandra TV, Kanamadi RD (2003) Biosorption of heavy metals. Res. J. Chem. Environ.7, 71-79.

Ahmad A, Mukherjee P, Senapati S, Mandal D, Khan ML, Kumar R, Sastry M (2003) Extracellular biosynthesis of silver nanoparticles using the fungus *Fusarium oxysporum.* Colloid Surface B. 28, 313-318.

Ahmad M, Lee SS. Yang JE, Ro HM. Lee YH, Ok YS (2012) Effects of soil dilution and amendments (mussel shell, cow bone, and biochar) on Pb availability and phytotoxicity in military shooting range soil. Ecotox. Environ. Safe. 79, 225-231.

Akar T, Celik S, Ari GA, Akar TS (2013) Nickel removal characteristics of an immobilized macro fungus: equilibrium, kinetic and mechanism analysis of the biosorption. J. Chem. Technol. Biotech. 88, 680-689.

Akhtar N, Iqbal M, Zafar SI, Iqbal J (2008) Biosorption characteristics of unicellular green alga *Chlorella sorokiniana* immobilized in loofa sponge for removal of Cr (III). J. Environ. Sci. 20, 231-239.

Aloma I, Lara MMA, Rodríguez IL, Blazquez G, Calero M (2012) Removal of nickel (II) ions from aqueous solutions by biosorption on sugarcane bagasse. J. Taiwan I. Chem. Eng. 43, 275-281.

Alslaibi TM, Abustan I, Ahmad MA, Foul AA (2013) Cadmium removal from aqueous solution using microwaved olive stone activated carbon. J. Environ. Chem. Eng. 1, 589-599.

Al-Wabel MI, Usman AR, El-Naggar AH, Aly AA, Ibrahim HM, Hesham M, Ibrahim, Elmaghraby S, Al-Omran A (2015) *Conocarpus* biochar as a soil amendment for reducing heavy metal availability and uptake by maize plants. Saudi J. Biol. Sci. 22 (4), 503-511

Amin F, Talpur FN, Balouch A, Surhio MA, Bhutto MA (2015) Biosorption of fluoride from aqueous solution by white—rot fungus *Pleurotus eryngii* ATCC 90888. Environ. Nanotechnol. Monitor. Manage. 3, 30-37.

Anirudhan TS, Sreekumari SS (2011) Adsorptive removal of heavy metal ions from industrial effluents using activated carbon derived from waste coconut buttons, J. Environ. Sci. 21, 1989–1998.

Arýca MY, Bayramoðlu G (2005) Cr (VI) biosorption from aqueous solutions using free and immobilized biomass of *Lentinus sajorcaju*: preparation and kinetic characterization. Colloid Surface A 253, 203-211.

Asci Y, Nurbaþ M, Açýkel YS (2010) Investigation of sorption/desorption equilibria of heavy metal ions on/from quartz using rhamnolipid biosurfactant. J. Environ. Manage. 91, 724-731.

Ay CO, Ozcan AS, Erdogan Y, Ozcan, A (2012) Characterization of *Punica granatum* L. peels and quantitatively determination of its biosorption behavior towards lead (II) ions and Acid Blue 40. Colloid Surface B. 100, 197-204.

Ayoub A, Venditti RA, Pawlak JJ, Salam A, Hubbe MA (2013) Novel hemicellulose–chitosan biosorbent for water desalination and heavy metal removal. ACS Sustain. Chem. Eng.1, 1102-1109.

Babu AG, Kim JD, Oh BT (2013) Enhancement of heavy metal phytoremediation by *Alnus firma* with endophytic *Bacillus thuringiensis* GDB-1. J. Hazard. Mater. 250, 477-483.

Bai RS, Abraham TE (2003) Studies on chromium (VI) adsorption–desorption using immobilized fungal biomass. Bioresour. Technol. 87, 17-26.

Bairagi H, Khan MMR, Ray L, Guha AK (2011) Adsorption profile of lead on *Aspergillus versicolor*: a mechanistic probing. J. Hazard. Mater. 186, 756-764.

Bansal M., Garg U, Singh D, Garg VK (2009) Removal of Cr (VI) from aqueous solutions using pre-consumer processing agricultural waste: A case study of rice husk. J. Hazard Mater. 162, 312-320.

Barka N, Abdennouri M, El Makhfouk M, Qourzal S (2013) Biosorption characteristics of cadmium and lead onto eco-friendly dried cactus (*Opuntia ficus indica*) cladodes. J Environ. Chem. Eng. 1, 144-149.

Bayramoglu G, Denizli A, Bektas S, Arica MY (2002) Entrapment of *Lentinus sajorcaju* into Ca-alginate gel beads for removal of Cd (II) ions from aqueous solution preparation and biosorption kinetics analysis. Microchem. J. 72, 63-76.

Bhargavi SD, Savitha J (2014) Arsenate resistant *Penicillium coffeae*: A potential fungus for soil bioremediation. Bull. Environ. Contam. Toxicol. 92, 369-373.

Birch L, Bachofen R (1990) Complexing agents from microorganisms. Experientia 46, 827-834.

Bolan NS, Kunhikrishnan A, Choppala GK, Thangarajan R,Chung JW (2012). Stabilization of carbon in composts and biochars in relation to carbon sequestration and soil fertility. Sci Total Environ. 424, 264-270.

Brady JM, Tobin JM (1995) Binding of hard and soft metal ions to *Rhizopus arrhizus* biomass. Enzyme Microbe. Technol. 17, 791- 796.

Cao X, Wahbi A, Ma L, Li, B, Yang Y (2009) Immobilization of Zn, Cu, and Pb in contaminated soils using phosphate rock and phosphoric acid. J. Hazard. Mater. 164, 555-564.

Chan J, Sanderson J, Chan W, Lai C, Choy D, Ho A, Leung R (1997) Prevalence of sleep-disordered breathing in diastolic heart failure. CHEST J. 111, 1488-1493.

Chen BY, Chen CY, Guo WQ, Chang HW, Chen WM, Lee DJ, Huang CC, Chang JS (2014) Fixed-bed biosorption of cadmium using immobilized *Scenedesmus obliquus* CNW-N cells on loofa (*Luffa cylindrica*) sponge. Bioresour. Technol. 160, 175-181.

Chen X, Chen, G. Chen L, Chen Y, Lehmann J, McBride,MB, Hay AG (2011) Adsorption of copper and zinc by biochars produced from pyrolysis of hardwood and corn straw in aqueous solution. Bioresour. Technol. 102, 8877-8884.

Chibuike GU, Obiora SC (2014) Bioremediation of hydrocarbon-polluted soils for improved crop performance. Int. J. Environ. Sci. 4(5), 840-858.

Chong HLH, Chia PS, Ahmad MN (2013) The adsorption of heavy metal by Bornean oil palm shell and its potential application as constructed wetland media. Bioresour. Technol. 130, 181-186.

Choppala GK, Bolan NS, Megharaj M, Chen Z, Naidu R (2012) The influence of biochar and black carbon on reduction and bioavailability of chromate in soils. J. Environ. Qual. 41, 1175-1184.

Çigdem, ARPA. Rýdvan, SAY, Satiroðlu N (2000) Heavy metal removal from aquatic systems by northern Anatolian smectites. Turk. J. Chem. 24, 209-215.

Cobbett CS (2000). Phytochelatins and their roles in heavy metal detoxification. Plant Physiol. 123, 825-832.

Crini G, Badot PM (2008) Application of chitosan, a natural aminopolysaccharide, for dye removal from aqueous solutions by adsorption processes using batch studies: A review of recent literature. Prog. Polym. Sci. 33(4), 399-447.

Daghistani H (2012) Bio-remediation of Cu, Ni and Cr from rotogravure wastewater using immobilized, dead, and live biomass of indigenous thermophilic *Bacillus* species. Int. J. Microbio. 10 (1) http://ispub.com/IJMB/10/1/13867.

Davis TA, Volesky B, Mucci A (2003). A review of the biochemistry of heavy metal biosorption by brown algae. Water Res. 37, 4311-4330.

Demirbas A (2008) Heavy metal adsorption onto agro-based waste materials a review. J. Hazard. Mater. 157, 220-229.

DeSchamphelaire L, Rabaey K, Boon N, Verstraete W, Boeckx P (2007) Minireview The potential of enhanced manganese redox cycling for sediment oxidation. Geomicrobiol. J. 24, 547-558.

Desta MB (2013). Batch sorption experiments: Langmuir and Freundlich isotherm studies for the adsorption of textile metal ions onto Teff Straw (Eragrostis tef) agricultural waste. J. Thermodyn. Article ID 375830, http://dx.doi.org/10.1155/2013/375830.

Diamadopoulos, Evan, Evita, Agrafioti., Dimitrios, Kalderis (2014). As (V), Cr (III) and Cr (VI) sorption on biochars and soil. EGU Gen Assem Conf Abstract. 16, 9816.

Ding Y, Jing D, Gong H, Zhou L, Yang X (2012) Biosorption of aquatic cadmium (II) by unmodified rice straw. Bioresour. Technol, 114, 20-25.

Dixit R, Malaviya D, Pandiyan K, Singh UB, Sahu A. Shukla R, Paul D. (2015) Bioremediation of heavy metals from soil and aquatic environment: an overview of principles and criteria of fundamental processes. Sustainability. 7, 2189-2212.

Durán N, Marcato PD, Alves OL, De Souza GI and Esposito E (2005) Mechanistic aspects of biosynthesis of silver nanoparticles by several *Fusarium oxysporum* strains. J Nanobiotechnol. 3, 1-7.

Environment Protection Authority of Australia. (2012) Classification and management of contaminated soil for disposal. Information Bulletin 105. Hobart, TAS 7001 Australia.

Environmental Protection Ministry of China (EPMC). (2014) National survey report of soil contamination status of China. Environmental Protection Ministry of China, Beijing, China.

European Environmental Agency (EEA), (2007) Progress in management of contaminated sites (CSI 015/LSI 003), http://www.eea.europa.eu/data-and-maps/indicators.

Fellet G, Marchiol L, Vedove DG, Peressotti A (2011) Application of biochar on mine tailings: effects and perspectives for land reclamation. Chemosphere 83, 1262–1297

Feng K, Sun H, Bradley MA, Dupler EJ, Giannobile WV, Ma PX (2010) Novel antibacterial nanofibrous PLLA scaffolds. J. Control Release. 146(3), 363-369.

Fomina M, Gadd GM (2002) Metal sorption by biomass of melanin producing fungi grown in clay containing medium. J. Chem. Technol. Biot. 78, 23–34.

Fomina M, Gadd GM (2014) Biosorption current perspectives on concept, definition and application. Bioresour. Technol. 160, 3-14.

Fourest E, Serre A, Roux JC (1996) Contribution of carboxyl groups to heavy metal binding sites in fungal wall. Toxicol. Environ. Chem. 54, 1-10.

Franco LDO, Maia RDCC, Porto ALF, Messias AS, Fukushima K, Campos-Takaki GMD (2004). Heavy metal biosorption by chitin and chitosan isolated from *Cunninghamella elegans* (IFM 46109) Braz. J. Microbiol. 35, 243-247.

Gadd G, Sayer J. (2000) Influence of Fungi on the Environmental Mobility of Metals and Metalloids. In: *Environmental Microbe-Metal Interactions* (Ed) Lovley DR, ASM Press, Washington, DC, pp 237-256.

Gadd GM (1990) Heavy metal accumulation by bacteria and other microorganisms. Experientia, 46, 834-840.

Gadd GM (1999) Fungal production of citric and oxalic acid: importance in metal speciation, physiology and biogeochemical processes. Adv. Microbial. Physiol. 41, 47-92.

Gadd GM (2007) Geomycology: biogeochemical transformations of rocks, minerals, metals and radionuclides by fungi, bioweathering and bioremediation. Mycol. Res. 111, 3–49.

Gadd GM, Gharieb MM, Ramsay LM, Sayer JA, Whatley AR, White C (1998) Fungal processes for bioremediation of toxic metal and radionuclide pollution. J. Chem. Technol. Biot. 71, 364-366.

Garbisu C, Alkorta I (1997) Bioremediation: Principles and future. J. Clean Technol Environ.Toxicol Occup. Med. T. 6, 351-366.

Garbisu C, Llama MJ, Serra JL (1997) Effect of heavy metals on chromate reduction by Bacillus subtilis. J. Gen. Appl. Microbiol. 43, 369-371.

Garbisu, C, Alkorta I (2003) Basic concepts on heavy metal soil bioremediation. Eur. J. Mineral Process. Environ. Prot. 3, 58-66.

Garbisu, C, Alkorta, I, Llama, MJ, Serra, JL. (1998) Aerobic chromate reduction by Bacillus subtilis. Biodegradation 9(2), 133-141.

Gaur N, Dehankhar R, (2009) Equilibrium modelling and spectroscopic studies for the biosorption of Zn+ 2 ions from aqueous solution using immobilized *Spirulina platensis*. Iran J. Environ. Health Sci. Eng. 6, 1-6.

Gautam SP, CPCB, New Delhi, August (2011) Hazardous metals and minerals pollution in India: sources, toxicity and management, a position paper.

Ghani NA, Sairi NA, Aroua MK, Alias Y, Yusoff R (2014) Density, surface tension, and viscosity of ionic liquids (1-ethyl-3-methylimidazolium diethylphosphate and 1, 3-dimethylimidazolium dimethylphosphate) aqueous ternary mixtures with MDEA. J. Chem. Eng. Data 59, 1737-1746.

GIZ, Contaminated sites in India: Indian Government supports commissioning of GIZ (2012) https://www.giz.de/en/mediacenter/6563.html.

Gochev V, Velkova Z, Stoytcheva M, Yemendzhiev H, Aleksieva Z, Krastanov A (2012) Biosorption of Cu (II) from aqueous solutions by immobilized mycelium of Trametes versicolor. Biotechnol. Biotech. Equipment 26, 3365-3370.

Gupta S, Babu BV (2010) Experimental investigations and theoretical modeling aspects in column studies for removal of Cr (VI) from aqueous solutions using activated tamarind seeds. J. Water Res. Protect. 2, 706-716.

Gupta VK, Pathania D, Agarwal S, Sharma S (2013) Removal of Cr (VI) onto *Ficus carica* biosorbent from water. Environ. Sci. Pollut. Res. 20, 2632-2644.

Hassan SW, Kassas HY (2014) Biosorption of cadmium from aqueous solutions using a local fungus *Aspergillus cristatus* (Glaucus group). Afr. J. Biotechnol. 11, 2276-2286

Hemambika B, Rani MJ and Kannan VR. (2011) Biosorption of heavy metals by immobilized and dead fungal cells: A comparative assessment. J. Ecol. Natl. Environ. 3, 168-175.

Hennebel T, De Gusseme B, Boon N. Verstraete W (2009) Biogenic metals in advanced water treatment. Trends Biotechnol. 27, 90-98.

Hobman JL, Brown NL (1997) Bacterial mercury-resistance genes. J. Metal Ions Biol. Syst. 34, 527.

Holan ZR, Volesky B (1995) Accumulation of cadmium, lead, and nickel by fungal and wood biosorbents. Appl. Biochem. Biotech. 53(2), 133-146.

Hossain MA, Ngo HH, Guo WS, Nguyen TV (2012) Biosorption of Cu (II) from water by banana peel based biosorbent: Experiments and models of adsorption and desorption. J. Water Sustain. 2, 87-104.

Hossain MA, Ngo HH, Guo WS, Setiadi T (2012) Adsorption and desorption of copper (II) ions onto garden grass. Bioresour. Technol.121, 386-395.

Hu MZC, Reeves M (1997) Biosorption of uranium by *Pseudomonas aeruginosa* strain CSU immobilized in a novel matrix. Biotechnol. Prog. 13, 60-70.

Huang H, Cao L, Wan Y, Zhang R, Wang W (2012) Biosorption behavior and mechanism of heavy metals by the fruiting body of jelly fungus (*Auricularia polytricha*) from aqueous solutions. . Microbiol. Biotechnol. 96, 829-840.

Inyang M, Gao B. Yao Y. Xue Y, Zimmerman AR, Pullammanappallil P, Cao X (2012) Removal of heavy metals from aqueous solution by biochars derived from anaerobically digested biomass. Bioresour. Technol. 110, 50-56.

Ippolito JA, Laird DA, Busscher WJ (2012) Environmental benefits of biochar. J Environ Qual. 41(4), 967-972.

Iqbal M., Edyvean, RGJ (2004) Biosorption of lead, copper and zinc ions on loofa sponge immobilized biomass of *Phanerochaete chrysosporium*. Miner. Eng. 17, 217-223.

Iram S, Shabbir R, Zafar H, Javaid M (2015) Biosorption and bioaccumulation of copper and lead by heavy metal-resistant fungal Isolates. Arab J. Sci. Eng. 40 (7), 1867-1873.

Jain M, Garg VK, Kadirvelu K (2013) Cadmium (II) sorption and desorption in a fixed bed column using sunflower waste carbon calcium–alginate beads. Bioresour. Technol, 129, 242-248.

Javaid A, Bajwa R, Manzoor (2011) Biosorption of heavy metals by pretreated biomass of Aspergillus niger. Pak. J. Bot. 43, 419-425.

Javaid A, Bajwa R, Shafique U, Anwar J (2011) Removal of heavy metals by adsorption on *Pleurotus ostreatus*. Biomass Bioenerg 35, 1675-1682.

Jayanthi M, Kanchana D, Saranraj P, Sujitha D (2014) Bioadsorption of chromium by *Penicillium* chrysogenum and Aspergillus niger isolated from tannery effluent. Int. J. Micr. Res. 5(1), 40-47.

Jiang TY, Jiang J, Xu RK, Li Z. (2012) Adsorption of Pb (II) on variable charge soils amended with rice-straw derived biochar. Chemosphere 89, 249-256.

Jimenez VLH, Mendez RJR (2014) Chemical and thermogravimetric analyses of raw and saturated agave bagasse main fractions with Cd (II), Pb (II), and Zn (II) ions: Adsorption Mechanisms. Ind. Eng. Chem. Res. 53, 8332-8338.

Joner E, Leyval C (2001) Time-course of heavy metal uptake in maize and clover as affected by root density and different mycorrhizal inoculation regimes. Biol. Fertil. Soils 33, 351-357.

Juarez MVM, Gonzalez CJF, Bouscoulet TME, Rodríguez AI (2012) Biosorption of mercury (II) from aqueous solutions onto fungal biomass. Bioinorganic Chem. App. Article ID 156190, http://dx.doi.org/10.1155/2012/156190

Kamnev A, Van Der Lelie D (2000) Chemical and biological parameters as tools to evaluate and improve heavy metal phytoremediation. Biosci. Rep. 20, 239-258.

Kausar A, Bhatti HN, MacKinnon G (2013) Equilibrium, kinetic and thermodynamic studies on the removal of U (VI) by low cost agricultural waste. Colloid Surface B. 111, 124-133.

Khoramzadeh E, Nasernejad B, Halladj R (2013) Mercury biosorption from aqueous solutions by sugarcane bagasse. J. Taiwan. Inst. Chem. Eng. 44, 266-269.

Krowiak WA, Szafran RG, Modelski S (2011) Biosorption of heavy metals from aqueous solutions onto peanut shell as a low-cost biosorbent. Desalination 265(1), 126-134.

Kumar SP, Ramalingam S, Abhinaya RV, Kirupha SD, Murugesan A, Sivanesan S (2012) Adsorption of metal ions onto the chemically modified agricultural waste. CLEAN– Soil Air Water 40, 188-197.

Kumar U, Acharya J (2012) Fixed Bed Column Study for the Removal of Copper from Aquatic Environment by NCRH. Res. J. Rec. Sci. 2, 9-12.

Kuroda K, Ueda M (2011) Yeast biosorption and recycling of metal ions by cell surface engineering.

In: Microbial Biosorption of Metals (Ed) Kotrba P, Mackova M, Macek T, Springer, Netherlands, pp 235–247.

Kuyucak N, Volesky B (1988) Biosorbents for recovery of metals from industrial solutions. Biotechnol. Lett. 10, 137-142.

Lakshmipathy R, Sarada NC (2013) Application of watermelon rind as sorbent for removal of nickel and cobalt from aqueous solution. Int. J. Miner. Process, 122, 63-65.

Ledin M (2000). Accumulation of metals by microorganisms—processes and importance for soil systems. Earth Sci. Rev. 51, 1-31.

Lee SS, Lim JE, El-Azeem SAA, Choi B, Oh SE, Moon DH, Ok YS (2013) Heavy metal immobilization in soil near abandoned mines using eggshell waste and rapeseed residue. Environ. Sci. Pollut. Res. 20, 1719-1726.

Li PS, Tao HC (2013) Cell surface engineering of microorganisms towards adsorption of heavy metals. Crit Rev Microbiol. 0, 1-10.

Lim JE, Ahmad M, Usman AR, Lee SS, Jeon WT, Oh, SE, Ok, YS (2013) Effects of natural and calcined poultry waste on Cd, Pb and As mobility in contaminated soil. Environ. Earth Sci. 69, 11-20.

Liu H, Guo L, Liao S, Wang G (2012) Reutilization of immobilized fungus *Rhizopus* sp. LG04 to reduce toxic chromate. J. App. Microbiol. 112, 651-659.

Liu H, Xian XU, Zhenhua WU, Guoxia WEI, Lei SUN (2015) Removal of heavy metals from aqueous solution using biochar derived from biomass and sewage sludge. Appl. Mech. Material. 768, 89-95.

Lloyd JR, Yong P, Macaskie LE (1998) Enzymatic recovery of elemental palladium by using sulfate-reducing bacteria. Appl. Environ. Microbiol. 64(11), 4607-4609.

Lovley DR, Lloyd JR (2000) Microbes with a mettle for bioremediation. Nat. Biotechnol. 18, 600-601.

Lupea M, Bulgariu L, Macoveanu M (2012) Biosorption of Cd (II) from aqueous solution on marine green algae biomass. Environ. Eng. Manage. J. 11, 607-615.

Martin S, Griswold W (2009) Human health effects of heavy metals. Environ. Sci. Technol. Briefs Citizens 15, 1-6.

Mazali IO, Alves OL (2005) Morphosynthesis: high fidelity inorganic replica of the fibrous network of loofa sponge (*Luffa cylindrica*). An Acad. Bras. Cien. 77, 25-31.

Maznah, W. W., Al-Fawwaz, A. T., & Surif, M. (2012) Biosorption of copper and zinc by immobilised and free algal biomass, and the effects of metal biosorption on the growth and cellular structure of *Chlorella* sp. and *Chlamydomonas* sp. isolated from rivers in Penang, Malaysia J. Environ. Sci. 24, 1386-1393.

Mehmood MS, Yasin T, Jahan MS, Mishra SR, Walters BM, Ahmad M, and Ikram M, (2013) Assessment of residual radicals in et γ.-sterilized shelf-aged UHMWPE stabilized with α-tocopherol. Polym Degrad Stabil. 98, 1256-1263.

Mehta SK., & Gaur JP (2005) Use of algae for removing heavy metal ions from wastewater: progress and prospects. Crit. Rev. Biotechnol. 25, 113-152.

Méndez A, Gomez A, Paz-Ferreiro J, Gasco G (2012) Effects of sewage sludge biochar on plant metal availability after application to a Mediterranean soil. Chemosphere 89:1354–1359

Mengistie AA, Rao TS, Rao AP (2012) Adsorption of Mn (II) ions from wastewater using activated carbon obtained from Birbira (*Militia ferruginea*) leaves. Global J Sci Frontier Res Chem. 12(1). 21-28.

Mezaguer M, Kamel N, Lounici H, Kamel Z (2013) Characterization and properties of *Pleurotus mutilus* fungal biomass as adsorbent of the removal of uranium (VI) from uranium leachate. J. Radioanal. Nucl. Chem. 295, 393-403.

Mishra A, Malik A (2013) Recent advances in microbial metal bioaccumulation. Crit. Rev. Environ. Sci. Technol. 43, 1162-1222.

Mishra S P (2014) Adsorption–desorption of heavy metal ions. Curr. Sci. 107, 601-612.

Morsy EM, El-Didamoney SMM (2013) *Mucor racemosus* as a biosorbent of metal ions from polluted water in Northern Delta of Egypt. Mycosphere 4, 1118-1131.

Mostaedi T M, Asadollahzadeh M, Hemmati A, Khosravi A (2013) Equilibrium, kinetic, and thermodynamic studies for biosorption of cadmium and nickel on grapefruit peel. J. Taiwan Inst. Chem. Eng. 44, 295-302.

Muraleedharan TR, Venkobachar C (1990) Mechanism of biosorption of Copper (Ii) by *Ganoderma* lucidum. Biotechnol. Bioengg. 35, 320-325.

Muzzarelli RA (1980) Immobilization of enzymes on chitin and chitosan. Enzyme Microbial Tech. 2, 177-184.

Nakajima A, Sakaguchi T (1993) Accumulation of uranium by basidiomycetes. Appl. Microbial Biotechnol. 38(4), 574-578.

Narayanaswamy T, Hemachandran, J, Thirumalai T, Sharma CV, Kannabiran K, David E (2013). Biosorption of heavy metals from aqueous solution by *Gracilaria corticata varcartecala* and *Grateloupia lithophila*. J. Coastal Life Med. 1, 102-107.

Neilands JB (1981) Microbial iron transport compounds (siderochromes). In: Inorganic Biochemistry Vol (1) (Ed) Eichorn GL, Elsevier, pp 167 – 202.

Nelson YM, Lion LW, Shuler ML, Ghiorse WC (2002) Effect of oxide formation mechanisms on lead adsorption by biogenic manganese (hydr) oxides, iron (hydr) oxides, and their mixtures. Environ. Sci. Technol. 36, 421-425.

Ngah WW, and Hanafiah MA, KM, (2008) Biosorption of copper ions from dilute aqueous solutions on base treatedrubber (*Hevea brasiliensis*) leaves powder kinetics, isotherm, and biosorption mechanisms. J. Sci. Technol. 20, 1168-1176.

Norris PR., Kelly DP (1979) Accumulation of metals by bacteria and yeasts. Dev. Ind. Microbiol. 20, 299-308.

Nwabanne JT, Igbokwe PK (2012) Adsorption performance of packed bed column for the removal of Lead using oil palm fibre. Int. J. App. Sci. Technology. 2(5). 106-115

O'Connell DW, Birkinshaw C, O'Dwyer TF (2008) Heavy metal adsorbents prepared from the modification of cellulose: A review. Bioresour. Technol. 99, 6709-6724.

Oves M, Khan MS, Zaidi A (2013) Biosorption of heavy metals by *Bacillus thuringiensis* strain OSM29 originating from industrial effluent contaminated north Indian soil. Saudi J. Biol. Sci. 20, 121-129.

Ow DW (1996) Heavy metal tolerance genes: prospective tools for bioremediation. Resour, Conserv. Recy. 18, 135–149

Park D, Yun YS, Park JM (2010) The past, present, and future trends of biosorption. Biotechnol. Bioprocess. Eng. 15(1), 86-102.

Park JH, Choppala GK, Bolan NS, Chung JW, Chuasavathi T (2011) Biochar reduces the bioavailability and phytotoxicity of heavy metals. Plant Soil 348(1-2), 439-451.

Pümpel T, Ebner C, Pernfuss B, Schinner F, Diels L, Keszthelyi Z, Wouters H (2001) Treatment of rinsing water from electrolysis nickel plating with a biologically active moving-bed sand filter. Hydrometallurgy 59, 383-393.

Rao RAK, Ikram S. (2011) Sorption studies of Cu (II) on gooseberry fruit (*Emblica officinalis*) and its removal from electroplating wastewater. Desalination 277, 390-398.

Raspanti E, Cacciola SO, Gotor C, Romero LC, García I (2009) Implications of cysteine metabolism in the heavy metal response in *Trichoderma harzianum* and in three *Fusarium* species. Chemosphere76, 48-54.

Rawat M, Rai JPN (2012) Adsorption of heavy metals by *Paenibacillus validus* strain mp5 isolated from industrial effluent–polluted soil. Bioremediation J 16, 66-73.

Raymond E, Dahan L, Raoul JL, Bang YJ, Borbath I, Lombard-Bohas C, Ruszniewski P (2011) Sunitinib malate for the treatment of pancreatic neuroendocrine tumors. New England J. Med. 364, 501-513.

Rich G, Cherry K (1987) Hazardous Waste Treatment Technology, Pudvan Publishing. Co., New York.

Rowell RM, Han JS, Rowell, JS (2000) Characterization and factors effecting fiber properties. Natural Polymers and Agrofibers Bases Composites. Embrapa Instrumentacao Agropecuaria, Sao Carlos, Brazil, 115-134.

Sadowski Maliszewska IH, Grochowalska B, Polowczyk I, Kozlecki T (2008) Synthesis of silver nanoparticles using microorganisms. Mater. Sci. - Poland 26, 419-424.

Saglam A, Yalcinkaya Y, Denizli A, Arica MY, Genc O, Bektas S (2002) Biosorption of mercury by carboxy methyl cellulose and immobilized *Phanerochaete chrysosporium*. Microchem. J. 71, 73-81.

Saravanan N, Basha CA, Kannadasan T, Manivasagan V, Babu NG (2012) Biosorption of Ni on biobeads and biofilms using immobilized Escherichia coli. Eur. J. Sci. Res. 81, 231-245.

Sastry M, Ahmad A, Islam KM, Kumar R (2003) Biosynthesis of metal nanoparticles using fungi and actinomycete. Curr. Sci. 85, 162-170.

Sati M, Verma M and Rai JPN (2014) Biosorption of Pb(ii) ions from aqueous solution onto free and immobilized cells of *Bacillus megaterium*. Int. J. Recent Sci. Res. 5, 1286-1292.

Sayer JA, Gadd GM (1997) Solubilization and transformation of insoluble inorganic metal compounds to insoluble metal oxalates by *Aspergillus niger*. Mycol. Res. 101, 653-661.

Sekhula MM, Okonkwo Jo, Zvinowanda CM, Agayel NN, Chaudhary AJ (2012) Fixed bed column adsorption of Cu (II) onto Maize Tazzel- PVA beads. J. Chem. Eng. Process. Technol. 3, 1-5.

Sepehr MN, Nasseri S, Yaghmaian MMAK (2005) Chromium bioremoval from tannery industries effluent by Aspergillus oryzae. Iran J. Environ. Health Sci. Eng. 2. 273-279.

Sharma R, Singh B (2013) Removal of Ni (II) ions from aqueous solutions using modified rice straw in a fixed bed column. Bioresour. Technol. 146, 519-524.

Sharma S, Malik A, Satya S, Mishra A (2011) Development of a biological system employing *Aspergillus lentulus* for Cr removal from a small scale electroplating industry effluent. Asia Pacific J. Chem. Eng. 6, 55-63.

Shazia I, Uzma GRS, Talat A (2013) Bioremediation of heavy metals using isolates of filamentous fungus *Aspergillus fumigatus* collected from polluted soil of Kasur, Pakistan. Int. Resour. J. Biol. Sci. 2. 1-6.

Sheng G, Yang, Y, Huang M, Yang K (2005) Influence of pH on pesticide sorption by soil containing wheat residue-derived char. Environ. Pollut. 134, 457-463.

Sheng PX, Ting YP, Chen JP, Hong L (2004) Sorption of lead, copper, cadmium, zinc, and nickel by marine algal biomass: characterization of biosorptive capacity and investigation of mechanisms. J. Colloid Interf. Sci, 275, 131-141.

Sheng XF, Xia JJ (2006) Improvement of rape (*Brassica napus*) plant growth and cadmium uptake by cadmium-resistant bacteria. Chemosphere 64, 1036-1042.

Shoaib A, Aslam N, Aslam N (2013) Trichoderma harzianum: Adsorption, desorption, isotherm and FTIR studies. J. Anim. Plant Sci. 23(5), 1460-1465.

Sober DL, Lakshmanan VI, McCready RGI, Dahya AS, (1996) Bioadsorption of U by fungal isolates. In: Biominet Proceedings (Ed) McCready RGL, CANMET Special Publications, 1986–96, pp 93–111.

Subbaiah MV, Yun YS (2013) Biosorption of Nickel (II) from aqueous solution by the fungal mat of *Trametes versicolor* (rainbow) biomass: equilibrium, kinetics, and thermodynamic studies. Biotech. Bioprocess. Eng. 18, 280-288.

Sud D, Mahajan G, Kaur MP (2008). Agricultural waste material as potential adsorbent for sequestering heavy metal ions from aqueous solutions–A review. Bioresourc. Technol. 99, 6017-6027.

Suh JH, Yun JW, Kim DS (1999). Effect of extracellular polymeric substances (EPS) on Pb2+ accumulation by *Aureobasidium pullulans*. Bioprocess. Eng. 21, 1-4.

Sulaymon AH, Ebrahim SE, Ridha MMJ (2013). Equilibrium, kinetic, and thermodynamic biosorption of Pb (II), Cr (III), and Cd (II) ions by dead anaerobic biomass from synthetic wastewater. Env. Sci. Pollut. Res. 20, 175-187.

Sun YM, Horng CY, Chang Fl, Cheng LC, Tian WX (2010). Biosorption of lead, mercury, cadmium ions by *Aspergillus terreus* immobilized in natural matrix. Pol. J. Microbiol. 59, 37-44.

Sutter HP, Jones EBG, Walchli O (1983). The mechanism of copper tolerance in *Poria placenta* (Fr.) Cke and *Poria caillantii* (Pers.) Fr. Mater Organismen 18, 243–263.

Tang J, Zhu W, Kookana R, Katayama A (2013). Characteristics of biochar and its application in remediation of contaminated soil. J. Biosci. Bioeng. 116, 653-659

Tobin JM, Cooper DG, Neufeld RJ (1990). Investigation of the mechanism of metal uptake by denatured *Rhizopus arrhizus* biomass. Enz. Microbiol. Technol. 12, 591– 595.

Tsezos M, Deutschmann AA (1990). An investigation of engineering parameters for the use of immobilized biomass particles in biosorption. J. Chem. Technol. Biotechnol. 48, 29-39.

Ullah I, Nadeem R, Iqbal M, Manzoor Q (2013). Biosorption of chromium onto native and immobilized sugarcane bagasse waste biomass. Ecol. Eng. 60, 99-107.

United States Environmental Protection Agency (US EPA) (2014). Cleaning up the Nation's Hazardous Wastes Sited. http://www.epa.gov/superfund.

Vargas KK, Lopez CM, Tellez RS, Bandala ER, Salas SJL (2012) Biosorption of heavy metals in polluted water, using different waste fruit cortex. Phys. Chem. Earth 37, 26-29.

Varshney R, Bhadauria S, Gaur MS (2011) Biosorption of Copper (II) from electroplating wastewaters by Aspergillus terreus and its kinetics studies. Water 2, 142-151.

Veglio F, Beolchini F (1997). Removal of metals by biosorption: a review Hydrometallurgy 44, 301-316.

Verma, DK, Hasan SH, Ranjan D, Banik R M (2014) Modified biomass of *Phanerochaete chrysosporium* immobilized on luffa sponge for biosorption of hexavalent chromium. Int J. Environ. Sci. Technol. 11, 1927-1938.

Vijayaraghavan K, Yun YS (2008). Bacterial biosorbents and biosorption. Biotechnol. Adv. 26, 266–291.

Villalobos M, Bargar J, Sposito G (2005). Mechanisms of Pb (II) sorption on a biogenic manganese oxide. Environ. Sci. Technol. 39, 569-576.

Vivas A., Biro B, Németh T, Barea JM, Azcon R (2006). Nickel-tolerant *Brevibacillus brevis* and arbuscular mycorrhizal fungus can reduce metal acquisition and nickel toxicity effects in plant growing in nickel supplemented soil. Soil Biol Biochem. 38, 2694-2704.

Volesky B (2007). Biosorption and me. Water Res. 41, 4017-4029.

Volesky B, Holan ZR (1995) Biosorption of heavy metals. Biotechnol Progr. 11, 235-250.

Wang H, Gao B, Wang S, Fang J, Xue Y, Yang K (2015) Removal of Pb (II), Cu (II), and Cd (II) from aqueous solutions by biochar derived from $KMnO_4$ treated hickory wood. Bioresour. Technol.197, 356-362.

Wang H, Liu ZM (2013) Biosorption of Cd (II)/Pb (II) from aqueous solution by biosurfactant-producing bacteria: Isotherm kinetic characteristic and mechanism studies. Colloid Surface B. 105, 113-119.

Wang J, Chen C (2014) Chitosan-based biosorbents: modification and application for biosorption of heavy metals and radionuclides. Bioresour. Technol. 160, 129-141.

Wang PC, Mori T, Komori K, Sasatsu M, Toda K, Ohtake H (1989) Isolation and characterization of an *Enterobacter cloacae* strain that reduces hexavalent chromium under anaerobic conditions. Appl. Environ. Microbiol. 55, 1665-1669.

White C. Shaman AK, Gadd GM (1998) An integrated microbial process for the bioremediation of soil contaminated with toxic metals. Nat. Biotechnol. 16, 572-575.

Wilde EW, Benemann JR (1993) Bioremoval of heavy metals by the use of microalgae. Biotechnol Adv. 11, 781-812.

Wu FC, Tseng RL, Juang RS (2010) A review and experimental verification of using chitosan and its derivatives as adsorbents for selected heavy metals. J. Environ. Manage. 91, 798-806.

Xu X., Cao X, Zhao L, Wang H, Yu H, Gao B. (2013) Removal of Cu, Zn, and Cd from aqueous solutions by the dairy manure-derived biochar. Environ. Sci. Pollut. Res. 20, 358-368.

Yadav SK (2010). Heavy metals toxicity in plants: an overview on the role of glutathione and phytochelatins in heavy metal stress tolerance of plants. S. Afr. J. Bot. 76, 167-179.

Yin P, Yu Q, Jin B, and Ling Z, (1999) Biosorption removal of cadmium from aqueous solution by using pretreated fungal biomass cultured from starch wastewater. Water Res. 33(8), 1960-1963.

Zamani AA, Shokri R, Yaftian MR, Parizanganeh AH (2013) Adsorption of lead, zinc and cadmium ions from contaminated water onto *Peganum harmala* seeds as biosorbent. Intl. J. Environ. Sci. Technol.10, 93-102.

Zouboulis Al, Matis KA, Loukidou M, Sebesta F (2003) Metal biosorption by PAN-immobilized fungal biomass in simulated wastewaters. Colloid Surface A. 212, 185-195.

4

Arbuscular Mycorrhizal Fungi in Quenching the Detrimental Effects of Heavy Metals for Sustainable Agriculture of Crop Plants

Sharma S. and Kapoor R.

Abstract

Contamination groundwater and soil due to heavy metals (HMs) has aggravated into a threat not only for people inhabiting the polluted areas but also for the vegetation of the region. Relying on the use of obligate symbionts, arbuscular mycorrhizal fungi (AMF) in remediation of contaminated sites can be a productive approach. These fungi due to high cation exchange capacity have the ability to adsorb heavy metals (HMs) on fungal tissues; they also possess enzymes responsible for reduction of metals to less toxic forms. In areas with high toxicity of various HMs such as lead, AMF extends its extraradical mycelium (ERM) beyond the depletion zone of plant roots increasing the availability of nutrients such as potassium, phosphorus, magnesium, and sulphur resulting in increased plant growth and yield. Production of phenol and organic acids by the fungus have been reported to increase iron diffusion, that is a co-factor for antioxidant enzyme superoxide dismutase. AMF sequester HMs in fungal structures and bind them in their chitin cell walls preventing their translocation to aerial plant parts. These symbionts also influence soil microbial community and soil phosphatase activity resulting in better nutrient acquisition by plants and improvement in soil health. Therefore, employing AMF could prove to be a pioneer in augmenting the usage of greener technologies towards remediation of sites contaminated with HMs. This chapter will highlight the potential of AMF in alleviation of HM toxicity, encouraging their use in amelioration of contaminated sites.

Keywords: Arbuscular mycorrhizal fungi, Heavy metal stress, Metal sequestration, Plant-symbiont interaction, Reactive oxygen species.

1. Introduction

In India where 70% population of the country resides in the rural belt and is dependent on agricultural practices as the major source of income, the agriculture sector constitutes a major portion of its economy. Vicious practice of employing pesticides in order to increase the crop yield has led to the buildup of heavy metals and their subsequent seepage in the water table causing a potential health hazard. The total land area of India is 329 million hectare of which 100 million hectare is characterized as non-cultivable wasteland - not feasible for growing food crops, and 140.60 million hectare is cultivated land (BAIF research foundation) BAIF research foundation 2006.Therefore, development of a technique for remediation of non-cultivable wasteland to cultivable land will have a positive impact in meeting the growing demand of food crops in the country.

Without a proper framework in place to conserve the flora and fauna of the country, the effect of rapid industrialization will result in a hefty price to be paid by the environment, and its implications will be reflected in the form of deteriorating natural resources. Even now in various towns and cities across the country whether it be small or large scale industries a proper waste management system is lacking. As a result of which various sources of harmful pollutants especially HMs, characterized in Table 2 are degrading the environment getting absorbed by soil, and ultimately reaching the water table. Therefore the need of the hour is the presence of an environmentally integrated stable technology employing the use of natural resources that instead of augmenting the carrying capacity of the ecosystem will focus on alleviating the pressure from an already degrading environment.

2. Arbuscular Mycorrhizal Fungi: Plant Patrons

Plants are known to associate with various microbial communities in the rhizosphere. Intriguing of all these interactions, of particular interest is a symbiotic mutualistic relationship between plants and fungi called mycorrhiza. Arbuscular mycorrhiza characterizes the most widespread terrestrial symbiosis occurring between 70-80% land plants and fungi belonging to monophyletic phylum Glomeromycota (Hibbett et al. 2007; Redecker et al. 2013; SchüBler et al. 2001). AMF depend on a living photoautotrophic partner in order to complete their life cycle and produce next generation of spores, therefore are obligate biotrophs (Parniske 2008).

The relationship between AMF and the plant with which they associate is based on shared compatibility as AMF result in extensive colonization of the host plant root without, or with weak elicitation of host defense responses (Harrisson 1997). This further emphasizes the fact that this mutualistic relationship has forged mechanisms to enable harmonized development of both plant and the symbiont to achieve a functional state (Harrisson *et al.* 1997).

AM symbiosis is especially beneficial in a scenario where nutrient availability for the plant is limited, resulting in better water absorption capabilities apart from other advantageous influences of the fungus (Leung *et al.* 2013). AMF are characterized by the development of arbuscules which are dichotomously branched intracellular structures formed inside the host root and are the main sites of nutrient exchange between fungi and plant, AMF supplies nutrients and water to plants and in turn obtains carbohydrates. AMF form an extensive network extending in the soil called extra radical mycelium (ERM), this structure allows the fungus to go beyond the depletion zone of the plant roots and acquire nutrients which is otherwise not possible for the plant, also increasing the surface area for absorption (Smith and Read 2008).

3. Role of AMF in Phytoremediation of Heavy Metal Contaminated Sites

Until now the use of various physical and chemical methods in treatment of sites contaminated with HMs have not yielded the desired results. These techniques are not appropriate for large areas of land, not to forget the financial constraints posed by employing these techniques apart from changes in soil chemistry (Wang *et al.* 2007a). In such a setting the technique of phytoremediation involving the use of plants for removing contaminants from the polluted sites might be productive. The phytoremediation technique besides being inexpensive and cost effective is also aesthetically attractive to the public eyes. Despite its merits the technique also has certain disadvantages associated with it:

- It is a very slow process as several years are needed to bring the metal contamination to halve in the contaminated sites (McGrath and Zhao 2003).

- Phytoremediating plants usually have difficulty in establishing themselves in heavily HM contaminated sites (Santana *et al.* 2015).

- It has limited efficiency as the contaminant in order to be accessible to the plant root must be present in its zone (Leung *et al.* 2013).

- Chelator application used in chemically assisted phytoextraction process can pose environmental risks and threat to soil microfauna and microorganisms (Romkens *et al.* 2002 and Wang *et al.* 2007a).

Table 1: Area wise characterization of industrial wastelands of India contaminated with heavy metals

Heavy metal	Contaminated sites in India	Major crops cultivated	Mining/ Industrial wastelands (Area in Sq. Kms.) {Dept. of Land Resources, Govt. of India, 2000}
Mercury (Hg)	Kodaikanal, Tamil Nadu	Tea, coffee, tapioca, rubber, coconut, sugarcane	120.46
	Gonjam, Orissa	Rice, maize, soybean, jute, potato	35.45
	Singrauli, Madhya Pradesh	Soybean, maize, sorghum, wheat, lentil	141.44
Arsenic (As)	West Bengal	Rice, potato, wheat, sugarcane, jute	47.34
	Ballia and other districts, Uttar Pradesh	Wheat, rice, pulses, sugarcane, oilseeds, potatoes	29.26
Copper (Cu)	Singbhum Mines, Jharkhand; Malanjkahnd, Madhya Pradesh	Rice, ragi, maize, wheat, redgram, niger	98.88
Chromium (Cr)	Ranipet, Tamil Nadu;Kanpur, Uttar Pradesh; Vadodra, Gujarat; Talcher, Orissa	Cotton, castor, onion, potato, bajra, wheat, soybean	49.66
Lead (Pb)	Ratlam, Madhya Pradesh; Bandalamottu Mines, Andhra Pradesh;	Rice, maize, ragi, pulses, jowar, small millets, castor.	24.4
	Vadodra, Gujarat;Korba, Chattisgarh	Rice, Wheat, maize, groundnut.	28.20

Therefore, solution to the above problem lies in improving the performance of plants involved in phytoremediation. Various reports have suggested the occurrence of AMF in HM polluted sites, and these fungi are also known to colonize hyperaccumulator plants growing in metal contaminated sites (Leung *et al.* 2007). Advantages of using AMF in conjuction with plants for remediation of HM contaminated sites are:

- Plants colonized by AMF are more robust, better growing, produce more biomass than plants not colonized by the fungi (Leung *et al.* 2006).

- They have boundless ability to survive in severely HM polluted soils (Leung *et al.* 2013).

- By improving plant nutrition AMF enhance the ability of plants to survive under various stresses in contaminated sites (Leung *et al.* 2013).

- They can results in either phytostabilization or phytoextraction of the contaminant.

- They contribute to the uptake of macronutrients as well as micronutrients by increasing the interface between plants and the soil (Smith and Read 2008).

- Fungal hyphae have the ability to extend into soil pores that are small for plant roots to enter (Leung *et al.* 2013).

- They penetrate beyond the nutrient depletion zone surrounding plant roots (Leung *et al.* 2013).

3.1 Strategies Adapted by AMF in Reducing HM Toxicity in Plants

The strategy adopted by AMF to impart tolerance to reduce HM toxicity in plants varies depending on type of HM, the level of contamination and the host plant. The mechanisms employed for specific metals have been discussed in the following sections individually. However, overall generalized mechanisms employed by AMF for reducing metal toxicity in plants are:

- Sequestration of HM in fungal structures such as vesicles, intra and extraradical hyphae and prevention of their translocation to upper plant parts (Christie *et al.* 2004; Wang *et al.* 2007b).

- Compartmentalization of HM in the vacoules (Kaldorf *et al.* 1999).

- Exclusion of HM by precipitation of polyphosphate granules (Andrade *et al.* 2008).

- Increase in the production of metal binding thiol peptides (Andrade *et al.* 2008).

- AMF can bind HM in their chitin cell wall leading to decreased translocation of HMs to the shoots of the plant (Rahmaty *et al.* 2008).

- AMF have the ability to form chelates that solubilize HMs and prevent their translocation to aerial plant parts (Levyal *et al.* 1997; Solis-Dominguez *et al.* 2011).

- AMF produce glomalin, an insoluble glycoprotein that binds to the HMs present in the soil and immobilizes them (Rahmaty *et al.* 2008).

Table 2: Sources of heavy metals contributing to environmental degradation

Heavy Metal	Sources of pollution	Reference
Mercury (Hg)	Thermal power plants, hospital waste (damaged thermometers, barometers, sphygmamometers), electrical appliances, chlor-alkali plants, fumigants; etc.	Mandal *et al.* (2014)
Mangnese (Mn)	Fertilizers.	Mandal *et al.* (2014)
Arsenic (As)	Fossil fuel burning, timber power plants, paints, pesticides, smelting processes, geogenic/natural processes, thermal power plants.	Zhao *et al.* (2009)
Vanadium (Vd)	Spent catalyst, sulphuric acid plant; etc.	Lone *et al.* (2008)
Cadmium (Cd)	Zinc smelting, paint sludge, incinerations and fuel combustion, waste batteries, e-waste.	Andrade *et al.* (2008)
Molybednum (Mo)	Spent catalyst and fertilizer.	Tjandraatmadja *et al.* (2010)
Nickel (Ni)	Battery industry, thermal power plants, smelting processes, mine tailings, alloys.	Jamal *et al.* (2002)
Zinc (Zn)	Electroplating, smelting, galvanization, dyes, fertilizers, mine tailings, timber treatment, paints; etc.	Christie *et al.* (2004)
Copper (Cu)	Smelting operations, mining, electroplating, fungicides, paints, pigments, timber treatment, fertilizers.	Santana *et al.* (2015), da Silva *et al.* (2005)
Lead (Pb)	e-waste, coal based thermal power plants, ceramics, bangle industry, paints, lead acid batteries, petrol additives, preservatives, gasoline additives, plastic recycling industry.	Chen *et al.* (2005), Lermen *et al.* (2015)
Chromium (Cr)	Industrial coolants, timber treatment, leather tanning, pesticides, dyes, chromium salt manufacturing, mining.	Lone *et al.* (2008), Mallick *et al.* (2010)
Uranium (U)	Chimney emissions, ash dumps from thermal power plants, mining and milling operations, nuclear waste, rock phosphate in fertilizers.	Joner *et al.* (2000), Rufykiri *et al.* (2003)
Selenium (Se)	Mining, coal fired power plants, petrochemical operations, procurement, processing and combustion of fossil fuels, fly ash and bottom ash from thermal power plants.	Lemly (1997a and 1997b).

4. Uranium (U)

4.1 Contamination in India

India is a growing and developing nation and therefore is committed to the present use and expansion of existing thermal and nuclear power plants. After thermal, hydroelectric, renewable sources, nuclear power is the fourth largest source of electricity in India (Nuclear Power Plants in India 2011). Waste generated from nuclear power plants has increased U content , that in high amount can be a potential threat to the environment (Chamberlain 2009). According to a report published in an international daily there has been upsurge of various deformities in children being born around the places close to thermal power plants. One such instance had been reported in Punjab wherein U level around plants growing in the vicinity was found to be fifteen times higher than the permissible limits set by World Health Organization (WHO). When coal is burned in thermal power plants the fly ash that is produced is reported to contain concentrated amount of U. The magnitude of this threat becomes multifold due to the fact that Punjab accounts for two-third of wheat in country's central reserves and 40% of its rice production (Chamberlain 2009). Strategies adopted for remediation of U contaminated soils have till now relied on the use of chemical and physical methods that are often costly and have the potential of changing the soil type and the environment. Therefore, the need is to channelize the untapped potential of phytoremediation and various rhizosphere inhabiting micro-organisms.

4.2 Strategies Adapted by AMF for Alleviation of U Contamination in Soil

There is lot of evidence in support of role of AMF in accumulation of U (Rufyikiri *et al.* 2003 and 2004). It has been reported that when AMF associates with a plant in U contaminated soil positive effects are seen with respect to plant growth. When supplied with high U concentration, plant biomass production and AMF colonization increases with the formation of intraradical structures characteristic of the mycorrhiza (Rufyikiri *et al.* 2004). Various mechanisms proposed, to contribute to the uptake and translocation of U by AMF are:

- Intracellular complex formation by polyphosphates (Wang *et al.* 2007a).

- High cation exchange capacity to adsorp U on negatively charged components of fungal tissues (Joner *et al.* 2000; Rufykiri *et al.* 2003).

- Production of glycoproteins like glomalin by extraradical hyphae that leads to U sequestration as these proteins absorb several metal cations (Gonzalez-Chavez *et al.* 2004).

Larger U concentration has been reported in mycorrhizal roots than in non-mycorrhizal roots. This could be attributed to the efficient uptake mechanism employed by AMF wherein after the translocation of U from extraradical to intraradical mycelium it is finally immobilized in the intraradical fungal structures (Rufyikiri et al. 2003).Vesicle abundance in HM contaminated sites has been found to be substantial in these fungi (Chen et al. 2005; Rufyikiri et al. 2004). Intraradical structures of the fungus with low metabolic activity store lipids and also have the capacity to sequester toxic metals and consequently reduce their chemotoxicity (Chen et al. 2006). The compartmentalization of toxic elements within the fungal structures not only protect the AMF, but also restrict the root to shoot translocation of U.

Uptake and translocation of U can also be influenced by the pH of the soil as reported by Rufyikri et al. (2003 and 2004). It was found by experimental analysis that translocation of uranyl phosphate at pH 4 was higher as compared to that at pH 5.5. This difference in translocation of the two U species is attributed to higher proton activity at pH 4 that leads to low adsorption of uranyl cations on the surface of fungal mycelium. In contrast to this assumption at pH 5.5 or higher, fewer number of protons lead to activation of binding sites so that positively charged cations are adsorbed at the surface of fungal hyphae and glycoproteins; and also at pH 5.5 phosphate species are dominant therefore complexation of U with polyphosphates occurs.

Studies to ascertain the role of AMF in remediation of U contaminated soils are scarce, pertaining to few species of AMF involved in remediation and the domain of these studies have been restricted, involving only in vitro and pot culture methods. Therefore, adequate field trials are required to optimize the use of these symbionts in remediation of U contaminated soils. The performance of AMF is influenced by various parameters such as soil composition and availability of other minerals. AMF strains isolated from U contaminated sites will throw light on some not yet known uptake and tolerance mechanisms of these symbionts.

5. Chromium (Cr)

5.1 Cr as a Contaminant in the Environment

Cr has several valence states that are unstable and are biologically short lived. However, trivalent Cr (III) and hexavalent Cr (VI) are the most stable forms. Cr (VI) is the most toxic form of Cr and is known to be associated with oxygen as chromate (CrO_4^{2-}) or dichromate ($Cr_2O_7^{2-}$) oxyanions (Shankera et al. 2005). In India, tanning industries contribute to about 2,000-32,000 tons of elemental Cr annually that escapes in the environment (Shankera et al. 2005). Sukinda valley in Orissa, contributes to 97% of India's chrome reserves and is the largest chromite ore mine in the world. In Sukinda mining area 7.6 million tonnes of solid waste is generated in the form of sub-grade ore, overburden material/ waste rock, rejected minerals, causing disastrous effects on the surrounding environment. Of these, waste rock is spread on the river beds of Bramhani river posing threat to the agricultural crops that are irrigated with the contaminated river water. In Sukinda valley, 60% drinking water and 70% surface water contains hexavalent Cr (Das amd Mishra 2008).

Presence of increased amount of Cr in soil can have deleterious effects on the plant and microbial community of the soil. Accumulation of Cr in various crop plants results in degradation of δ-aminolevulinic acid dehydratase, inhibition of chlorophyll biosynthesis (Choudhary *et al.* 2005). Cr causes deactivation of many enzymes by replacing Mg ions at their active sites. Due to its high oxidative potential Cr results in the production of reactive oxygen species (ROS) that cause oxidative damage in plants, and leads to peroxidation of membrane lipids, which cause degradation of photosynthetic pigments in plants (Mallick *et al.* 2010; Rahmaty *et al.* 2008). Cr mostly exists in the form of trivalent cation Cr^{3+} (Cr (III)) and divalent dichromate anion $Cr_2O_7^-$ Cr ((VI)) in soil (Davies *et al.* 2001). Of the two, Cr (VI) is dominant in plant roots (Smith *et al.* 2003), and crosses cell membrane, where phosphate-sulphate carrier transports Cr anions with higher transport index (Khan *et al.* 2000). On the other hand Cr (III) does not employ any specific membrane carrier for transportation, its biological activity depends on reduction and storage in different plant compartments and sequestration in the roots of the plants (Rahmaty *et al.* 2008).

5.2 Remediation of Cr Contaminated Sites Using AMF

The deleterious effects of Cr toxicity are significantly reduced in plants inoculated with *Glomus intraradices* (Ruscitti *et al.* 2011). Although the colonization and arbuscule formation are affected at high Cr concentration. Translocation of Cr from roots to shoots is relatively slow, and high amount of Cr is retained in the roots of the plants (Khan *et al.* 2000). The deleterious effects of Cr are more pronounced in the roots, affecting growth and membrane integrity, evaluated by electrolyte leakage, but Cr content in leaves does not cause significant damage to the membranes (Davies *et al.* 2002; Ruscitti *et al.* 2011). The reason for lower toxicity of Cr in shoots is attributed to the fact that it is mainly immobilized in the vacuoles of root cells. According to Shankar and Dubey (2004) when Cr(VI) passes through endodermis via symplast it is reduced to Cr(III) and retained in the root cortical cells resulting in reduced Cr(VI) concentration in the shoots. A vital aspect of this discussion is that the enzymes responsible for the reduction of Cr(VI) to Cr(III) are absent in higher plants, but are present in fungi, bacteria, and therefore AMF have an important role to play in reduction of Cr toxicity in plants (Cervantes *et al.* 2001; Vajpayee *et al.* 2000).

Rodriguez *et al.* (2004) suggested that mutualistic fungi provide assistance to host plants in order to perceive stress efficiently, this leads to activation of plant defense responses, not seen in case of non-symbiotic plants. AMF provide a protective barrier to plants avoiding the translocation of Cr to aerial parts (Ruscitti *et al.* 2011). Lower translocation of Cr to aerial parts of the plant allow proline metabolism to operate efficiently. High concentration of Cr in roots negatively affects the metabolism of this protective amino acid. Proline protects the fungal cells from abiotic stresses like heat, UV, drought. At high concentration of Cr, in the presence of AMF less degradation of chlorophyll and proteins occurs. AMF colonization increases the tolerance of plants (Gaur and Adholeya 2004; Ruscitti *et al.* 2011) towards high Cr toxicity in soils, by modifying the proline metabolism in the roots and shoots of the affected plants to make the stress response more efficient (Ruscitti *et al.* 2011). AMF decrease oxidative stress caused due to high Cr toxicity by reducing ROS production and lipid peroxidation (Rahamty *et al.* 2008). The increase in photosynthetic pigments, reduction in lipid peroxidation, lower activity of various antioxidant enzymes, better nutrient availability are few favorable outcomes of AMF association with plants in alleviation of toxic effects of Cr.

6. Lead (Pb) as a Contaminant and its Alleviation Using AMF

Pb is a cumulative, slow acting protoplasmic poison causing decrease in productivity of agricultural crops (Johnson and Eaton 1980). Pb accumulates in soil surface layers therefore to ascertain the amount of Pb directly available for plant uptake is difficult. It has the ability to bind organic matter present in the soil, and its uptake depends on cation exchange capacity, soil particle size, plant root surface area, root exudates and mycorrhization (Davies 1995). Toxic influence of Pb on plants include water imbalance, change in nutritional status, inhibition of enzyme activity, changes in membrane permeability, inhibition of enzymes containing sulphydryl (-SH) groups (Davies 1995).

The root/shoot ratio of Pb concentration is higher in AMF inoculated plants than in non-inoculated ones. Increased accumulation of Pb occurs in fungal structures like vesicles and hyphae. AMF with vesicle abundance are associated with plants growing in Pb contaminated sites (Wang *et al.* 2005). AMF colonized plants have been reported to accumulate 30% less Pb in shoots as compared to non-colonized plants, when supplied with high concentration of Pb (Lermen *et al.* 2015). This is reiterated by the fact that AMF mycelia have high metal sorption capacity (Joner *et al.* 2000). At Pb concentration phytotoxic for plants, AMF colonized soybean plants have shown 140% greater shoot dry matter production as compared to non-colonized plants (Lermen *et al.* 2015). Improved growth and greater yield in sites contaminated with Pb and other HMs occurs in AMF colonized plants. Increased availability of nutrients like phosphorus (P), nitrogen (N), potassium (K), magnesium (Mg) and sulphur (S) due to high volume of soil explored by extraradical hyphae of AMF colonized plants result in greater root absorption area (Gohre and Paskowski 2006; Lermen *et al.* 2015).

It has been substantiated by numerous reports that when AM symbiosis is formed, plant benefits by receiving adequate P supply. The AMF colonized plants growing in excess Pb contamination can reduce plant stress by maintaining higher P/Pb ratios in the shoots (Heggo and Angle 1990; Lermen *et al.* 2015). P is involved in detoxification of various HMs by forming molecules of phytates that have the ability to neutralize HMs present in excess (Heggo and Angle 1990). Presence of P provides metabolic energy in the form of ATP that aid in sequestering HMs in the cell.

7. AMF in Remediation of Copper (Cu) Contaminated Soils

Contamination of soil due to high Cu levels is toxic not only for the plants but also for the microbial communities growing in that soil. AMF occur not only in Cu contaminated areas but in various Cu mine spoils as well (da Silva *et al.* 2005; Griffioen *et al.* 1994). Several AMF strains tolerant to Cu have been reported (Wang *et al.* 2005). Deleterious effects of Cu on plants result in inhibition of photosynthesis and loss of chlorophyll a and b. This results in loss of vegetation, increased soil erosion of the affected area, ground and surface water contamination (Miotto *et al.* 2014). According to Gonsalez-Mendoza *et al.* (2013) excess Cu in *Canavalis ensiformis* (Jack bean) causes increase in minimal fluorescence (Fo). Cu toxicity results in destruction of photosystem II (PS II) reaction center and substitution of Mg by Cu in chlorophylls lead to increase in Fo (Cambrolle *et al.* 2015). Plants exhibiting PS II maximum quantum yield (Fv/Fm) close to 0.85 are considered to be healthy, but in the absence of AMF the ratio in Jack bean, decreased to 0.67 indicating stress conditions (Santana *et al.* 2015).

AMF inoculation of *Marsdeni volubilis* with *Glomus mosseae* increases iron (Fe) concentration in shoots. Exudation of phenol and organic acids by the fungus increases diffusion of Fe to the roots resulting in formation of Fe complexes (Sandhya *et al.* 2014). AMF colonization of plants helps in Fe uptake. Fe is a cofactor for superoxide dismutase (SOD), and plants with efficient Fe uptake show high activity of SOD (M'sehli *et al.* 2014). SOD is the first line of defense produced in response to reactive oxygen species (ROS) during increased stress conditions like HM toxicity. Increased activity of peroxidase (POD) in

AMF colonized plants is indicative of lower ROS production as compared to non-colonized plants (Santana *et al.* 2015). Cu is toxic to membranes. Increase in the activity of SOD and catalase (CAT) in Cu treated AMF colonized plants act as a protectant (Ahmed *et al.* 2010). Plants inoculated with AMF *Glomus mosseae* accumulate proline during Cu toxicity. Proline is an amino acid that accumulates in the presence of HM stress, and reduces free radicals formed during stress conditions (Castillo *et al.* 2011). Proline functions as a ROS scavenger or metal chelator during heavy metal toxicity.

Wang *et al.* (2007b) assessed the effect of AM fungus *Acaulospora mellea* on *Zea mays* (maize) supplied with high concentrations of Cu. The AMF used was isolated from a 100 year old coal mine spoil, where it evolved a strong tolerance towards heavy metals. Plants inoculated with AMF showed decreased concentration of Cu, resulting in minimal damage to the plants and leading to higher biomass. This is because improved P nutrition results in better nutrient uptake in AMF colonized plants (Smith *et al.* 2003). AMF increases soil pH especially under high Cu concentrations. Change in pH affects mobility of Cu in soil causing lower Cu concentrations in AMF colonized plants (Wang *et al.* 2007b). Change in concentration of various soil organic acids such as oxalic acid, citric acid, malic acid in AMF inoculated plants plays a key role in Cu availability to the plants (Wang *et al.* 2007a). All results corroborate that when AMF associates with a plant under increased Cu levels - Cu uptake efficiency, translocation and phytoextraction efficiency is lowered indicating reduced ability of Cu uptake from the soil. These symbionts have a potential in revegetation of Cu contaminated sites, but substantial studies need to be carried out to validate these results.

8. Role of AMF in Alleviating Cadmium (Cd) Toxicity in Plants

Cd is a non-essential element which can enter the food chain by accumulating in plants grown on Cd contaminated soils (Hutchinson *et al.* 2004). High Cd levels in soil are detrimental for plant growth reducing chlorophyll content in plants. Under conditions of high Cd concentration in soil, AMF colonized plants show high shoot dry weight, leaf area and increase in the content of photosynthetic pigments. However, in few instances plants grown in Cd contaminated soil and inoculated with AMF have shown reduced growth because of the direct toxic effects exerted on the physiology and metabolism of the plant (Andrade *et al.* 2008; Rivera-Becerril *et al.* 2002). In order to quantify the ability of the plant to accumulate Cd present in the medium Andrade *et al.* (2008) calculated the transfer factor (TF), it was found that AMF colonized plants have greater TF and greater Cd accumulating ability as compared to non-colonized plants. Several plants have the ability to naturally accumulate Cd in the roots of apoplast due to ionic interactions between components of the cell wall. The portion of the metal interacting, is complexed with phytochelatins and is sequestered in the vacuole (Cohen *et al.* 1998). In addition to the above mechanism employed by few plants the presence of AMF acts as a filter reducing the metal uptake by plants (Andrade *et al.* 2008 and 2005; Joner *et al.* 2000). Cd has the ability to influence concentration of polyvalent cations like Magnesium (Mg), Calcium (Ca) by competing with them for binding sites or for transporters, but their level is unchanged in AMF inoculated plants (Andrade *et al.* 2008). AMF colonized plants growing in high Cd concentration have higher P in shoots, increased content of photosynthetic pigments, low activity of enzyme guaiacol peroxidase (GPX) involved in cell protection against ROS (Andrade *et al.* 2008; Gohre and Paskowski 2006). Cd uptake and nutrient acquisition improves even if AMF colonization is low in some plants (Janouska *et al.* 2005).

According to Hutchinson *et al.* (2004) AMF inoculation decreases Cd concentration by avoiding its direct accumulation in plants rather than causing its translocation to aerial plant parts, signifying reduction in metal uptake by plants. Presence of AMF in soil increases soil phosphatase activity (Wang *et al.* 2005). The influence of AMF on phosphatase activity occurs by synthesis of soil enzymes or release of root exudates comprising soil enzymes by AMF colonized roots. Enhancement of soil phosphatase activity also accounts for better P acquisition by plants harboring AMF (Hu *et al.* 2013).

9. Arsenic (As)

9.1 As Menace in India

Severe As contamination of groundwater and soil is reported in Gangetic plains that share its boundary with Nepal and Bangladesh % the two heavily As contaminated sites of South Asia (Singh *et al.* 2015). In India, Bihar and West Bengal two states of the Middle Gangetic Plain) are the worst affected due to As toxicity (Singh *et al.* 2015). As basically seeps in the groundwater by uncontrolled use of fertilizers and other anthropogenic activities, this contaminated water is used for irrigation of crops in various parts of the two states, ultimately reaching the food chain. In one of the districts in Bihar wheat and banana are the major produce grown in As contaminated sites and are a threat not only to the people of the region but also to the consumers in various parts of the country where the produce is transported for consumption (Singh and Ghosh 2011).

9.2 Role of AMF in Alleviation of As Contaminated Sites

As is a HM occurring naturally that is toxic to plants even in minimal amount (Smith and Read 2008). It exists in oxidation states +V (arsenate), +III (arsenite), 0 (arsenic), and –III (arsine) (Jia et al. 2012). But in soil or aqueous environments arsenite and arsenate are the stable forms present (Bai et al. 2008). Organic forms of As mainly monomethylarsonic acid (MMA), dimethylarsinic acid (DMA) and trimethylarsine oxide (TMAO) are present in soil (Huang and Matzner 2006). Sensitivity of plants growing in As contaminated sites is intricately linked to the availability of P in soil (Wang et al. 2007b). Due to highly similar properties and chemical behavior of As (V) and phosphate (Pi), both compete for the same Pi transporter in plant plasma membrane, with Pi competing much more effectively (Mehrag and Macnair 1994) (Gomes et al. 2006; Gunes et al. 2009). Threats due to As toxicity further intensified due to its accumulation in agricultural lands and soil (Gonzalez et al. 2013). A number of studies have highlighted that arsenate can be removed partially from soil colloids by Pi (Gonzalez et al. 2013). Even when supplied with arsenate, the form dominant in plants is arsenite, suggesting plants have a mechanism for arsenate reduction (Zhao et al. 2009).

As (V) being similar to Pi replaces it in various essential biochemical reactions like photosynthesis, glycolysis, oxidative phosphorylation. Presence of As in plants have detrimental effects on ATP due to formation of ADP-As which is a highly unstable compound (Mehrag et al. 1994). As exposure in plants produces ROS such as hydrogen peroxide (H_2O_2), hydroxyl radical (OH^-), and superoxide radical (O_2^-) (Gunes et al. 2009). Increase in ROS production in the presence of As causes membrane damage resulting in electrolyte leakage, damage to DNA, proteins, and lipid peroxidation of membranes (Gunes et al. 2009). Various factors affect As uptake and its mobility in soil (Bai et al. 2008):

- Soil pH and redox status
- Physical constituents and composition of the soil
- Adsorption, desorption, transformation , and cycling of As in soil

In studies on *Zea mays* (maize), *Medicago trancatula* (barrel medic), *Triticum durum* (wheat) tolerance towards As enhanced along with better growth and improved nutrition in AMF inoculated plants (Bai et al. 2008; Zhang et al. 2015). In some AMF inoculated plants, high As accumulation in plant shoot occurs (Odanka et al. 1985), this mechanism can be exploited in As phytoextraction from contaminated soils (Zhang et al. 2015). AMF also influences speciation of As in contaminated sites. For instance it is reported by Zhang et al. (2015) that in plants inoculated with AMF there is reduction of As

(V) to As (III) and methylation of inorganic As occurs, into less toxic organic DMA. Increased growth and better yield of AMF colonized plants occurs along with enhancement of mineral nutrition (Xu *et al.* 2008). Membrane damage causes production of malondialdehyde (MDA), but its production is decreased in AMF inoculated plants.

Conclusion

According to world agricultural statistics (Food and Agriculture Organization Corporate Statistical Database 2010), India ranks second in the production of wheat and rice worldwide, and is also among the top five producers of 80% agricultural items like cotton, coffee and various other cash crops. India is the seventh largest agricultural exporter worldwide. These statistics are sufficient to point to the obvious fact that the agriculture sector constitutes the backbone of the Indian economy with agriculture and allied fields contributing to 13.7% of the Gross Domestic Product (GDP), 50% of the total workforce for the year 2013 (Asia Sentinel 2011).

Since for the growth of a country its industrialization cannot be tampered with so does the flora and fauna that have close ties with the culture and economy of the country. Therefore a natural and environmentally feasible technology for remediation of contaminated sites from industrial and other sources of pollution hampering the growth of plants and posing risk for human survival is the urgent need of the hour.

The need to act on the rising contamination of soils caused due to HM toxicity becomes aggravated. For the year 2014; West Bengal, Uttar Pradesh, Andhra Pradesh are leading producers of rice, whereas in wheat production states like Uttar Pradesh, Punjab, Madhya Pradesh have excelled, and Andhra Pradesh, Karnataka, Maharashtra are leading in maize production (Directorate of Economics and Statistics, Government of India 2014). Almost all these states have contamination of various HMs as mentioned in Table 1. For remediation of contaminated soils in these and other places of pollution, use of AMF could prove to be valuable, as highlighted in Fig 1.

In reports by various authors highlighted in this article AMF are of great importance in plants growing in HM contaminated sites contributing to lower metal uptake and increased yield of plants. According to Zhang *et al.* (2015) vesicular abundance presented constructive correlation with concentration of extractable Zn and Cd present in the soil. The mechanisms undertaken by AMF to alleviate metal toxicity in plants are neither well understood nor well documented, therefore more research focusing on this aspect could be productive. The effects exerted by AMF on metal tolerance cannot be generalized and vary depending on plant species, soil characters like pH, chemical composition. It is validated by various reports that for some AMF species colonization rate and spore number is low in soils heavily contaminated with HMs but AMF already growing in contaminated sites are better adapted to tolerate metal toxicity and contribute to increased plant growth (Lermen *et al.* 2015). AMF ecotypes from different habitats exhibit varying degree of tolerance and those isolated from HM contaminated sites are more tolerant.

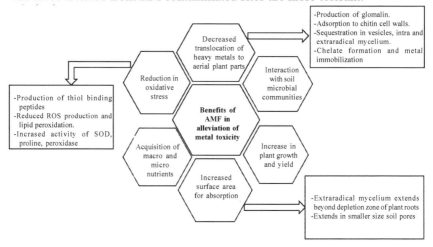

Fig. 1: Multi-faceted approach of AMF in alleviation of HM toxicity in plants

Furthermore, in the absence of sufficient information on availability of metals it becomes difficult to compare different studies pertaining to HM toxicity (Lermen *et al.* 2015). Another problem with the study of AMF is that most of the fungi detected in natural ecosystems pose difficulty in culturing and may have a restricted host range as compared to *Glomus mosseae* or *Glomus intraradices* that are intensively investigated and are easily cultured (Parniske 2008). The performance of these fungi in alleviation of metal toxicity in contaminated sites is enhanced by use of various amendments and by careful planning and understanding of the mechanisms employed by them. For instance Santana *et al.* (2015) reported the use of organic fertilizers in improving the process of phytoremediation due to presence of high concentration of HM binding functional groups present in these fertilizers, decreasing availability of HMs resulting in promotion of plant growth. According to literature survey conflicting viewpoints are predicted by various authors, some emphasize employment of AMF for of phytostabilization and not phytoextraction therefore constructive research needs to be conducted to bring light to the actual role played by these symbionts, leading to their extensive use in remediation strategies. In a study conducted by Solis-Domniguez *et al.* (2011) AMF interact with different groups of soil bacteria, and have the potential to modify the microbial community, which can indirectly exert a positive influence on plant growth and productivity. Therefore, employing these fungi in remediation of contaminated sites could prove to be productive without hampering the surrounding environment.

References

Ahmed H, Hasan A, Akhtar S, Hussain A, Abaidullah YG, Wahid A, Mahmood S (2010) Antioxidant enzymes as bio-markers for copper tolerance in safflower (*Carthamus tinctorius* L.). Afri. J. Biotech. 9, 5441-5444.

Andrade SAL, de Abre MF, Jorge RA, Silveira APD (2005) Cadmium effect on the association of jackbean (*Canavalia ensiformis*) and arbuscular mycorrhizal fungi. Sci. Agric. 62, 389-394.

Andrade SAL, Silveira APD, Jorge AR, de Abreu FM (2008) Cadmium accumulation in sunflower plants influenced by arbuscular mycorrhiza. Int. J. Phytorem. 10, 1-13.

Asia Sentinel (2011) India's rising nuclear safety concerns. http://www.asiasentinel.com/blog/indias-rising-nuclear-safety-concerns

Bai JF, Lin XG, Yin R, Zhang HY, Wang JH, Chen XM, Luo YM (2008) The influence of arbuscular mycorrhizal fungi on As and P uptake by maize (*Zea mays* L.) from As contaminated soils. Appl. Soil. Ecol. 38, 137-145.

BAIF Research Foundation (2006) Involvement of small farmers in cashew production. http://www.baif.org.in

Cambrollé J, Garcia JL, Figueroa ME, Cantos M (2015) Evaluating wild grapevine tolerance to copper toxicity. Chemosphere 120, 171-178.

Castillo OS, Dasgupta-Schubert N, Alvarado CJ, Zaragoza EM, Villegas HJ (2011) The effect of the symbiosis between *Tagetes erecta* L. (marigold) and *Glomus intraradices* in the uptake of Copper (II) and its implications for phytoremediation. Biotechnol. 29, 156-164.

Cervantes C, Garcia JC, Devars S, Corona FG, Tavera HL, Torres Guzman J (2001) Interactions of chromium with microorganisms and plants. FEMS Microbiol. Rev. 25, 335-347.

Chamberlain G (2009) India's generation of children crippled by nuclear waste.

Chen BD, Jakobsen I, Roos P, Zhu YG (2005) Effects of the mycorrhizal fungus *Glomus intraradices* on uranium uptake and accumulation by *Medicago truncatula* L. from uranium contaminated soil. Plant Soil 275, 349-359.

Chen BD, Zhu YG, Smith FA (2006) Effects of arbuscular mycorrhizal inoculation on uranium and arsenic accumulation by Chinese brake fern (*Pteris vittata* L.) from a uranium mining impacted soil. Chemosphere 62, 1464-1473.

Choudhury S, Panda SK (2005) Toxic effects, oxidative stress and ultrastructural changes in the moss *Taxithelium nepalense* (Schwaegr.) Broth under chromium and lead phytotoxicity. Water Air Soil Pollut. 157, 73-90.

Christie P, Li XL, Chen B (2004) Arbuscular mycorrhiza can depress translocation of zinc to shoots of host plants in soils moderately polluted with zinc. Plant Soil 261, 209-217.

Cohen CK, Fox TC, Garvin DF, Kochian LV (1998) The role of iron deficiency stress responses in stimulating heavy metal transport in plants. Plant Physiol. 116, 1063-1072.

da Silva GA, Trufem SF, Saggin Junior OJ, Maia LC (2005) Arbuscular mycorrhizal fungi in a semi-arid copper mining area in Brazil. Mycorrhiza 15, 47-53.

Das PA, Mishra S (2008) Hexavalent Chromium (VI): environment pollutant and health hazard. J. Environ. Res. Dev. 2, 3.

Davies BE (1995) Lead and other heavy metals in urban areas and consequences for health of their inhabitants. In: Environmental contaminants, Ecosystem and Human Health (Ed) Majumdar

SK, Miller EW and Brenner FJ, Pennsylvania Academy of Science, Easton PA, USA, pp 287-307.

Davies FT, Puryear JD, Newton JR, Egilla JN, Grossi JAS (2001) Mycorrhizal fungi enhance accumulation and tolerance of chromium in sunflower (*Helianthus annuus*) J. Plant Physiol. 158, 777-786.

Davies FT, Puryear JD, Newton RJ, Egilla JN, Grossi JAS (2002) Mycorrhizal fungi increase chromium uptake by sunflower plants: influence on tissue mineral concentration, growth and gas exchange. J. Plant Nutr. 25, 2389-2407.

Department of Land Resources, Ministry of Rural Development, Government of India (2000) District and category wise wastelands of India http://dolr.nic.in/wasteland.htm

Directorate of Economics and Statistics, Department of Agriculture and Cooperation, Ministry of Agriculture, Government of India (2014) Agricultural Statistics at a Glance.

Food and Agriculture Organization Corporate Statistical Database (FAOSTAT) 2010 data. http://faostat.fao.org/site/567/default.aspx#ancor

Gaur A, Adholeya A (2004) Prospects of arbuscular mycorrhizal fungi in phytoremediation of heavy metal contaminated soils. Curr. Sci. 86, 528-534.

Gohre V, Paskowski U (2006) Contribution of the arbuscular mycorrhizal symbiosis to heavy metal phytoremediation. Planta 223, 1115-1122.

Gomes-Junior RA, Moldes CA, Delite FS, Gratao PL, Mazzafera P, Lea PJ, Azevedo RA (2006) Nickel elicits a fast antioxidant response in *Coffea arabica* cells. Plant Physiol. Biochem. 44, 420-429.

González MD, Espadas GF, Escoboza GF, Santamaria JM, Zapata PO (2013) Copper stress on photosynthesis of black mangle (*Avicennia germinans*). Acad. Bras. Cienc. 85, 665-670.

Gonzalez-Chavez MC, Gonzalez C, Wright R, Nichols SE (2004) The role of glomalin, a protein produced by arbuscular mycorrhizal fungi, in sequestering potentially toxic elements. Environ. Pollut. 130, 317-323.

Griffioen WAJ, Ietswaart JH, Ernst WHO (1994) Mycorrhizal infection of an *Agrostis capillaries* population on a copper contaminated soil. Plant Soil 158, 83-89.

Gunes A, Pilbeam D, Inal A (2009) Effect of arsenic-phosphorus interaction on arsenic-induced oxidative stress in chickpea plants. Plant Soil 314, 211-220.

Harrisson J Maria (1997) The arbuscular mycorrhizal symbiosis: an underground association. Trends Plant Sci, Vol. 2, No 2.

Heggo A, Angle JS (1990) Effect of vesicular arbuscular mycorrhizal fungi on heavy metal uptake by soybeans. Soil Biol. Biochem. 22, 865-869.

Hibbett DS (2007) A higher level phylogenetic classification of the Fungi. Mycol. Res. 111, 509-547.

http://www.theguardian.com/world/2009/aug/30/india-punjab-children-uranium-pollution

Hu J, Wang H, Wu F, Wu S, Cao ZH, Lin X, Wong HM (2013) Arbuscular mycorrhizal fungi influence the accumulation and partitioning of Cd and P in bashful grass (*Mimosa pudica* L.) grown on a moderately Cd-contaminated soil. Appl. Soil Ecol. 73, 51-57.

Huang JH, Matzner E (2006) Dynamics of organic and inorganic arsenic in the solution phase of an acidic fen in Germany. Geochim. Cosmochim. Ac. 70, 2023-2033.

Hutchinson JJ, Young SD, Black CR, West HM (2004) Determining radiolabile soil cadmium by arbuscular mycorrhizal hyphae using isotopic dilution in a compartmental pot system. New Phytol. 164, 477-484.

Jamal A, Ayub N, Usman M, Khan GA (2002) Arbuscular mycorrhizal fungi enhance zinc and nickel uptake from contaminated soil by soybean and lentil. Int. J. Phytoremediation 4, 205-221.

Janouskova M, Pavlýkova D, Macek T, Vosatka M (2005) Arbuscular mycorrhiza decreases cadmium phytoextraction by transgenic tobacco with inserted metallothionein. Plant Soil

272, 29-40.

Jia Y, Huang H, Sun GX, Zhao FJ, Zhu YG (2012) Pathways and relative contributions to arsenic volatilization from rice plants and paddy soil. Environ. Sci. Technol. 46, 8090-8096.

Johnson MS, Eaton JW (1980) Environmental contamination through residual trace metal dispersal from a derelict lead-zinc mine. J. Environ. Qual. 9, 175-179.

Joner EJ, Briones R, Leyval C (2000) Metal-binding capacity of arbuscular mycorrhizal mycelium. Plant Soil 226, 227-234.

Kaldorf M, Kuhn AJ, Schroder WH, Hildebrandt U, Bothe H (1999) Selective element deposits in maize colonized by a heavy metal tolerance conferring arbuscular mycorrhizal fungus. J. Plant Physiol. 154,718-728.

Khan AG, Kuek C, Chaudhry TM, Khoo CS, Hayes WJ (2000) Role of plants, mycorrhizae and phytochelators in heavy metal contaminated land remediation. Chemosphere 41,197-207.

Lemly AD (1997a) Environmental implications of excessive selenium: a review. Biomed. Environ. Sci. 10, 415-435.

Lemly AD (1997b) Ecosystem recovery following selenium contamination in a freshwater reservoir.

Lermen C, Fabricio M, Zilda CG, da Silva PA, Eduardo JG, Dragunski DC, Albertona O (2015) Essential oil content and chemical composition of *Cymbopogon citratus* inoculated with arbuscular mycorrhizal fungi under different levels of lead. Ind. Crop. Prod 76, 734-738.

Leung HM, Wang ZW, Ye ZH, Yung KL, Peng XL, Cheung KC (2013) Interactions between arbuscular mycorrhizae and plants in phytoremediation of metal contaminated soils: a review. Pedosphere 23, 549-563.

Leung HM, Ye ZH, Wong MH (2006) Interactions of mycorrhizal fungi with *Pteris vittata* (As hyperaccumulator) in As contaminated soil. Environ. Pollut. 139, 1-8.

Leung HM, Ye ZH, Wong MH (2007) Survival strategies of plants associated with arbuscular mycorrhizal fungi on toxic mine tailings. Chemosphere 66, 905-915.

Leyval C, Turnau K, Haselwandter K (1997) Effect of heavy metal pollution in mycorrhizal colonization and function: physiological, ecological and applied aspect. Mycorrhiza 7,139-153.

Lone MI, He Z, Stoffella PJ, Xiao Y (2008) Phytoremediation of heavy metal polluted soils and water: Progresses and perspective. J. Zhejiang Univ. Sci. B. 9, 210-220.

M'sehli W, Houmani H, Donnini S, Zocchi G, Abdelly C, Gharsalli M (2014) Iron deficiency tolerance at leaf level in *Medicago ciliaris* plants. Am. J. Plant Sci. 5, 2541-2553.

Mallick S, Sinam G, Mishra RK, Sinha S (2010) Interactive effects of Cr and Fe treatments on plants growth, nutrition and oxidative status in *Zea mays* L. Ecotox. Environ. Safe. 73, 987-995.

Mandal A, Purakayastha TJ, Ramana S, Neenu S, Bhaduri D, Chakraborty K, MC Manna, Rao SA (2014) Status on phytoremediation of heavy metals in India-a review. Intl J. Bioresour. Stress Manage. 5, 553-560.

McGrath SP, Zhao FJ (2003) Phytoextraction of metals and metalloids from contaminated soils. Curr. Opin. Biotechnol.14, 277-282.

Meharg AA, Macnair MR (1994) Relationship between plant phosphorus status and the kinetics of arsenate influx in clones of *Deschampsia cespitosa* (L.) Beauv that differ in their tolerance to arsenate. Plant Soil 162, 99-106.

Miotto A, Ceretta CA, Brunetto G, Nicoloso FT, Girotto E, Farias J, Tiecher TL, De Conti L, Trentin G (2014) Copper uptake, accumulation and physiological changes in adult grapevines in response to excess copper in soil. Plant Soil 374, 593-610.

Nuclear power plants in India (2011). http://cea.nic.in/powersecreports/ExecutiveSummary/ 200812/27-33

Odanaka Y, Tsuchiya N, Matano O, Goto S (1985) Characterization of arsenic metabolites in rice plant treated with DSMA (disodium methanearsonate). J. Agr. Food Chem. 33; 757-763.

Parniske M (2008) Arbuscular mycorrhiza: the mother of plant root endosymbiosis. Nat. Rev. Microbiol. 6, 763-775.

Rahmaty R, Khara J (2008) Effects of vesicular arbuscular mycorrhiza *Glomus intraradices* on photosynthetic pigments, antioxidant enzymes, lipid peroxidation, and chromium accumulation in maize plants treated with chromium Turk. J. Biol. 35, 51-58.

Redecker D, Schüßler A, Stockinger H, Sturmer LS, Morton BJ, Walker C (2013) An evidence based consensus for the classification of arbuscular mycorrhizal fungi (Glomeromycota). Mycorrhiza 23, 515-531.

Rivera BF, Calantzis C, Turnau K, Caussanel JP, Belismov AA, Gianinazzi S, Strasser JR, Gianinazzi PV (2002) Cadmium accumulation and buffering of cadmium-induced stress by arbuscular mycorrhiza in three *Pisum Sativum* L. genotypes. J. Exp. Bot. 371, 1177-1185.

Rodriguez R, Redman R, Henson JM (2004) The role of fungal symbioses in the adaptation of plants to high stress environments. Mitig. Adapt. Strateg. Glob. Change 9, 261-272.

Romkens P, Bouwman L, Japenga J, Draaisma C (2002) Potentials and drawbacks of chelate enhanced phytoremediation of soils. Environ. Pollut. 116, 109-121.

Rufyikiri G, Declerck S, Thiry Y (2004) Comparison of 233U and 33P uptake and translocation by the arbuscular mycorrhizal fungus *Glomus intraradices* in root organ culture conditions. Mycorrhiza 14, 203-207.

Rufyikiri G, Thiry Y, Delvaux B, Declerck S (2003) Contribution of hyphae and roots to uranium uptake, accumulation and translocation by arbuscular mycorrhizal carrot roots under root-organ culture conditions. New Phytol. 158, 391-399.

Ruscitti M, Arango M, Ronco M, Beltrano J (2011) Inoculation with mycorrhizal fungi modifies proline metabolism and increases chromium tolerance in pepper plants (*Capsicum annuum* L.). Braz. J. Plant Physiol. 23, 15-25.

Sandhya A, Vijaya T, Sridevi A, Narasimba G (2014) Co-Inoculation studies of vesicular arbuscular mycorrhizal fungi (VAM) and phosphate solubilizing bacteria (PSB) on nutrient uptake of *Marsdenia volubilis* (T. Cooke). Ann. Plant Sci. 3, 765-769.

Santana AN , Ferreira AAP, Soriani HH, Brunetto G, Nicoloso TF, Antoniolli IZ, Jacques SJR (2015) Interaction between arbuscular mycorrhizal fungi and vermi-compost on copper phytoremediation in a sandy soil. Appl. Soil Ecol. 96 172-182.

Schüâler A, Schwarzott D, Walker C (2001) A new fungal phylum, the Glomeromycota: phylogeny and evolution. Mycol. Res. 105, 1413-1421.

Shankar P, Dubey SR (2004) Lead toxicity in plants. Braz. J. Plant Physiol. 17, 35-52.

Shankera KA, Cervantes C, Taverac LH, Avudainayagam S (2005) Chromium toxicity in plants. Environ. Int. 31, 739-753.

Singh KS (2015) Groundwater Arsenic Contamination in the Middle-Gangetic Plain, Bihar (India): The Danger Arrived. Int. Res. J. Environ. Sci. 4, 70-76.

Singh SK, Ghosh AK (2011) Entry of Arsenic into Food Material-A Case Study. World Appl. Sci. J. 13, 385-90.

Smith SE, Read DJ (2008) Mycorrhizal symbiosis, 3rd edition, Academic Press, Elsevier, New York.

Smith SE, Smith FA, Jakobsen I (2003) Mycorrhizal fungi can dominate phosphate supply to plants irrespective of growth responses. Plant Physiol. 133, 16-20.

Solis-Dominuguez FA, Vargas Valentine Alexis, Chorover Jon, Maier M Raina (2011) Effect of arbuscular mycorrhizal fungi on plant biomass and the rhizosphere microbial community structure of mesquite grown in acidic lead/zinc mine tailings. Sci. Total Environ. 409, 1009-1016.

Tjandraatmadja G, Pollard C, Sheedy C, Gozukara Y (2010) Sources of contaminants in domestic waste water nutrients and additional elements from household products. National Research Flagships, CSIRO.

Vajpayee P, Tripathi RD, Rai UN, Ali MB, Singh SN (2000) Chromium (VI) accumulation reduces chlorophyll biosynthesis, nitrate reductase activity and protein content of *Nymphaea alba*. Chemosphere 41, 1075-1082.

Wang FY, Lin XG, Yin R (2005) Heavy metal uptake by arbuscular mycorrhizas of *Elsholtzia splendens* and the potential for phytoremediation of contaminated soil. Plant Soil 269, 225-232.

Wang FY, Lin XG, Yin R (2007a) Inoculation with arbuscular mycorrhizal fungus *Acaulospora mellea* decreases Cu phytoextraction by maize from Cu contaminated soil. Pedobiologia 51, 99-109.

Wang FY, Lin XG, Yin R (2007b) Role of microbial inoculation and chitosan in phytoextraction of Cu, Zn, Pb and Cd by *Elsholtzia splendens*- a field case. Environ. Pollut. 147, 248-255.

Xu P, Christie P, Liu Y, Zhang J, Li X (2008) The arbuscular mycorrhizal fungus *Glomus mosseae* can enhance arsenic tolerance in *Medicago truncatula* by increasing plant phosphorus status and restricting arsenate uptake. Environ. Pollut. 156, 215-220.

Zhang X, Ren BH, Wu SL, Sun YQ, Lin G, Chen BD (2015) Arbuscular mycorrhizal symbiosis influences arsenic accumulation and speciation in *Medicago truncatula* L. in arsenic contaminated soil. Chemosphere 119, 224-230.

Zhao FJ, Ma F, Meharg AA, McGrath SP (2009) Arsenic uptake and metabolism in plants. New Phytol. 181, 777-794.

5

Bioremediation of Pesticide Contaminated Soil: A Cost Effective Approach to Improve Soil Fertility

Jyoti Bisht and Harsh, N.S.K.

Abstract

Excessive use and treatment of soil with pesticides under the adage, "if little is good, a lot more will be better" has played havoc with human and other life forms and can also cause populations of beneficial soil microorganisms to decline, which indirectly affect soil fertility. Soil constitutes a major environmental sink for many pesticides. These pesticides are either taken up from soil by plants and then pass into the bodies of invertebrates, water or air, directly or indirectly and affect public health and beneficial biota of the ecosystem. Bioremediation provides very attractive, eco-friendly and economic solution to many of our hazardous pollution problems. For both economic and ecological reasons, biological degradation has become an increasingly popular alternative for treatment of hazardous wastes.

Keywords: Bioremediation, Degradation, Fungi, Hazardous, Pesticides contamination, Soil fertility.

1. Introduction

One of the strategies to increase crop productivity is effective pest management because more than 45% of annual food production is lost due to pest infestation. Substantial food production is lost due to insect pests, plant pathogens, weeds, rodents, birds, nematodes and spoilage during storage. In tropical countries, crop loss is even more severe because the prevailing high temperature and humidity are highly conducive to rapid multiplication of pests. Chemical pesticides have contributed greatly to the increase of yields in agriculture by controlling insect pests and diseases and also towards checking the insect-borne diseases (malaria, dengue, encephalitis, etc.) in the human health sector. With the introduction of pesticides, farm practices have undergone revolutionary changes leading to incredible possibility that hunger can be vanished from earth. Although various methods are used to control pests, pesticides continue to be the major component of most of the pest control programmes and will probably remain so in near future.

2. Pesticides

Pesticides refer to chemical substances that alter biological processes of living organisms deemed to be pests, whether these are insects, moulds or fungi, weeds or noxious plants. They can be classified by considering the following criteria: target organisms, mode of action and chemical classes. According to the type of pests they control, the main classes of pesticides include herbicides (that kill weeds), fungicides (that kill fungi), insecticides (that kill insects) and rodenticides (that kill rodents). Depending upon the chemical nature, the major classes of pesticides include organochlorine, organophosphorus, carbamates, pyrethroids and neonicotiniods (US EPA 2006).

3. Pesticides Hazards: A Cause of Concern

3.1 Environmental Contaminations

The pesticide industry has become so integrated into our agricultural production and many farmers believe that productivity cannot survive without them. Unfortunately, high-volume pesticide application under the adage, "if little is good, a lot more will be better" has played havoc with human and other life forms much more than the intended targets. It has been estimated that only 0.1% of applied pesticides actually reach the target pests (Horrigan and Lawrence 2002) and more than 99.9% of used pesticides make their way into environment where they adversely affect public health and beneficial biota; and contaminate soil, water and atmosphere of the ecosystem (Mathew 2006). As a result, they are routinely detected in air, surface and ground water, sediment, soil, vegetable, and to some extent in foods.

Heavy treatment of soil with pesticides can cause populations of beneficial soil microorganisms to decline, which will indirectly affect the soil fertility. Sometimes pesticides have a negative impact on the available NPK from soil (Sardar and Kole 2005). Until 1962, pesticide use in agriculture and public health was indiscriminate. Only after the publication of "Silent Spring" by Rachel Carson in 1962, people's awareness towards the ill effects of pesticides increased.

Carson, more than anyone before, had pointed out the risks of pesticides (Carson 1962). The potential risk to human health makes the remediation of pesticide-contaminated sites a necessary and almost urgent undertaking. Most of the pesticides are banned under pesticide pollution and toxicity prevention act.

3.2 Direct Impact on Humans

Certain environmental chemicals, including pesticides termed as endocrine disruptors, are known to elicit their adverse effects by mimicking or antagonizing natural hormones in the body and it has been postulated that their long-term, low-dose exposure is increasingly linked to human health effects such as immune suppression, hormone disruption, diminished intelligence, reproductive abnormalities and cancer (Hurley *et al.* 1998). Additionally, many studies have indicated that pesticide exposure is associated with long-term health problems such as respiratory problems, memory disorders, dermatologic conditions, cancer, depression, neurological deficits, miscarriages and birth defects (WHO 2009).

3.3 Residues in Food

Of late, increased attention has been focused on chemical residues in food. It has been observed that a diet containing fresh fruit and vegetables far outweigh potential risks from eating very low residues of pesticides in crops (WHO 2009).

3.4 Reduction of Beneficial Species

Pesticides easily find their way into soils, where they may be toxic to arthropods, earthworms, fungi, bacteria, and protozoa. Small organisms are vital to ecosystems because they dominate both the structure and function of ecosystems. Like pest populations, beneficial natural enemies and biodiversity (predators and parasites) are adversely affected by pesticides (Aktar *et al.* 2009). It has been estimated that honeybee colonies have dropped in the US from 4.4 million in 1985 to <1.9 million in 1997 due to indirect and direct effects from pesticides (Horrigan and Lawrence 2002). Honeybees are involved in the pollination of an estimated $14 billion worth of US seeds and crops (Barrionuevo 2007).

3.5 Resistance in Target Pests

In addition to destroying natural enemy populations, the extensive use of pesticides has often resulted in the development and evolution of pesticide resistance in insect pests, plant pathogens and weeds (WHO 2009).

4. Bioremediation: A Strategy for Cost Effective Treatment of Pesticides

4.1 Terminology

The term bioremediation is defined as the use of biological agents like bacteria, fungi, actinomycetes and plants to reduce or eliminate hazardous substances from the environment. It uses naturally occurring bacteria and fungi or plants to degrade or detoxify hazardous substances. For bioremediation to be effective, microorganisms must enzymatically attack the pollutants and convert them to harmless products. As bioremediation can be effective only where environmental conditions permit microbial growth and activity, its application often involves the manipulation of environmental parameters to allow microbial growth and degradation to proceed at a faster rate. Bioremediation techniques are typically more economical than traditional methods such as incineration, and some pollutants can be treated on site, thus reducing exposure risks for clean-up personnel, or potentially wider exposure as a result of transportation accidents. Since bioremediation is based on natural attenuation the public considers it more acceptable than other technologies (Vidali 2001).

4.2 Mechanism of Biodegradation

Biodegradation or biotransformation of pesticides by soil microorganisms is predominantly an aerobic process. However, a variety of chemicals are decomposed also in the absence of molecular oxygen by reaction involving hydrolysis, dechlorinations or reductions, or hydrolytic ring cleavage. Microbial degradation has advantages because a large variety of compounds can be degraded completely under mild conditions compared to degradation using physical and chemical means. Microbes are the key players in bioremediation as they generate the enzymes that catalyze the degradative reactions. Microbes use organic substances as a source of carbon and energy and gain raw material for their multiplication and maintenance (Wallnofer and Engelhardt 1989).

4.3 Mycoremediation

Bacteria are very sensitive to fluctuating environmental conditions in soil since their growth requires films of water to form in soil pores. The attributes that distinguish filamentous fungi from other life-forms determine why they are good biodegraders (Paszczynski and Crawford 2000) are given below:

- The mycelial growth habit gives a competitive advantage over single cells such as bacteria and yeasts, especially with respect to the colonization of insoluble substrates.

- Fungi can rapidly ramify through substrates literally digesting their way along by secreting a battery of extracellular degradative enzymes.

- Hyphal penetration provides a mechanical adjunct to the chemical breakdown affected by the secreted enzymes. The high surface-to-cell ratio characteristic of filaments maximizes both mechanical and enzymatic contact with the environment.

- The extracellular nature of the degradative enzymes enables fungi to tolerate higher concentration of toxic chemicals

4.4 Potential of White Rot Fungi in Mycoremediation

The application of fungi for the cleanup of contaminated soil first came to attention in the mid-1980s, when the white rot fungus *Phanerochaete chrysosporium* was shown to metabolize an extremely diverse group of environmental contaminants (Trejo-Hernandez *et al.* 2001). Since then, there has been intense worldwide research to unravel the potential of white rot fungi in bioremediation. This ability of white rot fungi to degrade a wide spectrum of environmental pollutants sets them apart from many other microbes used in bioremediation. Later, this ability was demonstrated for other white rot fungi, including *Trametes versicolor* and *Pleurotus ostreatus* (Ghani *et al.* 1996). The lignin modifying enzymes (LMEs) of white rot fungi consists of a battery of enzymes that catalyze oxidation of xenobiotics in addition to their ability to degrade lignin. Free radical species generated during the degradation process (of either lignin or organopollutants) may serve as secondary oxidants, which may, in turn, mediate the oxidation of other compounds away from the active sites of the enzymes (Barr and Aust 1994).

Many fungi have been tested for their ability to degrade pesticides. More technically advanced research efforts are required for searching, exploiting new fungal species and improvement of practical application to propagate the use of fungi for bioremediation of harmful pesticides. This section includes the review of fungal degradation of pesticides, which is summarized in Tables 1 and 2.

Table 1: National status of fungal degradation of pesticides

Pesticides	Fungi	References
Endosulfan	*Aspergillus niger*	Mukherjee and Gopal 1994; Bhalerao 2012
	Mucor thermohyalospora	Shetty *et al.* 2000
	Aspergillus terreus and *Cladosporium oxysporum*	Mukherjee and Mittal 2005
	Trichoderma viride	Subashini *et al.* 2007
	Aspergillus nidulans	Dave and Dikshit 2011
	Aspergillus niger, Trichosporon sp. and *Verticillium dahaliae.*	Hussaini *et al.* 2013
Chlorpyrifos	*Aspergillus niger* and *Trichoderma viride*	Mukherjee and Gopal 1996
	Trichoderma viride and *T. harzianum*	Jayaraman *et al.* 2012
	Trichoderma harzinaum and *Rhizopus nodosus*	Harish *et al.* 2013
	Ganoderma australe, Trichosporon sp. and *Verticillium dahaliae*	Hussaini *et al.* 2013
Monocrotophos	*Aspergillus* sp.	Das and Anitha 2011
Lindane (γ-HCH)	*Phanerochaete chrysosporium* and *Trametes hirsuta*	Singh and Kuhad 1999
	Cyathus bulleri and *Phanerochaete sordid*	Singh and Kuhad 2000
β-cyfluthrin	*T.viride*5-2 and *T.viride* 2211 strains	Saikia and Gopal 2004
Malathion	*Rhizopus oligosporus*	Sri *et al.* 2011

Table 2: International Status of fungal degradation of pesticides

Pesticides	Fungi	References
Chlorpyrifos	*Phanerochaete chrysosporium*	Bumpus *et al.* 1993; Zhao *et al.* 2011
	Aspergillus terreus, *A. tamari*, *A. niger*, *Trichoderma harzianum* and *Penicillium brevicompactum*	Omar 1998
	Aspergillus sp., *Trichoderma harzianum* and *Penicillium brevicompactum*	Omar and Abdel-Sater2001
	Hypholoma fasciculare, Coriolu sversicolor	Bending *et al.* 2002
	Fusarium LK	Wang *et al.* 2005
	Trichosporon sp.	Xu *et al.* 2007
	Verticillium sp. DSP	Fanga *et al.* 2008
	Trametes versicolor	Gouma 2009
	Fusarium LK. ex Fx	Xie *et al.* 2010
	Cladosporium cladosporioides	Gao *et al.* 2012
Endosulfan	*Trichderma harzianum*	Katayama and Matsumura 1993
	Phanerochaete chrysosporium	Kullman and Matsumura 1996
	Fusarium ventricosum	Siddique *et al.* 2003

(Contd.)

Pesticides	Fungi	References
	Pleurotus pulmonarius	Rodriguez *et al.* 2006
	Chaetosartorya stromatoides, Aspergillus terricola and Aspergillus terreus	Hussain 2007
	Aspergillus sydowi	Goswami *et al.* 2009
	Mortierella sp. W8	Kataoka *et al.* 2010
	Phanerochaete chrysosporium, Trametes versicolor, T. hirsuta, Phlebia lindtneri, P. brevispora, Ceriporia lacerata, Pycnoporus coccineus, and *Phlebia* sp. MG-60	Kamei *et al.* 2011
Endosulfan sulfate	*Trametes hirsuta*	Kamei *et al.* 2011
Fonos	*Phanerochaete chrysosporium*	Bumpus *et al.* 1993; Pointing 2001
Turbofos	*Phanerochaete chrysosporium*	Bumpus *et al.* 1993; Pointing 2001
Dimethoate	*Aspergillus* sp. Z58	Liu and Zhong 2000
	Aspergillus niger	Liu *et al.* 2001
Malathion	*Rhizopus arrhizus*	Omar 1998
	Aspergillus sp.EM8 and *Penicillium* sp. EMT	Hasan 1999; Massoud *et al.* 2007
Fenitrothion	*Trichoderma viride, Mortierella isabellina* and *Saprolegnia parasitica*	Baarschers and Heitland 1986
Glyphosate	*Penicillium citrinum*	Pothuluri *et al.* 1998
	Pencillium notatum	Pothuluri *et al.* 1992
Lindane (β-HCH)	*Pleurotus eryngii, Polyporus ciliatus* and *Bjerkandera adusta*	Quintero *et al.* 2008
Lindane (δ-HCH)	*Bjerkandera adusta, Polyporus ciliatus, Lentinus tigrinus, Stereum hirsutum, Pleurotus eryngii* and *Irpex lacteus*	Quintero *et al.* 2008
Lindane (γ-HCH)	*Bjerkandera adusta, Polyporus ciliatus, Lentinus tigrinus* and *Irpex lacteus.*	Quintero *et al.* 2008
DDT	*Trametes* sp., *Polyporus* sp. and *Nigroporus* sp.	Siripong *et al.* 2009
	Trametes versicolor	Sari *et al.* 2013

5. Effect of Pesticides on Soil Fertility

Healthy soil is essential for the integrity of terrestrial ecosystems to remain intact or to recover from disturbances, such as drought, climate change, pest infestation, pollution, and human exploitation including agriculture. Pollutants, when incorporated in soil leads to adverse effect on soil quality. Easily degradable xenobiotics do not adversely affect the soil fertility, while the most persistent pollutants directly affect the soil quality. They inhibit the growth of soil microorganisms, which leads to affect the biogeochemical cycle of soil and ultimately the soil quality. The production of easily metabolized and less toxic transformation products may not be possible for all pesticides. Thus it is very important to assess the toxic effect of bioremediation products of pesticides in soil before their disposal in environment. Most of the time, bioremediation process provide an easy approach to recover soil fertility, which was not possible with conventional methods like degradation using physical and chemical methods. Various parameters like, microbial respiration, enzyme produced in soil, total microbial count, *etc.,* play important role to determine the soil quality (Ellert *et al.* 1997).

5.1 Soil Respiration

The oxidation of organic matter by aerobic organisms results in the production of carbon dioxide. Respiration may increase in response to an increase in microbial biomass or as a result of the increased activity of a stable biomass (Harris and Steer 2003). Mineralization studies involving measurements of total carbon dioxide production provide useful information on the biodegradability potential of pesticides in soil.

5.2 Soil Microbial Population

Microorganisms are vital for soil fertility and for degradation of organic material and pollutants, both in soil and sediments. Due to their function and ubiquitous presence, microorganisms can act as a relevant indicator of environmental pollution. Therefore, changes to the metabolic profiles of soil microbial communities could have potential use as early indicators of the impact of management or other perturbations on soil functioning and soil quality. Many workers have emphasized that pesticides in general, do not have much effect, except at concentrations greatly exceeding normal recommended field rates on soil microbial numbers and activities (Greaves 1987; Kale and Raghu 1989). There may be selective inhibition of some species, but others appear to rapidly replace the sensitive species, thus maintaining the metabolic integrity of the soil and resulting in no cumulative effects. It has been known that pesticides may stimulate, inhibit, or have no effect on microbial numbers (Grossbard 1976). Some workers have emphasized on the fact that insecticides in general, do not have much effect, except at concentrations greatly exceeding normal recommended field application rates on soil microbial numbers and activities (Greaves 1987; Kale and Raghu 1989). Sivasithamparam (1970) had found that there is no effects of insecticides on fungi whereas fungitoxic effects of many insecticides were demonstrated by Adust (1970) and Singh and Alka (1998), whereas stimulatory effects have also been confirmed (Naumann 1970).

Bisht (2014) also observed significant decline in bacterial population of soil amended with individual as well as with a mixture of endosulfan and chlorpyrifos (natural attenuation) and it was seen to remain at par with respect to the incubation periods. Chlorpyrifos caused a significant decline in the total fungal population when applied in soil, while endosulfan did not cause any inhibitory effect. Similar inhibitory effect of chlorpyrifos has been reported on soil bacteria, fungi, and actinomycetes (Pandey and Singh 2002; 2004; Shan et al. 2006; Xiaoqiang 2008). Effect of a pesticide on soil microorganisms depends not only on the chemical itself, but also on the pesticide concentration, soil type, and microbial composition in the soil under examination (Malkomes and Wohler 1983). Therefore, inconsistent trend or pattern of a given pesticide is often observed. In contrast to the results, significant stimulation of soil bacteria and fungi by chlorpyrifos has also been reported at lower concentrations (Tu 1991; Pozo et al. 1995; Pandey and Singh 2004; Shan et al. 2006).

5.3 Enzyme Production

Enzymatic activity is one of the good indicators of soil fertility, which provides reliable information on soil conditions. Decomposition of organic matter in soil releases nutrients such as nitrogen, phosphorus and sulphur required for plant and microbial growth. Various processes are involved in this nutrient cycling including reactions caused by inorganic catalysts and those catalyzed by intracellular enzymes (e.g. dehydrogenase; an oxidoreductase enzyme) and extracellular hydrolytic enzymes such as β-glucosidase, protease, acid phosphatase, alkaline phosphatase, urease and arylsulphatase (Table 3) (Acosta-Martinez et al. 2007).

Table 3: Soil enzymes as indicators of soil health

Soil enzyme	Enzyme reaction	Indicator of microbial activity
Dehydrogenase	Electron transport system	C-cycling
β-glucosidase	Cellobiose hydrolysis	C-cycling
Cellulase	Cellulose hydrolysis	C-cycling
Phenol oxidase	Lignin hydrolysis	C-cycling
Urease	Urea hydrolysis	N-cycling
Amidase	N-mineralization	N-cycling
Phosphatase	Release of PO_4	P-cycling
Arylsulphatase	Release of SO_4	S-cycling

Extracellular enzymes are derived from microorganisms, plant roots and soil animals. Their catalytic efficiency may be strongly influenced by the composition of the surrounding in which they act as catalysts. Pesticides, usually extraneous to soil component pools, are expected to affect the behavior of enzymes. Several investigations were devoted to study the effect of various pesticides on the activity of enzymes in soils from different origins. Persistent compounds including arsenic, DDT, and lindane caused long-term effects, including reduced microbial activity (Van-Zwieten *et al.* 2003), reduced microbial biomass, and significant decreases in soil enzyme activities (Ghosh *et al.* 2004; Singh and Singh 2005).

a) **Dehydrogenase:** Dehydrogenase enzyme activity is commonly used as an indicator of biological activity in soils. Dehydrogenase enzyme is known to oxidize soil organic matter by transferring protons and electrons from substrates to acceptors. These processes are part of respiration pathways of soil microorganisms and are closely related to the type of soil and soil air-water conditions (Kandeler *et al.* 1996). Dehydrogenase enzyme is often used as a measure of any disruption caused by pesticides, trace elements or management practices to the soil, as well as a direct measure of soil microbial activity (Garcia and Hernandez 1997).

Shetty and Magu (1998) reported that dehydrogenase activity is sensitive to insecticides. Inhibition of dehydrogenase activity by alachlor (Felsot and Dzantor 1995), DDT (Mitra and Raghu 1998) and pentachlorophenol (Meghraj *et al.* 1999) has also been reported. Bisht (2014) observed that chlorpyrifos caused a significant decline in dehydrogenase activity when applied in soil, and its higher concentrations caused more pronounced effect, while endosulfan did not cause any inhibitory effect. A mixture of both endosulfan and chlorpyrifos also caused inhibitory effect on soil dehydrogenase activity. This might be due to the toxic effect of chlorpyrifos. Similar observations were also made by Meghraj *et al.* (1998), Pandey and Singh (2004) and Singh and Singh (2005). Jastrzebslka (2011) also reported reduction in dehydrogenase activity in soil amended with chlorpyrifos. Pandey and Singh (2006) reported that use of chlorpyrifos for seed dressing in doses recommended by the manufacturer led to drop in activity of dehydrogenase. Nisha *et al.* (2006) and Defo *et al.* (2011) reported no measurable effect on the dehydrogenase activity at low concentration of endosulfan, while increased concentration caused inhibitory effect.

b) **Arylsulphatases:** Sulphur in different soil profiles are bound into organic compounds and are indirectly available to plants. In this regard, its availability will depend on the extracellular hydrolysis of these aromatic sulphate esters or intracellular oxidation of soluble organic matter absorbed by the microorganisms to yield energy and carbon skeletons for biosynthesis by which some SO_4-S is released as a by-product. All these processes are dependent on arylsuphatases enzymes. Arylsulphatases are typically widespread in nature as well as in soils. They are responsible for the hydrolysis of sulphate esters in the soil (Dodgson *et al.* 1982). Bisht (2014) reported that individual pesticide (chlorpyrifos and endosulfan) as well as their mixture when applied to soil caused a significant decline in arylsulphatase activity. Inhibitory effect of endosulfan on arylsulphatase activity has been reported earlier (Niemi *et al.* 2009; Girial *et al.* 2011). On the other hand, Kalyani *et al.* (2010) reported stimulatory effect of endosulfan on arylsulfatase when applied to soil in low concentration.

c) **Cellulases:** Cellulases are a group of enzymes that catalyze the degradation of cellulose, polysaccharides built up of b-1,4 linked glucose units. Studies have shown that activities of cellulases in agricultural soils are affected by several factors. These include temperature, soil pH, water and oxygen contents (abiotic conditions), the chemical structure of organic matter and its location in the soil profile horizon, quality of organic matter/ plant debris and soil mineral elements and the trace elements from fungicides (Arinze and Yubedee 2000).

d) **Phosphatases:** Phosphatases are a broad group of enzymes that are capable of catalysing hydrolysis of esters and anhydrides of phosphoric acid. Apart from being good indicators of soil fertility, phosphatase enzymes play key roles in P cycles of soil system (Dick *et al.* 2000). Bisht (2014) reported that individual pesticide (chlorpyrifos and endosulfan) as well as their mixture when applied to soil caused a significant decline in both acid and alkaline phosphatase. At higher concentration it caused more pronounced effect. Defo *et al.* (2011) reported no measurable effect on the phosphatase activity at a low concentration of endosulfan, while increased concentration caused inhibitory effect. Kalyani *et al.* (2010) reported stimulatory effect of endosulfan on phosphatase activity when applied in soil at low concentration. Increasing inhibitory effect on phosphatase enzyme activity with increased concentration of endosulfan was also recorded by Lal and Yadav (2000). Pozo *et al.* (1995) reported reduction of both acid and alkaline phosphatase activities in soil amended with chlorpyrifos.

6. Bioremediation: A Cost Effective Approach to Recover Soil Fertility

A common procedure to assess the outcome of bioremediation is to measure disappearance of lethal effects of toxic substances. Bisht (2014) reported that total microbial population was recovered at the end of bioremediation of pesticide contaminated soil. In pesticides contaminated soil augmented with fungi (*Trametes versicolor, T. hirsuta, Cladosporium cladosporioides* and *Penicillium frequentans*) as well as their consortium, microbial population and activity of soil enzymes reduced initially but increased significantly later on with respect to the incubation period. Schuster and Schroder (1990) reported that usually, after an initial toxic effect, mineralization of insecticides leads to an increase in the substrates for microbial growth, which might be the reason for increased microbial population and enzyme activities in bioaugmented soil, and the efficient fungi caused more pronounced increment.

Bisht (2014) reported high arylsulphatase activity in endosulfan amended soil, while both acid and alkaline phosphatase activities were high in chlorpyrifos amended soil. In addition, alkaline phosphatase activity was high in chlorpyrifos amended soil treated with *Cladosporium cladosporioides* and *Penicillium frequentans,* in which soil pH was in the alkaline range, while acid phosphatase was high in chlorpyrifos amended soil treated with *Trametes versicolor and T. hirsuta,* in which soil pH was in the acidic range. The soil fertility, however, was significantly improved in bioaugmented soil.

Soil contains both acid and alkaline phosphatases. Acid phosphatase predominates in acid soil while alkaline phosphatase in alkaline soil (Eivazi and Tabatabai 1977). Arylsulphatase enzyme is involved in mineralization of ester sulfate in soils and Phosphatase catalyzes the hydrolysis of a variety of organic phosphomonoesters and has been widely implicated in the degradation of organophosphorus pesticides (Kanekar *et al.* 2004). Increase in the activity of these enzymes in presence of respective pesticides also suggested the possible involvement of this group of enzymes in their degradation. Kalyani *et al.* (2010) and Bhalerao (2012) also suggested involvement of arylsulphatase in endosulfan degradation. Gao *et al.* (2012) has reported the involvement of phosphatase enzyme in degradation of chlorpyrifos.

7. Future Prospects and Scope

The damages caused to the environment and health due to the use and presence of pesticides dictate necessary development of technologies that guarantees their safe elimination from the environment. Cleanup of environmental pollution also presents a serious economic burden. However, use of indigenous or introduced microorganisms to decontaminate waste sites provides a very attractive, eco-friendly and economic solution to many of our hazardous pollution problems. This kind of treatment has a potential from a biotechnological point of view in order to have a methodology that is safer and more economic than the conventional treatments, as well as avoiding additional damages to the environment.

References

Acosta-Martinez V, Cruz L, Sotomayo-Rramirez D, Perez-Alegria, L (2007) Enzyme activities as affected by soil properties and land use in a tropical watershed. Appl. Soil Ecol. 35(1), 35-45.

Adust LJ (1970) The action of herbicides on the microflora of the soil, Proc. 10th Br. Weed Control Conf. pp 1036-1051.

Aktar W, Sengupta D, Chowdhury A (2009) Impact of pesticide use in agriculture: their benefits and hazards. Interdiscip. Toxicol. 2(1), 1-12.

Arinze AE, Yubedee AG (2000) Effect of fungicides on *Fusarium* grain rot and enzyme production in maize (*Zea mays* L.). Global J. Appl. Sci. 6, 629-634.

Baarschers WH, Heitland HS (1986) Biodegradation of Fenitrothion and Fenitrooxon by the fungus *Trichoderma viride*. J. Agric. Food Chem. 34, 707-709.

Barr DP, Aust SD (1994) Mechanisms white rot fungi use to degrade pollutants. Environ. Sci. Technol. 28, A78-A87.

Barrionuevo A (2007) Honeybees, Gone With the Wind, Leave Crops and Keepers in Peril. New York Times 27 February. New York, NY, http://select.nytimes.com/gst/abstract.html?res=F10B1FF8355A0C748EDDAB0894DF404482

Bending GD, Friloux M, Walker A (2002) Degradation of contrasting pesticides by white rot fungi and its relationship with ligninolytic potential. FEMS Microbiol. Lett. 212, 59-63.

Bhalerao TS (2012) Bioremediation of endosulfan-contaminated soil by using bioaugmentation treatment of fungal inoculant *Aspergillus niger*. Turkish J. Biol. 36, 561-567.

Bisht J (2014) Bioremediation of pesticide residues in soil using fungi. Ph.D. Thesis, Kumaun University Nainital, India.

Bumpus A, Kakar SN, Coleman RD (1993) Fungal degradation of organophosphorus insecticides. Appl. Biochem. Biotechnol. 39/40, 715-726.

Carson R (1962) Silent Spring. Boston: Houghton, USA.

Das SSM, Anitha S (2011) Mycoremediation of Monocrotophos, Int. J. Pharma Bio Sci. 2(1): 337-342.

Dave D, Dikshit AK (2011) Effect of different exogeneous compounds on biosorption of endosulfan. Am. J. Environ. Sci. 7(3), 224-236.

Defo MA, Njine T, Nola M, Beboua FS (2011) Microcosm study of the long term effect of endosulfan on enzyme and microbial activities on two agricultural soils of Yaounde-Cameroon, Afr. J. Agric. Res. 6(9), 2039-2050.

Dick WA, Cheng L, Wang P (2000) Soil acid and alkaline phosphatase activity as pH adjustment indicators. Soil Biol. Biochem. 32, 1915-1919.

Dodgson KS, White G, Fitzgerald JW (1982) Sulphatase enzyme of microbial origin. Afr. J. Biotechnol. 1, 156-159.

Eivazi F, Tabatabai MA (1977) Phosphatase in soils. Soil Biol. Biochem. 9, 167-172.

Ellert BH, Clapperton MJ, Anderson DW (1997) An ecosystem perspective of soil quality. In: Soil Quality for Crop Production and Ecosystem Health (Ed) Gregorich EG and Carter MR, Elsevier, Amsterdam, pp115-141.

Fanga H, Xianga YQ, Haoa YJ, Chua XQ, Pana XD, Yub JQ, Yua YL (2008) Fungal degradation of chloropyrifos by *Verticillium* sp. DSP in pure cultures and its use in bioremediation of contaminated soil and pakchoi, Int. Biodeter. Biodeg. 61(4), 294-303.

Felsot AS, Dzantor EK (1995) Effect of alachlor concentration and an organic amendment on soil dehydrogenase activity and pesticide degradation rate. Environ. Toxicol. Chem. 14(1), 23-28.

Gao Y, Chen S, Hu M, Hu Q, Luo J (2012) Purification and characterization of a novel chlorpyrifos hydrolase from *Cladosporium cladosporioides* Hu-01. PLoS ONE 7(6): e38137.

Garcia C, Hernandez T (1997) Biological and biochemical indicators in derelict soils subject to erosion. Soil Biol. Biochem., 29: 171-177.

Ghani A, Wardle DA, Rahman A, Lauren DR (1996) Interactions between 14C-labelled atrazine and the soil microbial biomass in relation to herbicide degradation. Biol. Fertil. Soils 21, 17-22.

Ghosh AK, Bhattacharyya P, Pal R (2004) Effect of arsenic contamination on microbial biomass and its activities in arsenic contaminated soils of Gangetic West Bengal, India. Environ. Int. 30, 491-499.

Girial PK, Sahab M, Halderb MP, Mukherjeeb D (2011) Effect of pesticides on microbial transformation of sulphur in soil. J Soil Sci. Environ Manage. 2, 97-102.

Goswami S, Vig K, Singh DK (2009) Biodegradation of á and â endosulfan by *Aspergillus sydoni*. Chemosphere 75, 883-888.

Gouma S (2009) Biodegradation of mixtures of pesticides by bacteria and white rot fungi. PhD Thesis, Cranfield University School of Health, Cranfield.

Greaves MP (1987) Side-effect testing: an alternative approach. In: Pesticide Effects on Soil Microflora (Ed) Somerville L and Greaves MP, Taylor and Francis, London, pp183-190.

Grossbard E (1976) Effects on the soil microflora In: Herbicides: Physiology, Biochemistry, Ecology, Vol 2, (Ed) Audus LJ, Academic Press, London, pp 99-148.

Harish R, Supreeth M, Chauhan JB (2013) Biodegradation of organophosphate pesticide by soil fungi. Adv. Bio Tech. 12(09), 1-5.

Harris J, Steer J (2003) Modern methods for estimating soil microbial mass and diversity: and integrated approach. In: The Utilization of Bioremediation to Reduce Soil Contamination: Problems and Solutions (Ed) Sasek V, Glaser JA and Baveye P, Springer Netherlands, pp 29-48.

Hasan HAH (1999) Fungi utilization of organophosphate pesticides and their degradation by *Aspergillus flavus* and *A. sydowii* in soil. Folia. Microbiol. 44, 77-84.

Horrigan L Lawrence RS (2002) How sustainable agriculture can address the environmental and human health harms of industrial agriculture. Environ. Health Perspectives, 110(5), 445-456.

Hurley PM, Hill RN, Whiting RJ (1998) Mode of carcinogenic action of pesticides inducing thyroid follicular cell tumours in rodents. Environ. Health Perspect. 106, 437.

Hussain S, Arshad M, Saleem M, Zahir ZA (2007) Screening of soil fungi for in vitro degradation of endosulfan, World J. Microbiol. Biotechnol. 23, 939.

Hussaini SZ, Shaker M, Iqbal MA (2013) Isolation of fungal isolates for degradation of selected pesticides. Bull. Env. Pharmacol. Life Sci. 2(4), 50-53.

Jastrzebslka E (2011) The effect of chlorpyrifos and teflubenzuron on enzymatic activity of soil. Polish J. Environ. Stud. 20(4), 903-910.

Jayaraman P, Kumar TN, Maheswaran P, Sagadevan E, Arumugam P (2012) *In vitro* studies on biodegradation of chlorpyrifos by *Trichoderma viride* and *T. harzianum*. J. Pure Appl. Microbiol. 6(3), 1-16.

Kale SP. Raghu K (1989) Relationship between microbial parameters and other microbial indices in soil. Bull. Environ. Contam. Toxicol. 43, 941–945.

Kalyani SS, Sharma J, Dureja P, Singh S (2010) Influence of endosulfan on microbial biomass and soil enzymatic activities of a tropical alfisol. Bull. Environ. Contam. Toxicol. 84, 351-356.

Kamei I, Takagi K, Kondo R (2011) Degradation of endosulfan and endosulfan sulfate by white-rot fungus *Trametes hirsuta*. J. Wood Sci. 57, 317-322.

Kandeler E (1996) Nitrate. In: Methods in Soil Biology (Ed) Schinner FO, Hlinger R, Kandeler E, Margesin R, Springer, Berlin, pp 408-410.

Kanekar PP, Bhadbhade BJ, Deshpande NM, Sarnaik SS (2004) Biodegradation of organophosphorus pesticides. Proc. Indian Natl. Sci. Acad. B 70(1), 57-70.

Kataoka R, Takagi K, Sakakibara F (2010) A new endosulfan degrading fungus, *Mortierella* species, isolated from a soil contaminated with organochlorine pesticides. J. Pestic. Sci. 35, 326-332

Katayama A, Matsumura F (1993) Degradation of organochlorine pesticides particularly endosulfan by *Trichoderma harzianum*. Environ. Toxicol. Chem. 12, 1059-1065

Kullman SW, Matsumura F (1996) Metabolic pathways utilized by *Phanerochaete chrysosporium* for degradation of the cyclodiene pesticide endosulfan. Appl. Environ. Microbiol. 62, 593-600.

Lal N, Yadav R (2000) Effect of endosulfan on activity and extracellular production of phosphatase by *Aspergillus fumigantus*. Indian J. Agric. Sci. 70(9), 627-629.

Liu Y, Zhong Y (2000) Degradation of organophosphate insecticide (dimethoate) by *Aspergillus* sp. Z58. Huanjing Kexue Xuebao 20, 95-99.

Liu Y, Ying CC, Xiong Y (2001) Purification and characterization of dimethoate degrading enzyme of *Aspergillus niger* ZH4256 isolated from sewage. Appl. Microbiol. Biotechnol. 67, 3746-3749.

Malkomes HP, Wohler B (1983) Testing and evaluating some methods to investigate soil effects of environmental chemicals on soil microorganisms. Ecotoxicol. Environ. Safety 7, 284-294.

Massoud AH, Derbalah AS, Belal EI-SB, EI-Fakharany, II (2007) Bioremediation of melathion in aquatic system by different microbial isolates. J. Pest Control Environ. Sci. 15(2), 13-28.

Mathews GA (2006) Pesticides: Health, Safety and the Environment. Black Well Publishing Ltd.

Meghraj M, Boul HL, Thiele JH (1999) Effects of DDT and its metabolites on soil algae and enzymatic activity. Biol. Fertil. Soils, 29 (2), 130-134.

Meghraj M, Singleton I, McClure NC (1998) Effect of pentachlorophenol pollution towards microalgae and microbial activities in soil from a former timber processing facility. Bull. Environ. Contam. Toxicol. 61 (1), 108-115.

Mitra J, Raghu K (1998) Long term DDT pollution in tropical soils: effect of DDT and degradation products on soil microbial activities leading to soil fertility. *Bull.* Environ. Contam. Toxicol. 60 (4), 585-591.

Mukherjee I, Gopal M (1994) Degradation of beta-endosulfan by *Aspergillus niger*. Toxicol. Environ. Chem. 46, 217-221.

Mukherjee I, Gopal M (1996) Degradation of chlorpyrifos by two soil fungi *Aspergillus niger* and *Trichoderma viride*. Toxicol. Environ. Chem. 57(1), 4.

Mukherjee I, Mittal A (2005) Bioremediation of endosulfan using *Aspergillus terreus* and *Cladosporium oxysporum*. Bull. Environ. Contam.Toxicol. 75, 1034-1040.

Naumann K (1970) Dynamics of the soil microflora following application of insecticides. Field trials on the effects of methyl parathion on the bacterial and actinomycetes population of soil, Zentbl. Bakt. Parasitkde Abt. 124, 743.

Niemi RM, Heiskanem I, Ahtiainen JH, Rahkonen A, Mantykoski K, Welling L, Laitinen P, Ruuttenen P (2009) Microbial toxicity and impact on soil enzymes activities of pesticides used in potato cultivation. Appl. Soil Ecol. 41, 293.

Nisha K, Shahi DK, Sharma A (2006) Effect of endosulfan and monocrotophos on soil enzymes in acid soil of Ranchi. Pestology 30(11), 42-44.

Omar SA (1998) Availability of phosphorus and sulfur of insecticide origin by fungi. J. Opthalmol., 9, 327-336.

Omar SA, Abdel-Sater A (2001) Microbial populations and enzyme activities in soil treated with pesticides. Water Air Soil Pollution, 127, 49-63.

Pandey S, Singh DK (2002) Residual impact of chlorpyrifos and quinalphos on soil bacterial and fungal population in the groundnut (*Arachis hypogaea* T.) field of Jaipur. National Conference: Soil Contamination and Biodiversity, Lucknow, February 8-10, pp 41

Pandey S, Singh DK (2004) Total bacterial and fungal population after chlorpyrifos and quinalphos treatments in groundnut (*Arachis hypogaea* L.) soil. Chemosphere 55, 197-205.

Pandey S, Singh DK (2006) Soil dehydrogenase, phosphomonoestraze and arginine deaminase activities in an insecticide treated groundnut (*Arachis hypogaea* L.) field. Chemosphere, 63, 869-880.

Paszczynski A, Crawford RC (2000) Recent advances in the use of fungi in environmental remediation and biotechnology. In: Soil Biochemistry, Vol. 10 (Ed) Bollag JM and Stotzky G, New York, Marcel Dekker, pp 379-422.

Pointing SB (2001) Feasibility of bioremediation by white-rot fungi. Appl. Environ. Microbiol. 57, 20-23.

Pothuluri JV, Chung YC, Xiong Y (1998) Biotransformationof 6-nitrochrysene. Appl. Environ. Microbiol. 64, 3106-3109.

Pothuluri JV, Heflich RH, Fu PP, Cerniglia CE (1992) Fungal metabolism and detoxification of fluoranthene. Appl. Environ. Microbiol. 58, 937-941.

Pozo C, Martinez-Toledo MV, Salmeron S, Rodelas B, Gonzalez-Lopez J (1995) Effect of chlorpyrifos on soil microbial activity. Environ. Toxicol. Chem. 14(2), 187-192.

Quintero JC, Moreira MT, Feijoo G, Lema JM (2008) Screening of white rot fungal species for their capacity to degrade lindane and other isomers of hexachlorocyclohexane (HCH). Cien. Inv. Agr. 35(2), 123-132.

Rodriguez DH, Sanchez JE, Nieto MG, Rocha FJM (2006) Degradation of endosulfan during substrate preparation and cultivation of *Pleurotus pulmonarius*, World J. Microbiol. Biotechnol. 22, 753-760.

Saikia N, Gopal M (2004) Biodegradation of â-cyfluthrin by fungi. J. Agric. Food Chem. 52 (5), 1220-1223

Sardar D, Kole RK (2005) Metabolism of Chlorpyriphos in relation to its effect on the availability of some plant nutrients in soil. Chemosphere, 61, 1273-1280.

Sari A, Tachibana S, Limin S (2013) Enhancement of ligninolytic activity of *Trametes versicolor* U97 pre-grown in agricultural residues to degrade DDT in soil. Water Air and Soil Pollution, 224, 1616.

Schuster E, Schroder D (1990) Side-effects of sequentially applied pesticides on non-target soil microorganisms: field experiments. Soil Biol. Biochem. 22(3), 367-373.

Shan M, Fang H, Wang X, Feng B, Chu XQ, Yu YL (2006) Effect of chlorpyrifos on soil microbial populations and enzyme activities. J. Environ. Sci. 18(1), 4-5.

Shetty PK, Magu SP (1998) *In vitro* effect of pesticide on carbon dioxide evolution and dehydrogenase activities in soil. J. Environ. Biol. 19 (2), 141-144.

Shetty PK, Mitra J, Murthy NBK, Namitha KK, Sovitha KN, Raghu K (2000) Biodegradation of cyclodiene insecticide endosulfan by *Mucor thermohyalospora* MTCC 1384. Curr. Sci.79, 1381-1383.

Siddique T, Okeke BC, Arshad M, Frankenberger WT (2003) Biodegradation kinetics of endosulfan by *Fusarium ventricosum* and a *Pandoraea* species. J. Agric. Food Chem. 51, 8015-8019.

Singh BK, Kuhad RC (1999) Biodegradation of Lindane (hexachlorocyclohexane) by the white-rot fungus *Trametes hirsutus*. Lett. App. Microbiol. 28, 238-241.

Singh BK, Kuhad RC (2000) Degradation of Insecticides Lindane (-HCH) by white-rot fungi *Cyathus bulleri* and *Phanerochaete sordida*. Pest Manage. Sci. 56(2), 142-146.

Singh J, Singh DK (2005) Dehydrogenase and phosphomonoesterase activities in groundnut (*Arachis hypogaea* L.) field after diazinon, imidacloprid and lindane treatments. Chemosphere, 60, 32-42.

Singh P, Alka (1998) Effect of pesticides on soil microorganisms. In: 39th AMI Conference, December 5-7, Jaipur.

Siripong P, Oraphin B, Sanro T, Duanporn P (2009) Screening of fungi from natural sources in Thailand for degradation of polychlorinated hydrocarbons. Am. Eur. J. Agric. Environ. Sci. 5(4), 466-472.

Sivasithamparam K (1970) Some effects of an insecticide (Dursban) and a weedicide (Linuron) on the microflora of a submerged soil. Risorgimento, 19, 339-346.

Sri PU, Rao KRSS, Rathinam KMS (2011) Biomineralisation of organophosphorus pesticide (Malathion) using fungal species *Rhizopus oligosporus* from cotton soils of Andhra Pradesh, India, 1: 039-050.

Subashini HD, Sekar S, Devi VRS, Rajam A, Malarvannan S (2007) Biodegradtion of pesticides residue using traditional plants with medicinal properties and *Trichoderma.* Res. J. Environ. Tox. 1(3), 124-130.

Trejo-Hernandez MR, Lopez-Munguia A, Ramirez RQ (2001) Residual compost of *Agaricus bisporus* as a source of crude laccase for enzymic oxidation of phenolic compounds. Process. Biochem. 36, 635-9.

Tu CM (1991) Effect of some technical and formulated insecticides on microbial activities in soil. J. Environ. Sci. Heal. B. 26(5-6), 557-573.

US EPA (2006) About Pesticides. http://www.epa.gov/pesticides/about/types.htm

Van-Zwieten L, Ayres MR, Morris SG (2003) Influence of arsenic co-contamination on DDT breakdown and microbial activity. Environ. Pollut., 124, 331-339.

Vidali M (2001) Bioremediation: An overview. Pure Appl. Chem. 73(7), 1163-1172.

Wallnofer PR, Engelhardt G (1989) Microbial degradation of pesticides. In: Chemistry of Plant Protection, Degradtion of Pesticides, Desiccation and Defoliation, Ach-Receptors as Targets (Ed) Haug G, Hoffmann H, Bowers WS, Ebing W, Fukuto TR, Martin D, Wegler R, Yamamoto I, Springer-Verlag Berlin Heidelberg, New York, pp 1-116.

Wang JH, Zhu LS, Wang J, Qin K (2005) Degradation characteristics of three fungi to chlorpyrifos. Chinese J. Appl. Environ. Bio. 11, 211-214.

WHO (World Health Organization) (2009) Health implications from monocrotophos use: a review of the evidence in India, pp 1-60.

Xiaoqiang CHU, Hua Fang, Xuedong Pan, Xiao Wang, Min Shan, Bo Feng, Yunlong YU (2008) Degradation of chlorpyrifos alone and in combination with chlorothalonil and their effects on soil microbial populations. J. Environ. Sci. 20, 464-469.

Xie H, Zhu L, Ma T, Wang J, Wang J, Su J, Shao B (2010) Immobilization of an enzyme from a *Fusarium fungus* WZ-I for chlorpyrifos degradation. J. Environ. Sci. (China), 22(12), 1930-5.

Xu G, Li Y, Zheng W, Peng X, Li W, Yan Y (2007) Mineralization of chlorpyrifos by co-culture of *Serratia* and *Trichosporon* spp. Biotechnol. Lett. 29, 1469-1473

Zhao L, Tang Y, Li Q, Li F, Shao X, Wang Y (2011) Studies on Degradation of the Pesticide of Chlorpyrifos by *Phanerochaete chrysosporium*. (iCBBE), 5th International Conference on Bioinformatics and Biomedical Engineering, Wuhan, China, 10 -12 May, pp 5119-5122.

6

Significance of Microbes for Humification Process

Manna, M.C., Asha Sahu, Patra, A.K. and Khanna, S.S.

Abstract

Within the time scale for soil formation, humans appeared to the planet within the last 1 million years. They shifted their lifestyle from hunter-gatherer to nomadic animal husbandry and finally to cultivation of domesticated crops only within the last 10,000 yrs. Earlier, these crops were cultured in a semi-nomadic style as land was plentiful, so when soil lost its productivity, people abandoned it in favor of new-land. Gradually, as land became scarcer and population exploded, agriculture evolved to continuous cropping of the same parcel of land. This discovered the major concern for soil health. As organic matter is key to soil health, humus is the last stage of organic matter decomposition. Humus is a complex aggregate of dark brown colored amorphous substances, which have originated during the decomposition of plant and animal residues by microorganisms, under aerobic and anaerobic conditions. Humus and living microorganisms are connected by numerous ways which must be appreciated in order to understand the origin and nature of humus. Our attempt is to narrate the role of organic matter and vital role of microbes in humification process.

Keywords: Humus, Microbes, Organic matter, Soil health.

1. Introduction

Healthy soil is the foundation of the healthy crops and in turn healthy people. Plants obtain nutrients in soil from two natural sources namely: organic matter and mineral matter. Soil organic matter (SOM) is the product of on-site biological decomposition that affects its chemical and physical properties and its health.

Its composition and breakdown rate affects the soil structure and porosity, moisture holding capacity of soils and the water infiltration rate, the biological activity and diversity of soil organisms, and plant nutrient availability. Most Indian soils contain 0.5–1.0 percent organic matter. However, even small amounts, organic matter is important. Soil is a living, dynamic ecosystem. Healthy soil is teeming with microscopic and larger organisms that perform many vital functions including converting dead and decaying matter as well as minerals to plant nutrients. Different soil organisms feed on different organic substrates. Their biological activity depends on the organic matter supply. Nutrient exchanges between soil, water and organic matter which are essential to soil fertility need to be maintained for sustainable food production purposes. Where the soil is exploited for crop production without restoring the nutrient contents and organic matter and maintaining a good structure, soil fertility declines and the balance in the agro-ecosystem is destroyed. SOM content is a function of organic matter inputs (residues and roots) and litter decomposition. It is dependent to moisture, temperature and aeration, physical and chemical properties of the soils as well as bioturbation (mixing by soil macrofauna), leaching by water and humus stabilization (organomineral complexes and aggregates). This chapter recognizes the importance of organic matter in agriculture, environment and climate change and significant role of microbes for humification process.

1.1 The Role of Organic Matter in Agriculture and Carbon Sequestration

Soil organic matter (SOM) has been called "the most complex and least understood component of soils" (Magdoff and Weil 2004). A broader definition of SOM proposed by Magdoff (1992), which consider SOM to be the diverse organic materials, such as living organisms, slightly altered plant and animal organic residues, and well-decomposed plant and animal tissues that vary considerably in their stability and susceptibility to further degradation. Soil organic matter is rich in nutrients such as nitrogen (N), phosphorus (P), sulfur (S), and micronutrients, and is comprised of approximately 50% carbon (C).

SOM is complex because it is heterogeneous (non-uniform) and, due to the biological factors under which condition it was formed, does not have a defined chemical or physical structure. It is not distributed evenly throughout the soil and breaks down at various rates by multiple agents that are influenced by the unique environmental conditions in which they are present. In fact, SOM and soil organisms are so interdependent that it is difficult to discuss one without the other. Soil organic matter accumulates to higher levels in cool and humid regions compared to warm and arid climates (Lal 2002). In addition, SOM associated with different soil textures (sand, silt, and clay), will differ in susceptibility to decomposition. Many studies have shown that SOM associated with the sand-size fraction is more susceptible to decomposition, and thus a higher turnover, than the silt- or clay-size fractions (Angers and Mehuys 1990, Dalal and Mayer 1986, Zhang *et al.* 1988). Soil organic matter has been directly and positively related to soil fertility and agricultural productivity potential. In most agricultural soils, organic matter is increased by leaving residue on the soil surface (as has been observed in Punjab soil), rotating crops with pasture or perennials, incorporating cover crops into the cropping rotation, or by adding organic residues such as animal manure, litter, or sewage sludge (Krull *et al.* 2004).

Continued addition of decaying plant residues to the soil surface contributes to the biological activity and the carbon cycling process in the soil. Breakdown of soil organic matter and root growth and decay also contribute to these processes. Carbon cycling is transformation of organic and inorganic carbon compounds by flora and fauna between the soil-plants-atmosphere continuum. Decomposition of organic matter is largely a biological process that occurs naturally. Its speed is determined by three major factors: soil organisms, the physical environment and the quality of the organic matter (Brussaard 1994). In the decomposition process, different products are released: carbon dioxide (CO_2), energy, water, plant nutrients and resynthesized organic carbon compounds. Successive decomposition of dead material and modified organic matter results in the formation of a more complex organic matter called humus (Juma 1998). This process is called humification. Humus affects soil properties. As it slowly decomposes, it colours the soil darker; increases soil aggregation and aggregate stability; increases the CEC (the ability to attract and retain nutrients); and contributes N, P and other nutrients. Soil organisms, including micro-organisms, use soil organic matter as food. As they break down the organic matter, any excess nutrients (N, P and S) are released into the soil in forms that plants can readily utilize. This release process is called mineralization. The waste products are also produced by microorganisms which are less decomposable than the original plant and animal material, but it can be used by a large number of organisms. By breaking down carbon structures and rebuilding new ones or storing the C into their own biomass, soil biota play the most important role in nutrient cycling processes and, thus, in the ability of a soil to provide the crop with sufficient nutrients to harvest a healthy product. The organic matter content, especially the humus, increases the capacity to store water and store (sequester) C from the atmosphere.

1.2 Following are the Important Role of SOM in Relation to Crop Productivity

(a) Nutrient release and retention

Humus is highly negatively charged because of carboxyl groups and thus capable of holding a large amount of cat ions. This highly charged humic fraction gives the SOM the ability to act as a slow release fertilizer. Over time, as nutrients are removed from the soil cat ion exchange sites, they become available for plant uptake. Several researchers have conducted trials to estimate the N release from SOM for plant growth. It has been concluded that soil with more SOM has less N requirement and *vice-versa*.

(b) Cation exchange capacity

The total sum of exchangeable cat ions (positively charged ions) that a soil can hold is termed as cat ion exchange capacity (CEC). CEC determines a soil's ability to retain positively charged plant nutrients, such as NH_4^+, K^+, Ca^{2+}, Mg^{2+}, Cu^{2+}, Fe^{2+}, Mn^{2+} and Zn^{2+}. As CEC increases for a soil, it is able to retain more of these plant nutrients and reduces the potential for leaching. Soil CEC also influences the application rates of lime and herbicides required for optimum effectiveness. The most important fraction for contributing to the CEC of a soil is the stable fraction (humus) of SOM.

Different soil textures have different CEC (Table 1). In most soils, organic matter contributes more to exchange capacity than the soil texture. The interaction of texture and organic matter components in soil has a tremendous influence on CEC potential (Coleman *et al.* 2004).

Table 1: The range of CEC for each soil texture and organic matter

Texture	CEC (cmol/kg)
Organic matter	40-200
Sand	1-5
Sandy loam	2-15
Silt loam	10-25
Clay loam/silty clay loam	15-35
Clay	25-60
Vermiculite	150

(c) Water holding capacity

It has been observed that soils with higher SOM are "fluffier" or have better "tilth" than soils with less SOM. This is because SOM is less dense than the mineral soil particles per unit of volume, and therefore provides greater pore space for water and air to be held. The result of increasing SOM is greater soil pore space, which provides an area for water to be stored during times of drought. A unique characteristic of the pore space in SOM is that the pores are found in many different sizes. The large pores do not hold water as tightly, and drained more readily. The medium and small-sized pores will hold water more tightly and for a longer period of time, so that during a dry period the soil retains moisture and a percentage of that water is made available over time for plant uptake. The benefit of leaving residue on the soil surface and increasing soil organic matter is that water infiltration is increased, soil crusting is decreased, and the soil can hold more of the water that infiltrates and will eventually make it available for plant use.

(d) Improve soil structure and reduce bulk density

Soil structure refers to the arrangement of individual soil mineral particles (sand, silt, and clay). Soil structure is stabilized by a variety of different binding agents. SOM is a primary factor in the development and modification of soil structure (Coleman *et al.* 2004).

While binding forces may be of organic or inorganic origins, the organic forces are more significant for building large, stable aggregates in most soils. Examples of organic binding agents include plant- and microbially-derived polysaccharides, fungal hyphae, and plant roots. Inorganic binding agents and forces include charge attractions between mineral particles and/or organic matter and freezing/thawing and wetting/drying cycles within the soil as well as compression and deformation forces. Both the stable and the active fraction of SOM contribute to and maintain soil structure and resist compaction.

(e) Increased biological activity

According to the father of soil pedology, Hans Jenny, a natural body of degraded mineral or organic material cannot be considered a "soil" without soil organisms. This emphasizes the importance of soil organisms in the study of soil science.

Soil microorganisms (bacteria and fungi), plants, and fauna (nematodes, springtails, mites, earthworms, and insects) are considered as the "life" in soils. While microorganisms only make up a small portion of the SOM (less than 5%) they are imperative to the formation, transformation, and functioning of the soil. In the soil, they conduct indispensable processes such as decomposition, nutrient cycling, and degradation of toxic materials, N fixation, symbiotic plant relationships, and pathogen control.

Jenny also said about soil fauna, "They break up plant material, expose organic surface areas to microbes, move fragments and bacteria-rich excrement around, up, and down, and function as homogenizers of soil strata" (Jenny 1980). Soil fauna play an important role in the initial breakdown of complex and large pieces of organic matter, making it easier for soil microorganisms to release carbon and plant nutrients from the material as they continue the process of decomposition.

(f) Organic matter and carbon sequestration

Global climate change is one of the areas receiving a great deal of attention and research effort. Soil organic matter plays a critical role in the global C cycle. The importance of soil in the C cycle is due to its role as both a major source and sink for C in the biosphere. The total soil C pool is three times greater than the atmospheric C pool and 3.8 times greater than the biotic C pool (Krull *et al.* 2004). The soil C pool contains about 1.7×10^{12} tons of organic C and about 8.3×10^{11} tons of inorganic C to a depth of 3.3 feet (Lal 2002).

Although the soil C cycle is complex, the concept of C sequestration for mitigating the release of greenhouse gases is relatively straightforward. Carbon stored in soils ties up C that would otherwise be released as green house gases to the atmosphere as carbon dioxide (CO_2) and methane (CH_4). Scientists are keenly interested in determining the extent to which atmospheric carbon can be diminished by storing C in soils.

2. Types of Organic Matter

Fractionating SOM into humic acid, fulvic acid, and humin has been known but this method of fractionation does not produce chemically discrete SOM fractions, and these, fractions are heterogeneous and non-reproducible. A more biologically and agriculturally meaningful method of describing soil organic matter is by dividing it into various "pools" which are sorted by how easily the material is decomposed (e.g. active or labile; slow or intermediate; and passive or stable). Pools, which have measurable organic matter components, are theoretically separate entities and are more concisely designated by fractions (Wander 2004). This method of SOM classification is far more commonly used now than the outdated measurements of humic and fulvic acid separation.

2.1 Active/Labile Fraction

Recently deposited roots and residues, dead organisms, or waste products, is the most biologically "active" fraction of the SOM, meaning that it serves as a food source for the living soil biological community. The younger fraction is also referred to as the "labile" SOM fraction, indicating that it is more readily decomposed than the passive/stable fraction. Generally, this fraction of the SOM is less than five years old. There are many ways to measure the active fraction but one of the most commonly used methods is to measure the particulate organic matter (POM). Particulate organic matter is defined as the microbially active fraction of soil organic matter. The reason that POM has become so frequently used is that it has been shown to have a strong response to management decisions, such as tillage, residue handling and levels, and crop rotation (Alvarez and Alvarez 2000, Carter 2002, Conteh *et al.* 1998, Franzluebbers *et al.* 2000).

2.2 Passive/Stable Fraction

During the process of decomposition many soil organisms assist in the process of decomposing plant and animal tissues and chemical transformations take place, creating new organic compounds in the soil. After years or even decades of these transformations, the original organic materials are converted into chemically complex, nutrient-poor compounds that few microbes can degrade. These compounds are referred to as "passive" or "stabilized" and can make up a third to a half of soil organic matter. Such passive or stabilized materials are what we commonly refer to as "humus" or the "stable fraction". This fraction does not contain many nutrients, and so is not directly important for soil fertility. However, the stable humus fraction of soil is very chemically reactive and contributes to the soil's net chemical charge, known as the cation exchange capacity and anion exchange capacity. In this way, humus directly/indirectly binds plant nutrients in the soil, preventing them from leaching, so that they are available for plant uptake. The stable fraction also modifies and "stabilizes" toxic materials so that they are less reactive and/or dangerous. Finally, the stable fraction enhances soil aggregation that reduces a soil's vulnerability to erosive forces and thereby soil loss by erosion is less.

3. Role of Microbes in Decomposition Process

The main components of organic matter are carbohydrates (*e.g.* cellulose), proteins, lipids and lignin. Their capacity to assimilate organic matter is dependent upon their ability to produce the enzymes needed for degradation of the substrate (Tuomela *et al.* 2000). The composition of the microbial community during composting is determined by several factors. Under aerobic conditions, temperature is the major selective factor for populations and determines the rate of metabolic activities. There are contradictory reports on the population of organisms during different phases (Table 2).

While some authors state that the total number of microorganisms does not significantly change during composting, others report higher numbers for the mesophilic stage. However, it is generally agreed that the composition of the community can vary during different phases of the composting (Ryckeboer *et al.* 2003). The diversity of prokaryotes and/or fungi during an entire composting process has been reported by very few workers. Ryckeboer *et al.* (2003) examined diversity and population densities of prokaryotes and fungi throughout the whole composting process of source-separated household wastes, i.e. from starting material to mature compost. Since starting material and process conditions determine the community composition to a large degree, it is difficult to make a generalized statement.

Table 2: Approximate numbers of microorganisms during different phases of composting (after Miller, 1993)

Organism	Number/ g^{-1} substrate
Bacteria in mesophilic stage	10^9-10^{13}
Bacteria in thermophilic stage	10^8-10^{12}
Actinomycetes, thermophilic stage	10^7-10^9
Actinomycetes, mesophilic stage	10^8-10^{12}
Fungi	10^5-10^8

4. Stages of the Composting Process

The composting process under optimal conditions can be divided into six phases/stages: (i) an initial (first) mesophilic phase (occurring at 10-42°C), which may last one week (ii) a thermophilic phase (at 45-70°C), lasting about 2 weeks, several weeks (particularly in food wastes) or even months (particularly in wood wastes); (iii) second mesophilic phase during which mesophile organisms, often dissimilar to those of the first mesophilic phase, recolonize the substrates; and (iv) the maturation phase which may last for two weeks (v) curing phase during which humification develops that may last about 2-3 weeks and (vi) stabilization phase which can last for 2-4 weeks (Fig. 1).

Different microbial communities predominate during the various composting phases, each of which being adapted to a particular environment. Physico-chemical environment is created in composting by primary decomposers, which attack the initial substrates and produce the metabolites that are suitable for secondary organisms. A rapid transition from mesophilic to thermophilic microflora is caused by the initial rapid increase of temperature. Often, however, a disruption of the process is observed at 42°C - 45°C. The inhibition of initial mesophilic microflora is caused by the high temperature 65°C - 75°C by the thermophilic organisms.

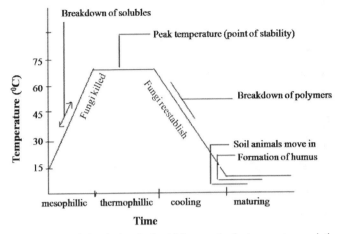

Fig. 1: Changes of chemical and microbial properties by temperature variations

4.1 Starting Phase - First Mesophilic Phase

Due to the heterogeneity of biodegradable material, little is known regarding the original composition of the waste microbial community. There are very few reports on microbial diversity present in organic waste material. During the initial phase, the substrates are at ambient temperature and the pH is usually slightly acidic. The dominant active degraders of fresh organic waste materials are mesophilic and/or thermotolerant fungi and bacteria (20-40°C). The proliferation of fungi and yeasts is stimulated by food wastes containing vegetable residues often have a low initial pH (4.5 to 5.0). These microorganisms rapidly break down soluble and easily degradable carbon sources, resulting in a pH drop due to the production of organic acids.

An increase in pH is favorable for bacteria that subsequently out-compete fungi within a few hours or days. The high surface/volume ratio of bacteria allows a rapid transfer of soluble substrates into the cell. Nutritionally also, bacteria are the most diverse group of compost organisms, which chemically degrade a variety of organic materials *via* a broad range of enzymes. Also, the average generation time of bacteria is much shorter than that of fungi which gives them a competitive advantage during such phases of the composting process which are characterized by rapid changes in substrate availability and other parameters like temperature, moisture (50-60%) and aeration by turning the material after a week.

Resultant to this factor, the number of bacteria (including actinomycetes) is usually much higher than the number of other microorganisms, e.g. fungi (if total numbers are comparable at all). It can be said that bacteria are responsible for most of the initial decomposition and heat generation in compost provided that the major growth requirements are met, Bacteria prefer a near-neutral pH and the optimal moisture content for their functioning is 50- 60% under aerobic condition.

Actinomycetes develop more slowly than most bacteria and fungi and are rather ineffective competitors when nutrient levels are high. A wide range of prokaryotes produce amylase which is also important during the initial phase for degradation of starch.

4.2 Thermophilic Phase

Thermophiles are reported to be good decomposers of cellulose. Cellulose degradation occurs at optimal temperature of around 65°C, implying that degradation is performed essentially by thermostable enzymes.

Thermophilic phase is characterized by decrease in moisture and temperature rise above 45°C. As a result, the substrates become more alkaline and the growth of actinomycetes in particular Streptomycetes strive increases. Thermophilic phase of composting is initiated by microorganisms metabolizing proteins, increasing liberation of ammonium and causing subsequent alkalinization. Degradation is faster in this phase as compared with the initial mesophilic phase.

Mesophilic microorganisms are inactivated or killed during the initial thermophilic phase (temperatures between 40-60°C), whereas the numbers and species diversity of thermophilic and/or thermo-tolerant bacteria, actinomycetes and fungi increase. However, during the thermophilic phase, overall diversity of bacterial species drops significantly. It may become necessary to spray the inoculant again.

4.2.1 Role of Different Microorganisms during Mesophilic and Thermophilic Phase

Various organisms such as bacteria, fungi and actinomycetes play an active role during various stages of compost production. The most dominant group of organisms which develop during different stages composting are listed in Table 3.

Table 3: Major organisms which develop during different stages of composting

Mesophilic stage	Thermophilic stage
Bacteria	**Bacteria**
Bacillus spp., *Cellumonas, Thiobacillus* spp.,	*Bacillus* sp., *Streptothermophilus*
Pseudomonas spp.	**Fungus**
Fungus	*Humicolla, Absidia, Chaetomium*
Aspergillus, Fusarium, Tricoderma, Mucor,	**Actinomycetes**
Helminthosporium	*Micromonosperma, Nocardia, Streptomyces*
	Termonospora, Thermopolyspora

a) Role of Actinomycetes

They cause the characteristic earthy smell of soil and compost by production of geosmine, which are sesquiterpenoid compounds. Actinomycetes compete with other organisms for nutrients and can inhibit microbial growth by production of antibiotics, enzymes or even by parasitism. In composting, they play an important role by degrading natural polymers and colonize organic material after bacteria and fungi have consumed easily degradable fractions. The enzymes produced by actinomycetes enable them to degrade tough debris such as woody stems, bark or newspaper, cellulose and hemicellulose originating from plant material. Most actinomycetes tolerate a higher pH than fungi, their optimum pH being 7-8. Under adverse conditions actinomycetes can survive as spores. Temperatures of 45 to 55°C, optimal for thermophilic actinomycetes for their growth, cause significant increase in their number and diversity.

b) Role of Bacteria

Endospore-forming bacteria, for e.g. *Bacillus* spp., are very active at temperatures around 50-60°C. At temperatures above 60°C, thermophilic bacteria dominate in the degradation process. Non-spore forming bacteria such as *Hydrogenobacter* spp. and *Thermus* spp. are the dominant active degraders in thermogenic composts at temperatures above 70°C, even up to 82°C.

c) Role of Fungi

The ability of fungi to degrade cellulose and lignin is higher than that of actinomycetes, and bacteria in general. Temperature is one of the most important factors affecting fungal growth. The majority of fungi are mesophilic (5°C to 37°C), with an optimum of 25-30°C. A low N content is a prerequisite for lignin degradation, although most fungi prefer a moderate level of N. Due to their extensive hyphal network, they can attack organic residues that are too dry, too acidic, or too low in nitrogen for bacterial decomposition. In comparison to actinomycetes, thermophilic fungi are generally less tolerant to high temperatures, their optimal temperature being 40 to 55 °C with a maximum at 60 to 62°C. At temperatures above 60°C, fungi are killed or transiently present as spores. Yeasts disappear during the thermophilic phase of composting, but when the temperature cools down to 54°C, they can be found again.

4.3 Cooling or Second Mesophilic Phase

Once the activity of the thermophilic organisms ceases due to depletion of substrates, turning the whole material is required, which may result in decreasing the temperature. Mesophilic organisms start to re-colonise the substrate, either originating from surviving spores, spreading through protected microniches, or from external inoculation. The bacterial numbers may decrease by 1 to 2 orders of magnitude during the start of mesophilic phase in comparison to the numbers present during the thermophilic phase (10^8-10^{11} g^{-1} dry wt), but their taxonomic and metabolic diversity increases. Metabolic studies have revealed that several isolates were not only simply organic oxidizers, but were involved in hydrogen-, ammonium-, nitrite- and sulphur-oxidation, nitrogen-fixation, sulphate-reduction, exopolysaccharide production, and nitrite production from ammonium under heterotrophic conditions. High numbers of diverse mesophilic and thermotolerant actinomycetes and yeasts reappear. Fall in temperature, lower water content and their ability to attack and/or degrade natural complex polymers (*e.g.* cellulose, hemicellulose, lignocellulose, lignin) also favour mesophilic and thermotolerant fungi during the cooling phase.

4.4 Maturation Phase

During the maturation phase the main activity which takes place is degradation of the more resistant compounds and getting them partly transformed into humus. These compounds are lignin, lignocellulose and other recalcitrant components of tree bark, yard wastes, agricultural wastes, etc. Paper may contain up to 20% of lignin. Most of the fungi, predominant cellulose and lignin degraders, are isolated during the maturation phase. During composting of solid wastes, the problem of the presence of undegraded cellulosic material in the compost at the end of the process is encountered. This is probably due to the inaccessibility of residual cellulose to enzymatic attack because of low water content or association with protective substances such as lignin.

4.4.1 Curing Stage

This is very important phase in composting process. Microbial population and species diversity involved in degradation of complex polymers such as cellulose, lignin, hemicelluloses etc. increase drastically during this phase. The microbial population at this stage varies from 10^8-10^9 cells/g Municipal Solid Waste (MSW)

Bacteria represent 80% of this population. Most of the population involved in the C-cycle had proteolytic, ammonifying, and aerobic cellulolytic capacities. Free living N- fixers, denitrifiers, sulfate reducers and sulfur oxidizers are important constituent of the total microbial population. A large variety of microbial mediated degradation and synthesis process operate during the maturity phase. The population of different groups of bacteria and fungi during maturity stage of MSW compost is given below in Table 4.

Table 4: Population and diversity of microbial species on MSW compost during Curing Stage

S.No	Types of microorganism	Species No.	Cell number/g MSWcompost
A. Bacteria			
1	Heterotrophic non sporeformers	18	10^7-10^9
2	Heterotrophic sporeformers	3	10^5-10^6
3	Hydrogen oxidizing	6	10^5-10^6
4	Sulfur-oxidizing (Obligatory)	2	10^3-10^4
5	Sulfur-oxidizing (Facultatively autotrophic)	7	10^4-10^7
6	N_2 Fixing	12	10^5-10^7
7	Nitrifying	4	10^4-10^6
8	Exopolysaccharide producers	10	10^6-10^8
B. Fungi		**14**	**10^4-10^5**

For better understanding, the following steps of FLOW CHART are a must for the production of quality compost. For preparing of good quality compost from MSW it is necessary for the organics to go through "Seven Stage of Composting".

MSW
↓

SEGREGATION
(100 mm sieve) Above 100 mm as Residual Derived Fuel (RDF) below 100 mm for compostable
↓

WINDROW FORMATION
(2.5 M Height Max., 3.5 m wide bottom Max)
↓

CULTURE TREATMENT
(At every 50 cm layer & 50-60% Moisture maintenance)
↓

FIRST MESOPHILIC STAGE
(Temp. below 45°C, 55% moisture & time 1 week) Turning for aeration
↓

THERMOPHILIC STAGE
Inoculation (Temp. 65-75 °C, Moisture at 50 % & time – 2 weeks) Turning after a week
↓

SECOND MESOPHILIC STAGE
(Cooling) (Temp. 45-50°C, moisture at 45% & time –1 week) Turning
↓

MATURATION STAGE
(CURING)
(Degradation of lignin, lignocellulose and other recalcitrant compounds, Time 1-2 weeks)
↓

SCREENING
(4mm seive)
↓

HUMIFICATION
(SYNTHESIS OF HUMUS)
Change of aliphatic compounds into aromatic compounds (high humic acid content & low fulvic acid content) high content of nutrient but no pollutants, time 1 week, & moisture around 20-25%)
↓

QUALITY COMPOST
(AS PER FCO STANDARDS)
To achieve this essentials are inoculants and aeration. We have Indigenous culture and Effective Microorganisms (EM) culture as inoculants and turning for aeration is by Pocklains.

5. Synthesis of Humus During Composting Process

Humus is the end product of composting in which organic compounds of natural origin are partially transformed into relatively stable humic substances. Humus is a black to dark brown component of organic matter which has high molecular weight, very high CEC and is a store house of plant nutrients. The pathway from organic matter to humus formation involves a number of degradative and condensation reactions. Initially, the mixture of wastes, whether agricultural or MSW contain a number of mono, di- and polysaccharides. The monosaccharides (glucose, fructose, mannose etc) are easily biodegradable and least resistant. The intermediate biodegradable compounds are disaccharides or oligosaccharides such as sucrose, melibiose, maltose, lactose and cellobiose while the organic polymers such as cellulose, hemicelluloses and chitin (indigestible); starch, glycogen, gum (digestible) and lignin (α-guaicyl-glycerol, β- coniferyl-ether) are most resistant to decomposition.

During humification, aliphatic compounds of waste material changes into aromatic compounds upon microbial action. Lignin is first degraded by extracellular enzymes to smaller units, which are then absorbed into microbial cells where they are partly converted to phenols and quinones. The substances are discharged together with oxidizing enzymes into the system where they get polymerized.

In general, immature compost contains high levels of fulvic acids and low levels of humic acids. As the decomposition proceeds, the fulvic acid fraction either decreases or remains unchanged while humic acids are produced. Mature humified compost is characterized by (1) a high content of stable organic matter rich in humic acid containing aromatic moieties (2) refeeding of soils with humus into soil microbes, (3) high nutrient supply capacity (4) support of better plant health, and (5) minimum content of pollutants.

The degree of humification is generally accepted as a criterion of compost maturity. The parameters used in this regard are the per cent of humic acid, humic acid/fulvic acid ratio (HA/FA) and humification index

$\left[\dfrac{HA + FA}{TOC} \right] * 100$ Non-destructive methods such as 13C-NMR and FTIR (Fourier

transform IR) spectroscopy have been used for the analysis of humus (Almendros *et al.* 1992). The increments of aromatic structures are widely used to monitor the degree of humification by spectroscopic methods.

The functional groups responsible for the cation exchange capacity (CEC) of decomposed material are mainly the carboxyl, phenolic and enolic groups. Therefore, CEC of the organic materials is also helpful for estimating the degree of humification. The humification index and CEC/TOC ratio increases with increase in decomposition. The amount of humic acid in Table 5 varied throughout the days of decomposition suggesting that at initial stage where the activity of microorganism have just started, for example at initial stage (0-7 days), the amounts of HA (mainly from the manures themselves) was high at 48.9 mg/g due to partially decomposed mixed manure (animal wastes, *Leucaena leucocephala*, ricebran, fish formulated juice, etc) used in this study. During thermophilic stage (42 days), where activities of microorganisms are high, HA is reduced as some of it might be used as source of food/energy in the biological processes. This is supported by the highest C:N ratio of HA at the beginning which decreases during the thermophilic stage. Because the amount of HA is changing over the period of mixing, some researchers believe that the amount of HA could not be the only good parameter to monitor compost maturity. At the maturity stage (42-120 days), the amounts of HA raises again as the formation of large HA molecules takes place from many small molecules. As the production of humus usually requires 100-120 days, it is incorrect to claim that quality of compost can be produced within 7-10 days, as is sometimes claimed to be.

Table 5: Formation of humic acid (HA) and changes in C:N ratio during decomposition

Day	HA(mg/g)	C:N
0	48.9	12.6
7	29.7	11.1
14	19.5	7.4
21	14.1	6.6
28	25.1	6.9
35	28.0	6.7
49	36.3	7.7
77	45.2	7.7
91	31.4	7.8
105	46.0	7.7
119	60.1	7.5

Source: Sanmanee *et al.* (2011)

6. Evaluation of the Humification Process

The humification processes have been evaluated in terms of humification index, CEC, CEC/TOC, lignin content, etc. at four stages of composting by Bernal *et al.* (1998) that is, at the initial stage (7 days), at the thermophilic phase (42 days of composting), at the end of active phase (42-84 days) and at the maturation phase (105-133 days of composting) from different composts using wide range of wastes. The humic acid like fraction (HA; FA and HI) generally increased during composting although in some mixtures this tendency was not so clear and even a slight decreases are observed (Table 6).

Since the humification parameters HA, FA and HI of composts have very different values depending of the origin of the waste, no threshold value can be established to describe the maturity of all kinds of composts. The ratio of CEC/TOC decreases with increase in maturity. However, the threshold values of CEC established (>3.5) to indicate a good degree of maturity but cannot be generalized for city garbage compost because the content of inert materials is very high in it.

Table 6: Characteristics of the organic matter of the wastes mixture at different composting times

Composting stage	C:N ratio	HA (%)	FA (%)	HI (%)	CEC/TOC
SC compost*					
Initial stage	21.1	7.18	3.75	16.4	1.22
Thermophilic stage	14.2	7.42	3.80	18.6	2.39
End of active phase	9.9	7.75	3.11	21.5	2.80
Maturity phase	9.4	7.9	2.57	22.2	3.50
PCO compost*					
Initial stage	15.0	3.47	3.72	8.5	2.28
Thermophilic stage	11.5	3.85	2.98	10.7	2.71
End of active phase	9.7	3.67	3.13	10.9	3.04
Maturity phase	9.7	2.94	2.65	8.7	3.15
SCO compost*					
Initial stage	21.1	2.83	2.95	7.0	1.88
Thermophilic stage	11.0	1.81	2.44	5.4	3.72
End of active phase	10.1	1.95	2.06	6.5	4.41
Maturity phase	9.4	1.36	1.84	4.6	4.54
SM compost*					
Initial stage	11.0	2.59	8.14	6.3	0.94
Thermophilic stage	9.1	3.71	6.54	10.1	1.46
End of active phase	9.6	5.67	3.66	18.6	1.92
Maturity phase	8.6	4.73	2.39	17.7	2.76
SMO compost*					
Initial stage	31.1	8.94	5.36	18.9	0.73
Thermophilic stage	18.4	6.25	4.06	15.3	1.34
End of active phase	13.8	7.34	4.03	17.7	1.63
Maturity phase	11.8	6.47	3.85	16.4	2.28
PPB compost*					
Initial stage	24.1	3.90	2.28	9.7	1.17
Thermophilic stage	13.3	5.28	2.10	14.9	1.85
End of active phase	11.3	6.59	2.14	19.7	2.60
Maturity phase	11.0	5.62	1.88	18.5	-

*SC= sewage sludge + cotton waste; PCO=Poultry manure + cotton waste+ olive mill waste water; SCO= Sewage sludge + cotton waste+ olive-mill wastewater; SM= sewage sludge+ maize straw; SMO= sewage sludge+ maize straw+ olive oil-mill waste water; PPB= pig slurry + poultry manure+ sweet sorghum bagasse.

Infrared spectroscopy can be used to predict the various functional groups present in humic acid which indicates the quality and maturity of compost. The energy uptake by the matured compost can be recorded by a wave number range (4000-400 cm^{-1} mid infra red area). The indicator bands reflect the changes of compost maturity and stability. However, this technique is very costly.

7. The Role of Humus in the Organic Cycle in Nature

After decomposition the plant and animal residues do not become completely mineralized. A certain part of these residues is more or less resistant to microbial decomposition and remains for a period of time in an undecomposed or in a somewhat modified state, and may even accumulate under certain conditions. This resistant material is dark brown to black in color and possesses certain characteristic physical and chemical properties; it is usually called humus. As a result of the formation and accumulation of this humus, a part of the elements essential for organic life, especially carbon, nitrogen, phosphorus, sulfur, and potash, become locked up and removed from circulation. In view of the fact that the most important of these elements; namely, carbon, combined nitrogen, and available phosphorus, are present in nature in only limited concentrations, their transformation into an unavailable state, in the form of humus, tends to serve as a check upon plant life. On the other hand, since humus can undergo slow decomposition under certain favorable conditions, it tends to supply a slow but continuous stream of the elements essential for new plant synthesis. Humus thus serves as a reserve and a stabilizer for organic life on this planet. The actual concentration of organic matter in the form of humus in the soil and in the sea far exceeds that present in all the living forms of plants and animals. It is sufficient to call attention to the fact that the humus content of the soil is considerably greater than the total amount of organic matter present in all the crops harvested in a given year from all the fields, orchards, and gardens, or that available in the form of reserve foodstuffs. The large quantities of humus present in peat and in brown and hard coal, which far exceed the supplies of organic matter in our forests, represent organic accumulations during many thousands of years. The amount of organic matter found in seas, rivers, and lakes, whether in true solution, in colloidal suspension, or in the bottom material, also exceeds many times (Krogh 1934) the organic matter content of all plant and animal life in those waters.

8. Characteristics of Humus

Humus is a natural body; it is a composite entity, just as are plant, animal, and microbial substances; it is even much more complex chemically, since all of these materials contribute to its formation. Humus possesses certain specific physical, chemical, and biological properties which make it distinct from other natural organic bodies. Chemically, humus consists of certain constituents of the original plant material resistant to further decomposition; of substances undergoing decomposition; of complexes resulting from decomposition, either by processes of hydrolysis or by oxidation and reduction; and of various compounds synthesized by microorganisms (Schreiner and Dawson 1927).

Humus, in itself or by interaction with certain inorganic constituents of the soil, forms a complex colloidal system, the different constituents of which are held together by surface forces; this system is adaptable to changing conditions of reaction, moisture, and action of electrolytes. The numerous activities of the soil microorganisms take place in this system to a large extent. It is now definitely recognized that humus has resulted from the decomposition of plant and animal bodies, mainly through the agency of microorganisms although the possibility of certain chemical reactions taking place in the process is not excluded. Humus has, therefore, certain specific properties which distinguish it from other natural bodies. These properties can be briefly summarized as follows:

a) Humus possesses a dark brown to black color.

b) Humus is practically insoluble in water, although a part of it may go into colloidal solution in pure water. Humus dissolves to a large extent in dilute alkali solutions, especially on boiling, giving a dark colored extract; a large part of this extract precipitates when the alkali solution is neutralized by mineral acids. Certain constituents of humus may also dissolve in acid solutions and be precipitated at the isoelectric point, which is at a pH of about 4.8.

c) Humus contains a somewhat larger amount of carbon than do plant, animal, and microbial bodies; the carbon content of humus is usually about 55 to 56 per cent, and frequently reaches 58 per cent.

d) Humus contains considerable nitrogen, usually about 3 to 6 per cent. The nitrogen concentration may be frequently less than this figure; in the case of certain high moor peats, for example, it may be only 0.5-0.8 per cent. It may also be higher, especially in subsoils, frequently reaching 10 to 12 per cent.

e) Humus contains the elements carbon and nitrogen in proportions which are close to 10: 1; this is true of many soils and of humus in sea bottoms. This ratio varies considerably with the nature of the humus, the stage of its decomposition, the nature and depth of soil from which it is obtained, and climatic and other environmental conditions under which it is formed.

f) Humus is not in a static, but rather in a dynamic condition, since it is constantly formed from plant and animal residues and is continuously decomposed further by microorganisms.

g) Humus serves as a source of energy for the development and growth of various groups of microorganisms, and, during decomposition, gives off a continuous stream of carbon dioxide and ammonia.

h)   Humus is characterized by a high capacity of base-exchange, of combining with various other inorganic soil constituents, of absorbing water, and of swelling and by other physical and physico-chemical properties which makes it a highly valuable constituent of substrates which support plant and animal life.

i)   Humus determines the soil health.

9. Process of Humification

"Humification" is a specific process of humus formation. The earlier investigators believed that humus formation takes place in nature by a specific process of "humification," whereby the plant residues are bodily transformed into dark colored substances, or humus. The first definite suggestion concerning the nature of this process is traced to Wallerius (1761), who stated that humus is formed from decomposing vegetation. As long as the plant residues underwent continued decomposition, the product was called "humus"; when they were saturated with water, the product gave rise to peat. This idea was expended by de Saussure in 1804, who emphasized that humus originates from vegetable matter through the combined action of air and water. These simple conceptions prevailed among many chemists and, as late as recent years, one finds definite statements to the effect that the decomposition of organic residues in nature and the formation of humus are largely processes which involve simple oxidation and condensation. In many instances, even when the role of microorganisms in the decomposition of plant and animal substances in soils and in composts was taken into consideration, the processes were designated by such vague generalized terms as "decay," "fermentation," and "putrefaction." Liebig spoke of the production or humus in the decomposition of vegetable matters by the action of acids and alkalies; he further stated that "woody fibre in a state of decay is the substance called humus." Rosenberg-Lipinsky (1879) defined humus as "a mass of brown, partly soluble, partly insoluble, partly acid and partly neutral products of decomposition, which, in the presence of air; water and heat, undergo further decomposition, giving carbon dioxide, water and ammonia." According to Hilgard plant material added to the soil has first to be "humified" before the nutrient elements contained therein become available for plant growth. Humus was thus considered to be a substance produced as an intermediary product during the decomposition of organic residues; it was also looked upon as that portion of soil organic matter which can readily become available for plant nutrition (Ladd 1901). Oden (1919), among the more recent investigators, also believed that the process of humus formation in nature is purely chemical, similar to the formation of artificial humic acids." Maillard proposed a purely chemical theory of humus formation, namely, the interaction of the sugars produced in the hydrolysis of polysaccharides with the amino acids formed from the proteins to give rise to humus; this process was believed to be one of "dehydration and humification." As late as 1930, humus formation, beginning with the first stages of browning, was considered to be physicochemical and non-biological in nature: during this process of "humification," oxygenated water and benzol-ring compounds which act as antiseptics are formed; the rays of the sun and temperature were believed to be the chief agents in this process; a definite relation was said to exist between plant pigmentation and humification, since the maternal substance (aromatic bodies) are the same and the energy of formation is the same. These ideas and others of a similar nature are largely speculative and not based upon sufficient experimental evidence. Wieler stated that all plants contain "humic acids." Fraps and Hamner (1910) and Gortner (1916), using ammonium hydroxide solution, obtained from fresh plant materials an extract similar to "humus"; after the plant residues were added to the soil, the concentration of this "humus" complex was found to diminish; they suggested, therefore, that no specific "humification" process takes place in the soil but that "humus" is actually added to the soil in the plant remains. These results seem to be quite contrary to the prevalent theories concerning the formation of humus and its resistance to decomposition, a phenomenon responsible for its persistence in soil and its accumulation in peat bogs. The current ideas of "humification" considered as a specific process of humus formation were thus found to be, in most instances, as vague as, if not more so than, those concerning the chemical nature of humus itself. No attention was usually paid to the fact that the chemical constituents of the plant and animal residues may be decomposed at different rates, by various organisms, giving a variety of products. Decomposition of the plant material was frequently looked upon as a single process. The complex chemical nature of the materials was not fully recognized, nor was the fact that in the microbial transformation of different organic complexes many reactions are involved which lead to the formation of numerous new compounds. In many instances, one finds only broad generalizations as to the role played by microorganisms in the formation and accumulation of the dark colored organic matter, namely, humus.

10. Role of Humus in Soil Health

Humus performed threefold function in the soil:

1. Physical: It modifies the soil color, texture, structure, moisture-holding capacity, and aeration.

2. Chemical:

- It Influences the solubility of certain soil minerals

- Forms compounds with certain elements, such as iron, which renders them more readily available for plant growth.

- Increases the buffering properties of the soil.

3. Biological:

- It serves as a source of energy for the development of microorganisms, as well as by making the soil a better medium for the growth of higher plants;

- It also supplies a slow but continuous stream of nutrients for plant life.

Humus exerts various other important effects upon plant growth which are still awaiting explanation; although some of these effects are believed to be injurious in nature, most are definitely established to be highly beneficial. The various attempts to explain these effects by the formation of plant toxins, on the one hand, and of "plant stimulating substances" or "auximones" (Mockeridge 1924; Saeger 1925; Wolfe 1926), on the other, are still a matter of dispute. Of course, all reference to the "spirit of the soil" or the "reserve soil force" represents speculative generalizations not based upon sound experimental evidence.

11. Compounds of Humus

Humus or humified organic matter is the remaining part of organic matter that has been used and transformed by many different soil organisms. It is a relatively stable component formed by humic substances, including humic acids, fulvic acids, hymatomelanic acids and humins (Tan 1994). It is probably the most widely distributed organic carbon-containing material in terrestrial and aquatic environments. Humus cannot be decomposed readily because of its intimate interactions with soil mineral phases and is chemically too complex to be used by most organisms. It has many functions. One of the most striking characteristics of humic substances is their ability to interact with metal ions, oxides, hydroxides, mineral and organic compounds, including toxic pollutants, to form water-soluble and water-insoluble complexes. Through the formation of these complexes, humic substances can dissolve, mobilize and transport metals and organics in soils and waters, or accumulate in certain soil horizons. This influences nutrient availability, especially those nutrients present at microconcentrations only (Schnitzer 1986). Accumulation of such complexes can contribute to a reduction of toxicity, *e.g.* of aluminium (Al) in acid soils (Tan and Binger 1986), or the capture of pollutants in the cavities of the humic substances (Vermeer 1996).

Humic and fulvic substances enhance plant growth directly through physiological and nutritional effects. Some of these substances function as natural plant hormones (auxins and gibberillins) and are capable of improving seed germination, root initiation, uptake of plant nutrients and can serve as sources of N, P and S (Tan 1994; Schnitzer 1986). Indirectly, they may affect plant growth through modifications of physical, chemical and biological properties of the soil, for example, enhanced soil water holding capacity and CEC, and improved tilth and aeration through good soil structure (Stevenson 1994).

About 35–55 percent of the non-living part of organic matter is humus. It is an important buffer, reducing fluctuations in soil acidity and nutrient availability. Compared with simple organic molecules, humic substances are very complex and large, with high molecular weights. The characteristics of the well-decomposed part of the organic matter, the humus, are very different from those of simple organic molecules. While much is known about their general chemical composition, the relative significance of the various types of humic materials to plant growth is yet to be established.

Humus consists of different humic substances:

- *Fulvic acids:* the fraction of humus that is soluble in water under all pH conditions. Their colour is commonly light yellow to yellow-brown.

- *Humic acids:* the fraction of humus that is soluble in water, except for conditions more acid than pH 2. Common colours are dark brown to black.

- *Humin:* the fraction of humus that is not soluble in water at any pH and that cannot be extracted with a strong base, such as sodium hydroxide (NaOH). Commonly black in colour.

The term acid is used to describe humic materials because humus behaves like weak acids. Fulvic and humic acids are complex mixtures of large molecules. Humic acids are larger than fulvic acids. Research suggests that the different substances are differentiated from each other on the basis of their water solubility. Fulvic acids are produced in the earlier stages of humus formation. The relative amounts of humic and fulvic acids in soils vary with soil type and management practices. The humus of forest soils is characterized by a high content of fulvic acids, while the humus of agricultural and grassland areas contains more humic acids. Soil humic substances comprise a physically and chemically heterogeneous mixture of naturally occurring, biogenic, relatively high molecular weight, yellow to black coloured, amorphous, polydispersed organic polyelectrolytes of mixed aliphatic and aromatic nature, formed by secondary synthesis reaction (humification) during the decay process and transformation of biomolecules that originated from dead organisms and microbial activity.

12. Theories of Humus Formation in Soils

Several pathways exist for the formation of humic substances during the decay of plant and animal remains in soil, the main ones being shown in the picture: The classical theory, popularized by Waksman, is that humic substances represent modified lignins (pathway 1) but the majority of present-day investigators favor a mechanism involving quinones (pathway 2 and 3). In practice all four pathways must be considered as likely mechanisms for the synthesis of humic and fulvic acids in nature, including sugar-amine condensation (pathway 4).These four pathways may operate in all soils, but not to the same extent or in the same order of importance. A lignin pathway may predominate in poorly drained soils and wet sediments (swamps, etc.) whereas synthesis from polyphenols may be of considerable importance in certain forest soils.

12.1 Pathway 1 - The Lignin Theory (Waksman)

Plant lignin as the main source of soil humic substances (HS), with the involvement of amino compounds produced by microbial synthesis. For many years it was thought that humic substances were derived from lignin (pathway 1). According to this theory, lignin is incompletely utilized by microorganism and the residuum becomes part of the soil humus. Modification in lignin include loss of methoxyl (OCH_3) groups with the generation of o-hydroxyphenols and oxidation of aliphatic side chains to form COOH groups. The modified material is subject to further unknown changes to yield first humic acids and then fulvic acids. This pathway, illustrated on the picture (Fig. 2), is exemplified by Waksman's lignin-protein theory.

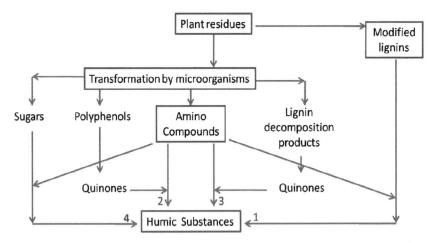

Fig. 2: Mechanisms for the formation of soil humic substances (*Source:* Stevenson 1982)

The following evidence were cited by Waksman (Fig. 3) in support of the lignin theory of humic acid formation:

- Both lignin and humic acid are decomposed with considerable difficulty by the great majority of fungi and bacteria.
- Both lignin and humic acid are partly soluble in alcohol and pyridine.
- Both lignin and humic acid are soluble in alkali and precipitated by acids.
- Both lignin and humic acid contain OCH_3 groups.
- Both lignin and humic acid are acidic in nature.
- When lignins are warmed with aqueous alkali, they are transformed into methoxyl-containing humic acids.
- Humic acids have properties similar to oxidized lignins.

Although lignin is less easily attacked by microorganisms than other plant components, mechanisms exist in nature for its complete aerobic decomposition. Otherwise undecomposed plant remains would accumulate on the soil surface and the organic matter content of the soil would gradually increase until CO_2 was depleted from the atmosphere. The ability of soil organisms to degrade lignin has been under estimated in some quarters and its contribution to humus has been exaggerated.

In normally aerobic soils lignin may be broken down into low-molecular-weight products prior to humus synthesis. On the other hand, the fungi that degrade lignin are not normally found in excessively wet sediments. Accordingly, it seems logical to assume that modified lignins may make a major contribution to the humus of peat, lake sediments, and poorly drained soils.

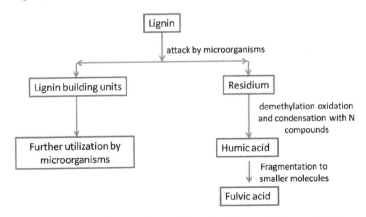

Fig. 3: The lignin theory of humus formation (*Source:* Waksman 1932)

12.2 Pathway 4 - Sugar-Amine Condensation

Reducing sugars and amino acids formed as by-products of microbial metabolism are assumed to be only precursor of HS (Fig. 4). The notion that humus is formed from sugars (pathway 4) dates back to the early days of humus chemistry. According to this concept reducing sugars and amino

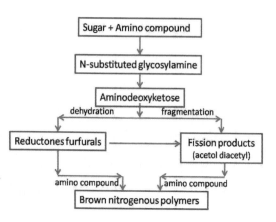

Fig 4: Sugar-amine condensation (*Source:* Stevenson 1982)

acids, formed as by-products of microbial metabolism, undergo non-enzymatic polymerization to form brown nitrogenous polymers of the type produced during dehydratation of certain food products at moderate temperatures. A major objection to this theory is that the reaction proceeds rather slowly at the temperatures found under normal soil conditions. However, drastic and frequent changes in the soil environment (freezing and thawing, wetting and drying), together with the intermixing of reactants with mineral material having catalytic properties, may facilitate condensation. An attractive feature of the theory is that the reactants (sugars, amino acids etc.) are produced in abundance through the activities of microorganisms.

The initial reaction in sugar-amine condensation involves addition of the amine to the aldehyde group of the sugar to form the n-substituted glycosylamine. The glycosylamine subsequently undergoes to form the N-substituted-1-amino-deoxy-2-ketose. This is subject to: fragmentation and formation of 3-carbon chain aldehydes and ketones, such as acetol, diacetyl etc.; dehydration and formation reductones and hydroxyl methyl furfurals.

All of these compounds are highly reactive and readily polymerize in the presence of amino compounds to form brown-colored products.

12.3 Pathways 2 and 3 - The Polyphenol Theory (Current Concept)

"Polyphenol theory" involves polyphenol and quinones, either derived from lignin or synthesized by microorganisms (Fig. 5). In pathway 3 lignin still plays an important role in humus synthesis, but in a different way. In this case phenolic aldehydes and acids released from lignin during microbiological attack undergo enzymatic conversion to quinones, which polymerize in the presence or absence of amino compounds to form humic like macromolecules.

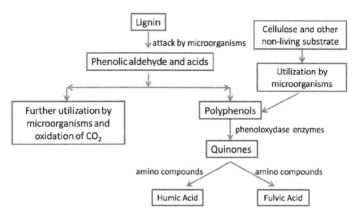

Fig. 5: Polyphenol theory (*Source:* Stevenson 1994)

Pathway 2 is somewhat similar to pathway 3 except that the polyphenols are synthesized by microorganisms from nonlignin C sources (*e.g.,* cellulose). The polyphenols are then enzymatically oxidized to quinones and converted to humic substances. As noted earlier, the classical theory of Waksman is now considered obsolete by many investigators. According to current concepts quinones of lignin origin, together with those synthesized by microorganisms, are the major building blocks from which humic substances are formed.

The formation of brown-colored substances by reactions involving quinones is not rare event, but is a well-known phenomenon that takes place in melanine formation, such as in the flesh of ripe fruits and vegetables following mechanical injury and during seed coat formation.

Possible sources of phenols for humus synthesis include lignin, microorganisms, uncombined phenols in plants and tannins. Of these, only the first two have received serious attention.

Flaig's concept of humus formation is:

1. Lignin, freed of its linkage with cellulose during decomposition of plant residues, is subjected to oxidative splitting with the formation of primary structural units (derivatives of phenylpropane).

2. The side-chains of the lignin-building units are oxidized, demethylation occurs, and the resulting polyphenols are converted to quinones by polyphenoloxidase enzymes.

3. Quinones arising from the lignin (and from other sources) react with N-containing compounds to form dark-colored polymers.

The role of microorganisms as sources of polyphenols has been emphasized by Kononova (1962). She concluded that humic substances were being formed by cellulose-decomposing myxobacteria prior to lignin decomposition.

The stages leading to the formation of humic substances were postulated to be:

1. Fungi attack simple carbohydrates and parts of the protein and cellulose in the medullary rays, cambium, and cortex of plants residues.

2. Cellulose of the xylem is decomposed by aerobic myxobacteria. Polyphenols synthesized by the myxobacteria are oxidized to quinones by polyphenoloxidase enzymes, and the quinones subsequently react with N compounds to form brown humic substances.

3. Lignin is decomposed. Phenols released during decay also serve as source materials for humus synthesis.

In general, all four pathways may operate for the synthesis of HS in all soils but not to the same extent, one pathway usually being prominent. Humic acid may originate from plant-lignin and microbially synthesized polyphenols. Fulvic acid may arise preferentially from sugar-amine condensation mechanism.

According to Martin and Haider (1971), microscopic fungi of the imperfecti group play a significant role in the synthesis of humic substances in soil. Their studies have shown that fungi such as *Aspergillus sydowi, Epicoccum nigrum, Hendersonula toruloidea, Stachybotrys atra* and *S. chartarum* degrade lignin as well as cellulose or other organic constituents and in the process synthesize appreciable amounts of humic acid like polymers. Phenolic units making up the polymer originated from lignin, as well through synthesis by the fungi. The quantities of humic acid synthesized by fungi can be appreciable. For example, Martin *et al.* (1972) found that as much as one-third of the substances synthesized by *H. toruloidea*, including the biomass, consisted of humic acid (Fig. 6). Furthermore, a humic acid-type polymer could be recovered from the mycelium tissue by extraction with 0.5 N NaOH. The production of humic substances by microorganisms may be partly an extracellular process. Following synthesis, the polyphenols are secreted into the external solution, where they are enzymatically oxidized to quinines, which subsequently combined with other metabolites (e.g. amino acids and peptides) to form humic polymers (Stevenson 1982).

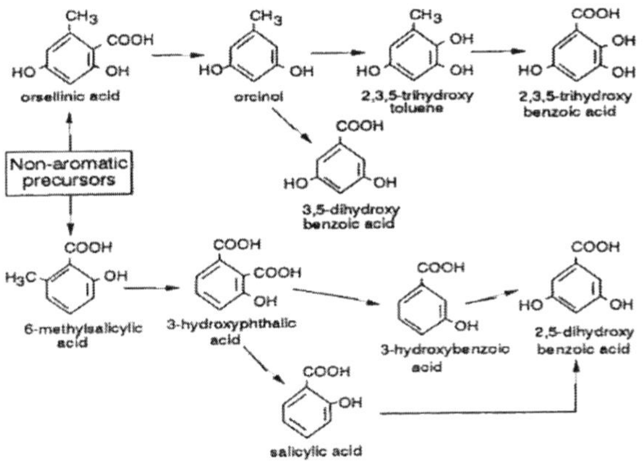

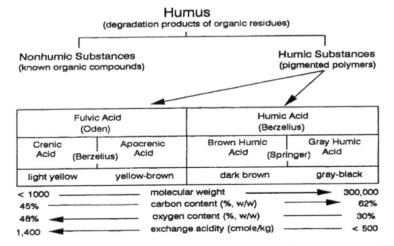

Fig. 6: Synthesis and possible transformations of polyphenols by *H. toruloidea* (*Source:* Martin et al. 1972)

13. Humic Substances as a System of Polymers

A useful concept that has evolved over the years, popularized several decades ago by the Russian scientist Kononova, is that the various humic fractions represent a system of polymers that vary in a systematic way in elemental content, acidity, degree of polymerization and molecular weight. The proposed interrelationships are shown in Fig. 7. No sharp difference exist between the two main fractions (humic and fulvic acids) or their subgroups. The humic fraction (material not extracted with alkali) is not represented, but this component may consist of one or more of the following:

Fig. 7: Classification and chemical properties of humic substances (*Source:* Stevenson 1994)

a) Humic acids so intimately bound to mineral matter that the two cannot be separated.

b) Highly condensed humic matter that has a high carbon content (>60%) and is thereby insoluble in alkali.

c) Fungal melanins, which have properties similar to humic acids and that are partially insoluble in alkali

d) The complete degradation of lignin occurs through the synergistic action of several groups of organisms. The main mechanism of microbial attack is the splitting of bonds at random in the aliphatic and or aromatic portions of the molecule.

The demethylation and the ring cleavage of lignin during attack by lignolytic fungi are shown in Fig. 8.

A particularly popular theory at this time is that in many terrestrial soils humic and fulvic acids are formed primarily through pathways 2 and 3, and that the processes include:

a) Degradation of all plant components into simpler monomers

b) Metabolism of the monomers accompanied by an increase in the soil biomass

c) Repeated recycling of biomass C and N with new cell synthesis

d) Concurrent polymerization of reactive monomers (i.e. of lignin origin and newly synthesized) into high molecular weight polymers (Stevenson 1994)

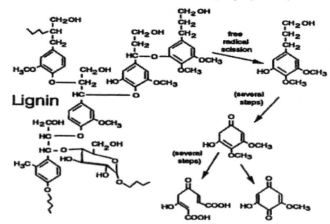

Fig. 8: Chain scission of lignin and production of phenols and other products during the microbial degradation of lignin (*Source:* Stevenson and Cole 1999)

From the above figure it can be seen that humic substances consist of a heterogeneous mixture of compounds, with each fraction (humic acid, fulvic acid, etc.) being made up of molecules of different sizes. In contrast to humic acids, low molecular weight fulvic acids contain higher oxygen but lower C contents, and they contain considerably more acidic functional groups, particularly COOH. Another important difference is that practically all the oxygen in fulvic acids can be accounted for in known functional groups (COOH, OH, C=O); a high portion of the oxygen in humic acids occurs as a structural component of the nucleus (e.g., in ether or ester linkages).

14. Chemical Properties and Structures of Humic and Fulvic Acids

Humic substances consist of a heterogenous mixture of compounds with each fraction (humic acid, fulvic acid, etc) being made up of molecules of different sites. In contrast to humic acids, the low molecular weight fulvic acids have higher oxygen but lower C contents, and they contain considerably more acidic functional groups, particularly –COOH. Various "type" structures have been proposed for humic and fulvic acids, but none of them can be considered as entirely satisfactory. The representative structural unit for a humate molecule shown in Fig. 9 conforms to many of the properties of humic and fulvic acids, notably a mixture of aliphatic and aromatic properties and a high content of oxygen containing functional groups.

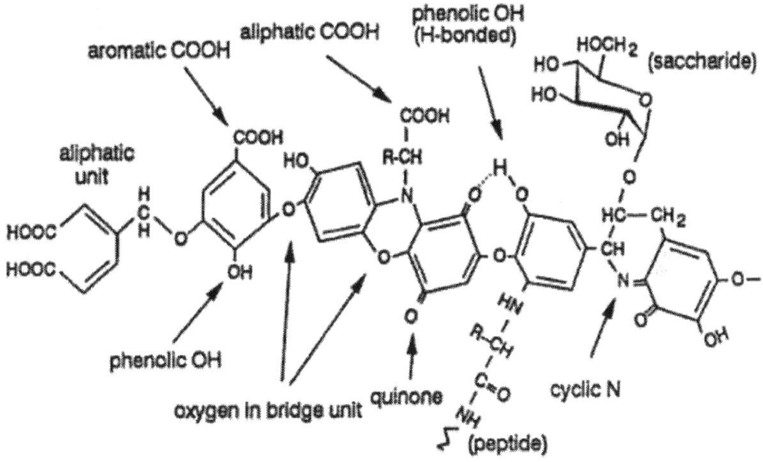

Fig 9: Representative structure unit of a humate molecule showing the presence of free and bound phenolic OH groups, quinones, oxygen and nitrogen as bridge units and –COOH groups variously placed on the molecule (*Source:* Stevenson 1994)

Conclusion

Soil organic carbon is a keystone as soil quality indicator and is the most consistently reported soil attribute, being inextricably linked to other physical, chemical, and biological soil quality indicators. But, despite of several research efforts, our understanding of the functions that SOM afford to soil quality, soil health and crop productivity still remain primarily descriptive in nature and matter of concern. The puzzle of SOM becomes even more complex given the varied responses of soil organisms to their environment and to our management efforts. It appears that increasing SOM has a host of benefits from both an agricultural and environmental standpoint.

References

Almendros G, Gonzalez-Vila FJ, Martin F, Frnd R, Ludemann HD (1992) Solid state NMR studies of fire-induced changes in the structure of humic substances. Sci. Total Environ. 117, 63–74.

Alvarez R, Alvarez CR (2000) Soil organic matter pools and their associations with carbon mineralization kinetics. Soil Sci. Soc. Am. J. 64, 184-189.

Angers DA, Mehuys GR (1990) Barley and alfalfa cropping effects on carbohydrate contents of a clay soil and its size fractions. Soil Biol. Biochem. 22, 285-288.

Bernal MP, Paredes C, Sánchez-Monedero MA, Cegarra J (1998) Maturity and stability parameters of composts prepared with a wide range of organic wastes. Bioresour. Technol. 63, 91-99.

Brussaard L (1994) Interrelationships between biological activities, soil properties and soil management. In: Soil Resilience and Sustainable Land Use (Ed) Greenland DJ and Szabolcs I, Wallingford, UK, CAB International, pp 309–329.

Carter MR (2002) Soil Quality for sustainable land management: Organic matter and aggregation interactions that maintain soil function. Agron. J. 94, 38-47.

Coleman DC, Crossley DA, Hendrix PF (2004) Fundamentals of Soil Ecology, 2nd Ed. Elsevier, Inc.

Conteh A, Blair GJ, Rochester IJ (1998) Soil organic carbon fractions in a Vertisol under irrigated cotton production as affected by burning and incorporating cotton stubble. Aust. J. Soil Res. 36, 655-667.

Dalal RC, Mayer RJ (1986) Long-term trends in fertility of soils under continuous cultivation and cereal cropping in southern Queensland: III. Distribution and kinetics of soil organic carbon in particle-size fractions. Aust. J. Soil Res. 24, 293-300.

Franzluebbers AJ, Stuedemann JA, Schomberg HH, Wilkinson SR (2000) Soil organic C and N pools under long-term pasture management in the Southern Piedmont USA. Soil Biol. Biochem. 32, 469-478.

Fraps GS, Hamner NC (1910) Studies on the ammonia-soluble organic matter of the soil. Texas Agr. Exp. Sta. Bull. 129.

Hilgard E W (1912) Soils, McGraw Hill Book Company, New York.

Jenny H (1980) The Soil Resource, Origin and Behavior, Ecological Studies 37, Springer-Verlag, New York.

Juma NG (1998) The pedosphere and its dynamics: a systems approach to soil science, Volume 1. Edmonton, Canada, Quality Color Press Inc.

Krogh A (1934) Conditions of life in the ocean. Ecol. Monogr. 4: 421-439.

Krull E, Skjemstad J, Baldock J (2004) Functions of Soil Organic Matter and the Effect on Soil Properties: A Literature Review. Report for GRDC and CRC for Greenhouse Accounting. CSIRO Land and Water Client Report. Adelaide: CSIRO Land and Water.

Ladd EF (1901) Humus and soil nitrogen. North Dak. Agr. Exp. Sta. Bull. 32. 1898, 47.

Lal R (2002) Soil carbon dynamics in cropland and rangeland. Environ. Pollut. 116, 353-362.

Lundegardh H (1924) Der Kreislauf der Kohlensaure in der Natur. Jena.

Magdoff F (1992) Building soils for better crops: Organic matter management. Univ. of Nebraska Press, Lincoln.

Magdoff F, Weil RR (2004) Soil Organic Matter in Sustainable Agriculture. CRC Press, Boca Raton, USA.

Maillard LC (1913) Genese des matierei,s prote'iques et des matieres humiques. Paris.

Martin JP, Haider K (1971) Microbial activity in relation to soil humus formation. Soil Sci. 111, 54-63.

Martin JP, Haider K, Wolf D (1972) Synthesis of phenols and phenolic polymers by Hendersonula toruloidea in relation to soil humus formation. Soil Sci. Soc. Am. Proc. 36, 311-315.

Miller FC (1993) Ecological process control of composting. In: Soil Microbial Ecology (Ed) Metting FB Jr, Marcel Dekker, New York, pp 529-536.

Mockeridge FA (1924) The occurrence and nature of plant growth-promoting substances in various organic manurial composts. Biochem. J. 14, 432-450.

Oden S (1919) Die huminsauren, chemische, physikalische und bodenkundliche forschung. Kolloidchem . Beihefte. 11, 75-260.

Parker EC (1920) Field Management and Crop Rotation. Copyright 1915, Webb Publishing, St. Paul, MN, USA.

Ryckeboer J, Mergaert J, Coosemans J, Deprins K, Swings J (2003) Microbiological aspects of biowaste during composting in a monitored compost bin. J. Appl. Microbiol. 94(1), 127-137.

Saeger A (1925) The growth of duckweeds in mineral nutrient solutions with and without organic extracts. J. Gen. Physiol. 7, 517-526.

Sanmanee N, Panishkan K, Obsuwan K, Dharmvanij S (2011) Study of compost maturity during humification process using UV-spectroscopy. World Acad. Sci. Eng. Technol. 56, 403-405.

Schnitzer M (1986) The synthesis, chemical structure, reactions and functions of humic substances. In: Humic Substances: Effect on Soil and Plants (Ed) Burns RG, dell'Agnola G, Miele S, Nardi S, Savoini G, Schnitzer M, Sequi P, Vaughan D and Visser SA, Congress on Humic Substances, March 1986, Milan, Italy.

Schreiner O, Dawson RP (1928) The chemistry of humus formation. Proc. First Intern. Congr. Soil Sci. 3, 255-263.

Schroeder H (1919) Die jahrliche Gesamtproduktion der grunen Pflanzendecke der Erde. Die Naturwiss. 7, 23-29.

Soil Science Society of America (1987) Glossary of Soil Science Terms. SSSA, Madison, WI.

Stevenson FJ (1982) Humus Chemistry. John Wiley, New York.

Stevenson FJ (1994) Humus Chemistry: Genesis, Composition, Reactions. 2nd edition, Wiley Interscience, New York, USA.

Stevenson FJ, Cole MA (1999) Cycles of Soil: Carbon, Nitrogen, Phosphorus, Sulfur, Micronutrients. 2nd Edition, Wiley-Interscience, New York.

Tan KH, Binger A (1986) Effect of humic acid on aluminium toxicity in corn plants. Soil Sci. 14, 20–25.

Tan KH (1994) Environmental Soil Science. New York, USA, Marcel Dekker Inc.

Tuomela M, Vikman M, Hatakka A, Itavaara M (2000) Biodegradation of lignin in a compost environment: A review. Bioresour. Technol. 72(2), 169-183.

Vermeer AWP (1996) Interactions between humic acid and hematite and their effects on metal ion speciation. Wageningen University, The Netherlands (Ph.D. thesis).

Wallerius JG (1761) Agriculturae fundamenta chemica spez. De Humo. Diss. Upsala.

Wander M (2004) Soil organic matter fractions and their relevance to soil function. In: Soil Organic Matter in Sustainable Agriculture (Ed) Magdoff F and Weil RR, CRC Press, pp 67-102.

Wolfe HS (1926) The Auximone Question. Bot. Gaz. 81, 228-231.

Zhang H, Thompson ML, Sandor JA (1988) Compositional differences in organic matter among cultivated and uncultivated Argiudolls and Hapludalfs derived from loess. Soil Sci. Soc. Am. J. 52, 216-222.

7

Biological Recovery of Metals from Electronic Waste Polluted Environment Using Microorganisms

Narayanasamy, M., Dhanasekaran, D. and Thajuddin, N.

Abstract

Electronic waste or e-waste is one of the emerging problems in developed and developing countries worldwide. The constantly changing today's world of technology has led to the serious problem of e-waste. The previous studies show that India has generated 0.6 million tons of e-waste in 2014 which may increase to 0.7 to 0.9 million tons by 2015–2016. E-waste constitutes multiple components some of which are toxic that can cause serious health and environmental issues if not handled properly. The challenge is to develop innovative and cost-effective solutions to decontaminate polluted environments, to make them safe for human habitation and consumption, and to protect the functioning of the ecosystems which support life. Biological approach is currently applied to recovery of metals from contaminated soil, groundwater, surface water, and sediments including air. There are different methods used in recovery of metals from e-waste using the microorganisms includes bioleaching, biosorption, bioaccumulation, biotransformation and biomineralization. These technologies have become attractive alternatives to conventional cleanup technologies due to relatively low capital costs and their inherently aesthetic nature. Therefore, these technologies need to be applied to decontaminate e-waste from the soil-water environment. The present chapter summarizes the hazardous effects of e-waste, in Indian and global scenario and innovative biological technologies using microorganisms to recover metals from e-waste in environment.

Keywords: Bioaccumulation, Bioleaching, Biosorption, Environmental hazard E-waste, Microorganisms and toxic metals.

1. Introduction

Electronic waste describes discarded electrical or electronic devices such as computers, office electronic equipment, entertainment device electronics, mobile phones, television sets and refrigerators. The used electronics which are destined for reuse, resale, salvage, recycling or disposal are also considered as e-waste. Consumer-oriented growth combined with rapid product obsolescence and technological advances are posing a new environmental challenge – the growing threat of 'electronic waste' or 'e-waste' that consists of obsolete electronic devices (Monika and Kishore 2010). Proper management and safe disposal of electronic waste has become an emerging issue worldwide. According to UNEP (United Nations Environment Programme), some 20 to 50 million metric tons of electronic waste are generated per year across the globe. Electronic waste includes obsolete electrical and electronic items which are categorized into six categories such as (a) monitors (10%); (b) televisions (10%); (c) computers, telephones, and their peripherals (15%); (d) DVD/VCR players, CD players, radios and hi-fi sets (15%); (e) refrigerators (20%); and (f) washing machines, air conditioners, vacuum cleaners etc (30%). Most of the electronic waste is a complex mixture of precious metals (Ag, Au, and Pt); base metals (Cu, Al, Ni, Si, Zn and Fe); toxic metals (Hg, Be, Cd, Cr (VI), As, Sb and Bi) along with halogens and combustible substances such as plastics and flame retardants (Robinson 2009). Now, electronic waste may be measured as a 'secondary ore' or 'artificial ore' for the concentrations of precious metals richer than natural ores, which makes their recycling important from both economic and environmental perspectives. However, complex composition and increasing volumes of e-waste, along with difficulties in treating it, are causes of concern (Tsydenova and Bengtsson 2011). So, recycling of electronic waste is an important subject not only from the point of waste management but also from the recovery aspect of valuable materials. However, recycling of electronic waste is still quite limited due to the heterogeneity of the materials present in the products and the complexity of the production equipment. Pyrometallurgical (Cui et al. 2003) mobile phones have a lifespan of less than 2 years in developed countries. When the millions of computers and other electronic devices around the world become obsolete yearly, they leave behind lead, cadmium, mercury and other hazardous wastes. EPA (Environmental Protection Agency) recently reported that over 3.19 million tons of e-waste was discarded in U.S alone and e-waste has been identified as the fastest growing component of solid municipal waste stream. Only 18% was recycled and the remaining was incinerated or land filled. Land filling poses challenges, due to scarcity of landfill space as well as the concern for pollution caused by leaching of toxic heavy metals into the environment through groundwater and rainwater.

Fortunately, biological process could be an environmentally sound and economical alternative. Using bioleaching techniques, the efficiency of recovery of metals can be increased, as revealed in copper and gold mining where low grade ores are biologically treated to obtain metal values, which are not accessible by conventional treatments (Agate 1996). Bioleaching is considered to be one of the most promising technologies without too much capital investment, labour need, and energy consumption (Hazard Mater *et al.* 2008). Bioleaching of different metals from numerous ores have been well documented. However, using microorganisms to leach metals from electronic waste materials is a new field of study with few publications in the literature (Brandl *et al.* 2001).

Among major groups of bacteria, the most commonly used are acidophilus and chemolithotrophs like *Acidithiobacillus ferrooxidans, Acidithiobacillus thiooxidans, Leptospirillum ferrooxidans* which use CO_2 as carbon source and inorganic compounds (Fe^{2+}, reduced S) as an energy source. These are microorganisms most widely considered group of microorganism in terms of bioleaching applications due to their ability to facilitate metal dissolution through a series of bio-oxidation and bioleaching reactions (Brandl *et al.* 2001). Other organisms including thermophiles *Sulfobacillus thermosulfidooxidans*, *Bacillus stearothermophilus, Metallospherasedula* and heterotrophic fungi including *Aspergillus niger* and *Penicillium simplicissimum* have also been used to effectively dissolve various metallic fractions from e-wastes. The ability of microorganisms (bacteria, fungi and algae) to leach and mobilize metals from solid materials occurs *via* three mechanisms redoxolysis, acidolysis and complexolysis (Krebs *et al.* 1997). In the direct mechanism for bacterial metal leaching *via* redox reaction, metals are solubilized by enzymatic reactions through physical contact between the microorganisms and the leaching materials, which lead to the destruction of the mineral. In the indirect mechanism of redox reaction, oxidation of ions such as Fe^{2+} to Fe^{3+} by bacteria and subsequently the Fe^{3+} (as oxidizing agent), in turn dissolves the metal chemically from solid material. The second mechanism of metal solubilization is through the formation of organic or inorganic acids. Examples include the production of citric acid or gluconic acid by *A. niger* and *P. simplicissimum,* and sulphuric acid by *A. ferrooxidans* and *A. thiooxidans*. The acid supplies the protons and contributes to the solubilization process. The third mechanism of metal extraction is through complexation. This mechanism involves complex formation between metabolites produced by the microorganisms and the metal ions, which can increase their mobility.

2. E-Waste

E-waste is a term used to cover all items of electrical and electronic equipment (EEE) and its parts that have been discarded by its owner as waste without the intent of reuse. It is also referred to as WEEE (Waste Electrical and Electronic Equipment), electronic waste or e scrap in different regions. E-waste includes a wide range of products, – almost any household or business item with circuitry or electrical components with power or battery supply. Basically, EEE can be calcified in to following six categories and therefore also e-waste:

a) Temperature exchange equipment. Also more commonly referred to as, cooling and freezing equipment. Typical equipments are refrigerators, freezers, air conditioners and heat pumps.

b) Screen/ monitors, typical equipment comprises of televisions, monitors, laptops, notebooks, and tablets.

c) Lamps. Typical equipment comprises of straight florescent lamp, compact florescent lamps, florescent lamps, high intensity lamps and LED lamps.

d) Large equipment. Typical equipment comprises of washing machines, clothe dryers, dish washing machines, electric stoves, large printing machines, copying equipment and photovoltaic panels.

e) Small equipment. Typical equipment comprises of vacuum cleaners, microwaves, ventilation equipment, toasters, electric kettles, electric shavers, scales, calculators, radio sets, video cameras, electrical and electronic toys, small electrical and electronic tools, small medical devices, small monitoring and control instruments.

f) Small IT and telecommunication equipment. Typical equipment comprises of mobile phones, GPS, pocket calculators, routers, personal computers, printers and telephones.

For each category, its original function, weight, size, material composition differ. These end of-life attributes determine that each category has different waste quantities, economic values, as well as potential environmental and health impacts through inappropriate recycling. Consequently, the collection and logistic means and recycling technology are different for each category in the same way as the consumer's attitude in disposing of the electrical and electronic equipment.

3. Worldwide Disposal of e-Waste

The global quantity of e-waste generation in 2015 was around 43.8 million tonnes (Mt); approximately 4 billion people were covered by national e-waste legislation, though legislation does not necessarily come together with enforcement. Driven by these national laws, around 7.5 Mt of e-waste was reported as formally treated by national take-back systems (Scenario 1). Not all e-waste laws have the same scope as the comprehensive scope in this report. In total, 0.8 Mt of e-waste is thrown into the waste bin in the 28 EU Member States (Scenario 2). The amount of e-waste that is disposed of in waste bins is unknown for other regions. The quantities of the collection outside formal take-back systems (Scenarios 3 and 4) are not documented systematically. However, they are likely to be gap between e-waste generated and the e-waste in the waste bin. Official data for the trans-boundary movement of e-waste (mostly from developed to developing countries) are unknown. The global quantity of e-waste in 2015 is comprised of 1.4 Mt lamps, 3.2 Mt of Small IT, 6.8 Mt of screens and monitors, 7.6 Mt of temperature exchange equipment (cooling and freezing equipment), 12.8 Mt large equipment, and13.8 Mt of small equipment. The amount of e-waste is expected to grow to 49.8 Mt in 2018, with an annual growth rate of 4 to 5 per cent (Table 1).

Table 1: Global quantity of e-waste generated

Year	E-waste generated (Mt)	Population (billion)	E-waste generated (kg/inh.)
2010	33.8	6.8	5.0
2011	35	6.9	5.2
2012	37.8	6.9	5.4
2013	39.8	7.0	5.7
2014	41.8	7.1	5.9
2015	43.8	7.2	6.1
2016	45.7	7.3	6.3
2017	47.8	7.4	6.5
2018	49.8	7.4	6.7

Source: (Balde *et al.* 2015)

4. Increasing Amount of e- waste

Product obsolescence is becoming more rapid since the speed of innovation and the dynamism of product manufacturing / marketing has resulted in a short life span (less than two years) for many computer products. Short product life span coupled with exponential increase at an average 15% per year will result in doubling of the volume of e-waste over the next five to six years.

5. Toxic Components of e-Waste

E-waste are known to contain certain toxic constituents in their components such as lead, cadmium, mercury, polychlorinated bi-phenyls, etched chemicals, brominated flame retardants etc., which are required to be handled safely. The recycling practices were found to be more in informal sectors leading to uncontrolled release of toxic materials into the environment as a result of improper handling of such materials (Table 2).

6. Environmentally Sound Management of e-waste in India

The growth of e-waste has significant economic and social impacts. The increase of electrical and electronic products, consumption rates and higher obsolescence rate leads to higher generation of e-waste. The increasing obsolescence rate of electronic products also adds to the huge import of used electronics products. The e-waste inventory based on this obsolescence rate in India for the year 2012 has been estimated to be 8, 00,000 tonnes which is expected to exceed 44, 00,000 tonnes by 2018. There is no large scale organized e-waste recycling facility in India and there are two small e-waste dismantling facilities functioning at Chennai and Bangalore, while most of the e-waste recycling units are operating in un-organized sector.

7. The Hazardous Wastes (Management and Handling) Rules, 2015

The Hazardous Waste (Management and handling) Rule, 2015, defines "hazardous waste" as any waste which by reason of any of its physical, chemical, reactive, toxic, flammable, explosive or corrosive characteristics causes danger or likely to cause danger to health or environment.

8. E-waste is a Global Challenge

The e-waste problem is global concern because of the nature of production and disposal of waste in a globalized world. According to the recent studies, up to 50– 80% of the WEEE, generated in developed markets, is being shipped to developing countries for reuse and recycling (Widmer *et al.* 2005) often against the international laws. E-waste, if not managed properly, can have very fatal effects on the environment as well as the living beings in the vicinity. In most of the developing and under-developed countries, e-waste is dumped directly in to the soil without any treatment; often due to weak environmental regulations and financial problems. This can induce toxicants into the soil, making it barren and often leaching into the ground water and contaminating it (Bastiaan *et al.* 2010). Hence, management of this ever piling WEEE is a big matter of concern in developed as well as in developing countries.

Table 2: Effect of E-waste toxic metals on human health

Components	Constituents	Affected body parts
Printed circuit boards (PCBs)	Lead, cadmium and berillium	Nervous system, kidney and liver
Mother boards	Lead oxide, barium and cadmium	Lungs, skin
Cathode ray tubes (CRTs)	Mercury	Heart, liver and muscles
Switches and flat-screen monitors	Cadmium	Brain, skin
Computer batteries	Polychlorinated biphenyls	Kidney, liver
Capacitors and transformers	Brominated flame-retardant casings cable	-
Printed circuit boards, Cable insulation/coating Plastic housing	Plastic polyvinyl chloride Bromine	- Immune system

Source: (Sharma Pramila *et al.*, 2012)

Table 3: Occurrence of metals in e-waste

Types of electronic waste	Metal abundance (in %)				Metal present (in ppm)			
	Pb	Ni	Al	Fe	Cu	Pd	Au	Ag
Printed wiring board	2.6	2	7	12	16	-	0.04	-
Printed circuit boards	3	29	2	7	12	16	-	0.04
Printed wiring board	3	2	7	12	16	-	110	280
PC board scrap	1.5	1	5	7	20	110	250	1000
PC main board scrap	2.2	1.1	2.8	4.5	14.3	12.4	506	6.36
TV board scrap	1	0.3	10	28	10	10	17	280
Mobile phone scrap	0.3	0.1	1	5	13	210	350	1340
Portable audio scrap	0.14	0.03	1	23	21	4	10	150
DVD player scrap	0.3	0.05	2	62	5	4	15	115
Calculator scrap	0.1	0.05	5	4	3	5	50	260
TV scrap	0.2	0.04	1.2	-	3.8	27.1	27.1	27.1
PC scrap	6.3	0.85	14.17	20.47	6.93	3	16	189

Source: (Bastiaan *et al.* 2010)

9. E-waste Recycling and Recovery

The efficient recycling of electronic waste is very necessary for profitable recovery of materials and sustainable environment which is still regarded as a major challenge for today's society. Recycling of e-waste Printed Circuit Boards (PCBs) is an important subject not only from the treatment of hazardous waste point of view but also from the recovery of the valuable materials point of view (Guo *et al.* 2008; Hall *et al.* 2007). The e-waste PCB recycling is difficult particularly because of its heterogeneity in organic materials, metals and glass fibers (Table 3). Hence, in this review, major focus is on hazards caused by such e-waste, followed by various recycling and recovery techniques of metal and non-metallic fractions of e-waste PCB and their reusability.

10. Recycling Techniques

PCBs are one of the most important components of electronic equipment. These PCBs encompass majority of the valuable metals and also most of the toxic components in the e-waste. PCBs waste recycling includes three processes namely, pretreatment, physical recycling and chemical recycling (Sohaili *et al.* 2012). Starting with pretreatment stage, it includes disassembling of the reusable and toxic parts using shredding or separation and followed by physical recycling process. Finally material is recovered by chemical recycling process that includes gasification and pyrolysis (Sohaili *et al.* 2012). There are various traditional and some modern methods to recover the valuable metallic and non-metallic fractions from PCBs. The following section will consist of various physical and chemical recycling processes for recycling of metallic and non-metallic fractions from e-waste PCBs (Fig. 1).

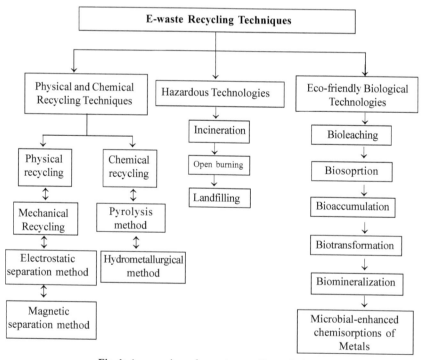

Fig. 1: An overview of e-waste recycling techniques

10.1 Chemical Recycling Methods

10.1.1 Pyrolysis Method

Pyrolysis is a chemical recycling technique, extensively used for recycling synthetic polymers including polymers that are mixed with glass fibers. Pyrolysis of such polymers gives gases, oils and chars and these products can further be used as chemical feedstock or fuels (Guo et al. 2008). The printed circuit boards are heated to a temperature, high enough to melt down the solder, used to bind the electrical components to the circuit board, in the presence of oxygen. After pyrolysis blackish metal substance is left behind (Mankhand et al. 2012). This blackish metal substance on leaching gives good yield of copper.

10.1.2 Hydrometallurgical Method

This process is majorly used for profitable recycling of metallic fraction (Bernardes et al. 1997). In this method, metal contents are dissolved into leaching solutions such as strong acids and alkalis. This is followed by electro refining of desired metals (Bernardes et al. 1997; Kinoshita et al. 2003). This technique is considered to be more flexible and energy saving, hence cost effective. Widely used leachants are aqua regia, nitric acid, and sulfuric acid and cyanide solutions (Kinoshita et al. 2003). In case of non-metallic substrates, metals leach out in the resulting solution, from the substrate (Li et al. 2004). Electrochemical processing can be done to recover metals, in case of metallic substrates (Li et al. 2004). Thus, a pure metal recovered is sold without any further processing while the remaining non-metallic substrates still need to be treated thermally prior to reusing or dumping in landfills. The major disadvantage of this method is the corrosive and poisonous nature of the liquid being used (Kinoshita et al. 2003).This process also leaves high totally dissolved solids (Eswaraiah et al. 2008).

10.2 Physical Recycling Methods

10.2.1 Mechanical Recycling

It is a physical recycling method. In this method, the disassembled samples are first cut into specific sizes depending upon the milling needs. Then the pieces are put through a milling process resulting into fine pulverized PCB powder. This powder is subjected to eddy current separators that separates the metal by their eddy current characteristics (Zhang et al. 1999). Finally the pulverized samples are subjected to density separation process (Zhang et al. 1997). Depending upon the density and particle size, stratification can be seen in the liquid column (Rohwerder et al. 2003).

10.2.2 Electrostatic Separation Method

In electrostatic separation technologies, electric force acting on charged or polarized bodies is used for the separation of granular materials (Shapiro *et al.* 2005). These technologies have been applied for recycling of metals and plastics from industrial wastes (Luga *et al.* 2001). Electrostatic separation technologies can be used to recycle Cu, Al, Pb, Sn, Fe and certain amount of noble metals and plastic from scrapped PCBs (Zhang *et al.* 1998).

10.2.3 Magnetic Separation Method

Magnetic separators with low intensity drum separators are widely used for the recovery of ferromagnetic metals from non-ferrous metals and other non-magnetic wastes (Hanafi *et al.* 2012). The disadvantage of magnetic separation is agglomeration of the particles. This agglomeration causes the magnet to also pull the non-metal materials which agglomerate with the ferrous materials (Sohaili *et al.* 2012).

10.3 Hazardous Technologies

10.3.1 Incineration

Incineration is the process of destroying waste through burning. Because of the variety of substances found in e-waste, incineration is associated with a major risk of generating and dispersing contaminants and toxic substances. The gases released during the burning and the residue ash is often toxic (Sasse *et al.* 1998). This is especially true for incineration or co-incineration of e-waste with neither prior treatment nor sophisticated flue gas purification. Studies of municipal solid waste incineration plants have shown that copper, which is present in PCBs and cables, acts a catalyst for dioxin formation when flame-retardants are incinerated. These brominated flame retardants when exposed to low temperature (600-800°C) can lead to the generation of extremely toxic polybrominated dioxins (PBDDs) and furans (PBDFs). PVC, which can be found in e-waste in significant amounts, is highly corrosive when burnt and also induces the formation of dioxins. Incineration also leads to the loss valuable of trace elements which could have been recovered had they been sorted and processed separately (Shapiro *et al.* 2005).

10.3.2 Open Burning

Since open fires burn at relatively low temperatures, they release many more pollutants than in a controlled incineration process at a municipal solid waste (MSW) plant. Inhalation of open fire emissions can trigger asthma attacks, respiratory infections, and cause other problems such as coughing, wheezing, chest pain, and eye irritation (Chen *et al.* 2000). Chronic exposure to open fire emissions may lead to diseases such as emphysema and cancer. For example, burning PVC releases hydrogen chloride, which on inhalation mixes with water in the lungs to form hydrochloric acid. This can lead to corrosion of the lung tissues, and several respiratory complications. Often open fires burn with a lack of oxygen, forming carbon monoxide, which poisons the blood when inhaled. The residual particulate matter in the form of ash is prone to fly around in the vicinity and can also be dangerous when inhaled (Zhu *et al.* 2013).

10.3.3 Landfilling

Landfilling is one of the most widely used methods of waste disposal. However, it is common knowledge that all landfills leak (Luga *et al.* 2001). The leachate often contains heavy metals and other toxic substances which can contaminate ground and water resources. Even state-of-the-art landfills which are sealed to prevent toxins from entering the ground are not completely tight in the long-term. Older landfill sites and uncontrolled dumps pose a much greater danger of releasing hazardous emissions. Mercury, Cadmium and Lead are among the most toxic leachates. Mercury, for example, will leach when certain electronic devices such as circuit breakers are destroyed (Yamawaki *et al.* 2003). Lead has been found to leach from broken lead-containing glass, such as the cone glass of cathode ray tubes from TVs and monitors. When brominated flame retarded plastics or plastics containing cadmium are landfilled, both polybrominated diphenyl ethers (PBDE) and cadmium may leach into soil and groundwater (Rotter *et al.* 2002). Similarly, landfilled condensers emit hazardous PCB's. Besides leaching, vaporization is also of concern in landfills. For example, volatile compounds such as mercury or a frequent modification of it, dimethylene mercury can be released. In addition, landfills are also prone to uncontrolled fires which can release toxic fumes (Huang *et al.* 2007). Significant impacts from landfilling could be avoided by conditioning hazardous materials from e-waste separately and by landfilling only those fractions for which there are no further recycling possibilities and ensure that they are in state-of-the-art landfills that respect environmentally sound technical standards (Yokoyama *et al.* 1995).

11 Eco-Friendly Biological Technologies

11.1 Micro-Remediation Methods

Micro-remediation is defined as the use of the microorganisms to eliminate, contain or transform the contaminants to non-hazardous or less-hazardous form in the environment through the metabolisms of microorganisms (Mulligan *et al.*

2001).There are 6 major mechanisms in micro-remediation of toxic metals which are (1) Bioleaching; (2) Biosorption; (3) Bioaccumulation; (4) Biotransformation; (5) Biomineralization; and (6) Microbial enhanced chemisorption of metals.

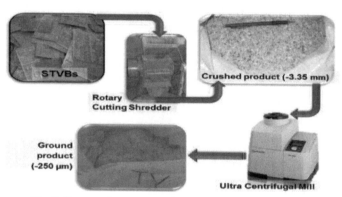

Fig. 2: Process of e-waste powder preparation (Bas et al., 2012)

11.1.1 Bioleaching

Bioleaching uses the natural ability of the microorganisms to transform metals present in the waste in a solid form to a dissolved form. There are two forms of bioleaching: direct leaching and indirect leaching. Direct leaching makes use of the organic acids produced by the microbes, to oxidize the insoluble toxic metals, turning them into soluble ions. In Indirect leaching metal oxidizing bacteria are used that oxidize the metal surrounding the microbe. In most cases, the an ion of the target metal compounds is oxidized, giving out free metal ions in aqueous medium (Tichy *et al.* 1998).

a) **Evolution of Bioleaching:** Bioleaching technology was first investigated in the mid 1980s at King's College, London, England for the elimination of sulphur from coal. Technology, still in its infancy, migrated to Perth, Australia, in the late 1980s, where it was funded privately. In 1994 a public company, Gold Mines of Australia, build the first BacTech bioleach plant at the Youanmi Mine in Western Australia. In 1998 a public company, Allstate Mining, licenses and installed the 2nd BacTech bioleach plant to process refractory arsenic ore from the Beaconsfield Mine in Tasmania (Australia). The mine is now in its 16th year of continuous operation. In 2000 a Chinese company, Shandong Tarzan Biogold Co. Ltd. ("Biogold"), licenses and installs a bioleach plant capable of treating 100 tonnes of concentrate per day from mines both in China and abroad.

b) **Importance of Bioleaching:** Natural resources can be categorized as recyclable and non-recyclable. Metallic elements are included in the first category whereas oil, coal, natural gas, etc. fall in the second category. Recyclable resources are of value as the element, and are permanent and neither decompose nor cease to exist, but merely transform from one form into another. Non-recyclable resources lose their value as they decompose *via* oxidation reactions. Many of the existing high-grade mineral deposits are successively depleted due to exponential growth in the demand for metals. In view of this, there is a need to find alternative metal resources from lower-grade ores and other metal-containing solid wastes in order to meet the demand for metal by mining activities and avert the shortage of certain metals in the future. Bioleaching, a relatively new developed method, may provide an alternative to conventional techniques in the recycling of the solid wastes, thus allowing for the recovery of the metal elements (Tables 4, 5, 6 and 7).

c) **Principles of Microbial Metal Leaching:** Mineralytic effects of bacteria and fungi on minerals are based mainly on three principles, namely acidolysis, complexolysis, and redoxolysis. Microorganisms are able to mobilize metals by (1) the formation of organic or inorganic acids (protons); (2) oxidation and reduction reactions; and (3) the excretion of complexing agents. Sulfuric acid is the main inorganic acid found in leaching environments. It is formed by sulfur-oxidizing microorganisms such as *Thiobacillus* spp. A series of organic acids are formed by bacterial (as well as fungal) metabolism resulting in organic acidolysis, complex and chelate formation. A kinetic model of the coordination chemistry of mineral solubilization has been developed which describes the dissolution of oxides by the protonation of the mineral surface as well as the surface concentration of suitable complex forming ligands such as oxalate, malonate, citrate, and succinate. Proton-induced and ligand-induced mineral solubilization occurs simultaneously in the presence of ligands under acidic conditions.

In many cases it was concluded that the "direct" mechanism dominates over the "indirect" mostly due to the fact that "direct" was equated with "direct physical contact". This domination has been observed for the oxidation of covelite or pyrite in studies employing mesophilic *T. ferrooxidans* and thermophilic *Acidianus brierleyi* in bioreactors which consisted of chambers separated with dialysis membranes to avoid physical contact (Chiang *et al.* 2007). However, the attachment of microorganisms on surfaces is not an indication for the existence of a direct mechanism. The term "contact leaching" has been introduced to indicate the importance of bacterial attachment to mineral surfaces (Clark *et al.* 1996). The following equations describe the "direct" and "indirect" mechanism for the oxidation of pyrite (Sand *et al.* 1999).

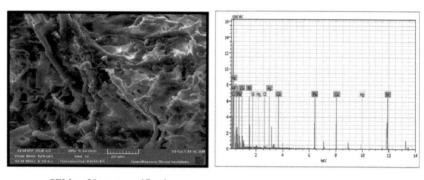

SEM on 20 μm magnification   EDX on 20 μm magnification

Fig. 3: The SEM micrograph shows e-waste interacted with *Aspergillus niger* and EDX shows e-waste metals leach out and reduction concentration of e-waste treated with *A.niger*.

Aspergillus niger was grown in standard sucrose medium and used for bioleaching of metals from e-waste. The one step and two step bioleaching experiments were performed using different concentration of e-waste. At lower concentration (0.1 %) e- waste was effective in the leaching of precious metal when compared with higher concentration. The two step bioleaching experiment was effective with high bioleaching, since the production medium had high fungal metabolites like organic acids. The organic acid plays a main role in bioleaching of metals and thus its presence in the broth was confirmed by Fourier Transform Infrared Spectroscopy (FT-IR) analysis and also by reduction of pH in the growth medium. The leaching of metals in the production medium with e-waste (0.1 %) was compared with control. The quantity of Ag (0.06), Cu (0.09) and Fe (0.06 mg /g) increased from initial concentration of Ag (0.29), Cu (0.23) and Fe (0.31 mg /g). However the toxic metals in e-waste reduced up to Hg (0.04), Br (0.59) and Si (1.07 mg /g) from initial concentration of Hg (0.00), Br (0.001) and Si (0.002) by detoxification process of *Aspergillus niger*. The above finding was confirmed by SEM with EDX, AAS and XRD.

Table 4: Bioleaching of toxic metals from e-waste (Stephen & Macnaughtont, 1999)

Organism used	Type	Name of toxic metals removed
Acidithiobacillus thioxidans	Bacteria	Arsenic, Lead
Micrococcus roseus	Bacteria	Cadmium
Thiobacillus ferrooxidans	Bacteria	Arsenic, Lead
Aspergillus fumigatus	Fungus	Arsenic
Aspergillus niger	Fungus	Cadmium, Lead

Table 5: Examples of industrial waste treated by bioleaching

Type of Waste	Metal values	Microorganisms	Reference
Fly ash	Al, Zn, Cu, Cd	*Acidithiobacillus* sp *Aspergillus niger*	Seidel *et al.* 2001
Sewage sludge	Cu, Mn, Zn, Ni At	*At. thiooxidans*	Pathak *et al.* 2009
Sediment	Cr, Cu, Zn At	*At. thiooxidans*	Fang *et al.* 2009
Tannery sludge	Cr At	*At. thiooxidans*	Wang *et al.* 2007
Electronic scrap	Cu, Ni, Sn, Al, Zn	*Acidithiobacillus* sp *Sulfobacillus*, *Aspergillus niger*	Ilyas *et al.* 2010 & Yang *et al.* 2009
Spent battery	Co, Li, Ni, Cd	*Acidithiobacillus* sp	Mishra *et al.* 2008 & Zhao *et al.* 2008
Spent refinery catalyst	Co, Ni	*Acidithiobacillus* sp	Beolchini *et al.* 2010
Spent petroleum catalyst	V, Ni, Mo	*Acidithiobacillus* sp	Mishra *et al.* 2008 & Mishra *et al.* 2009
Spent fluid cracking catalyst	Al, Mo, V	*Aspergillus niger*	Saanthiya *et al.* 2005 & Saanthiya *et al.* 2006
Waste electric device	Au	*Chromobacterium violaceum*	Faramarzi *et al.* 2004 & Kita *et al.* 2006
Jewelry waste/Automobile catalyst	Ag, Pt, Au	*Chromobacterium violaceum*, *Pseudomonas fluorescens*, *Bacillus megaterium*	Brandl *et al.* 2008

Table 6: Examples of electronic waste treated with bacterial leaching

E-waste	Leached metal	Microorganisms used	Reference
Lithium batteries	Li, Co	*Acidithiobacillus ferrooxidans*	Dong *et al.* 2008
Fly ashes	Zn, Al, Cd, Cu, Ni, Cr, Pb, Mn, Fe	*Thiobacillus thiooxidans Thiobacillus ferrooxidans*, *Aspergillus niger*	Bosshard *et al.* 1996
Tannery	Cr	*Acidithiobacillus thiooxidans*	Yuan-shan *et al.* 2007.
Sewage sludge	Cu, Ni, Zn, Cr	Iron oxidizing bacteria	Pathak *et al.* 2009.
Used cracking catalysts, hydro-processing catalysts	Al, Ni, Mo, V, Sb	*Aspergillus niger; Acidithiobacillus thiooxidans*	Santhiya *et al.* 2005.
BOF slag from steelmaking, slag from copper production	Zn, Fe, Cu, Ni	*Acidithiobacillus* sp *Leptospirillum* sp	Vestola *et al.* 2010
Jewelers waste,automobile catalytic converter, electronic scrap	Ag, Au, Pt	*Chromobacterium violaceum, Pseudomonas fluorescens, Pseudomona, Plecoglossicida*	Brandl *et al.* 2008.
Electronic scrap	Cu, Ni, Al, Zn	*Acidithiobacillus ferrooxidans Acidithiobacillus thiooxidans*	Brandl *et al.* 2001.

Table 7: The level of various metal bioleaching from e-waste

Species of microorganisms	Level of leached metal	Reference
A. ferrooxidans +A. thiooxidans	Cu, Ni, Al, Zn >90%	Chi *et al*. 2011
Acidithiobacillus ferrooxidans	Cu 99%	Guo *et al*. 2009
Acidithiobacillus ferrooxidans	Cu 99%	Jinuweiwany *et al*. 2009
Acidithiobacillus thiooxidans	Cu 74.9%	
A. ferrooxidans + A. thiooxidans	Cu 99.9%	
Sulfobacillus thermosulfidooxidans	Ni 81%, Cu 89%, Al 79% Zn 83%	Iiyas *et al*. 2010
Chromobacterium violaceum,	Au 68.5%	Faramarzi., 2004
Aspergillus niger,	Cu, Sn 65%Al, Ni, Pb,	
Penicillium simplicissimum	Zn >95%	Bosshard., 2001
Thermosulfidooxidans sulfobacilllus	Cu 86%, Zn 80%, Al 64%,	Ilyas., 2010
+ Thermoplasma acidophilum	Ni74%	

11.1.2 Biosorption

Biosorption refers to the concentrating and binding of soluble contaminants to the surface of cellular structure, it does not require active metabolism, in this case the soluble contaminants are ionized toxic metals (Volesky *et al.* 1991). There are several bacterial and fungal species are used in the biosorption of cadmium, chromium, lead and uranium from e-waste (Table 8).

Table 8: Biosorption of toxic metals from e-waste (Hu *et al.* 1996; Atkinson *et al.* 1998; Ahalya *et al.* 2003)

Organism used	Microbial Type	Name of toxic metals removed
Bacillus sphaericus	Bacteria	Chromium
Myxococcus xanthus	Bacteria	Uranium
Pseudomonas aeruginosa	Bacteria	Uranium
Streptoverticillium cinnamoneum	Bacteria	Lead
Rhizopus arrhizus	Fungus	Uranium
Saccharomyces cerevisiae	Fungus	Cadmium

11.1.3 Bioaccumulation

It is defined as the absorption of contaminants within the organism, which are transferred into a biomass within the cellular structure and concentrated there, this process requires active metabolism (Juwarkar *et al.* 2010). For organic contaminants, there are sometimes chemical reactions in the cell cytoplasm to convert them to other compounds; however, the metals entering the cell cytoplasm will not undergo any reaction but sequestered instead (Demirba *et al.* 2001).The cadmium, chromium, lead and uranium from e-waste are bioaccumulated in *Aspergillus niger, Bacillus megaterium, Bacillus circulans, Micrococcus luteus*, etc. (Table 9).

Table 9: Bioaccumulation of toxic metals from e-waste (Demirba, 2001; Srinath *et al.* 2002; Malik, 2004; Juwarkar and Yadav, 2010)

Organism used	Type	Name of toxic metals removed
Bacillus circulans	Bacteria	Chromium
Bacillus megaterium	Bacteria	Chromium
Deinococcus radiodurans	Bacteria	Uranium
Micrococcus luteus	Bacteria	Uranium
Aspergillus niger	Fungus	Chromium, Lead
Monodictys pelagica	Fungus	Chromium, Lead

11.1.4 Biotransformation

Biotransformation refers to the process in which a substance is changed from one chemical form to another chemical form by chemical reactions; in the case of toxic metals, the oxidation state is changed by the addition or removal of electrons, thus their chemical properties are also changed (Juwarkar *et al.* 2010). There are 2 ways for biotransformation process. Direct enzymatic reduction, in which multivalent toxic metal ions are reduced by accepting electrons from the enzymes in the exterior of the cell (Table 10). Indirect reduction can be used to reduce and immobilize multivalent toxic metal ions in sedimentary and subsurface environment by actions of metal-reducing or sulfate reducing bacteria (Tabak *et al.* 2005).

Table 10: Biotransformation of toxic metals from e-waste (Lovley and Coates, 1997; Malik, 2004)

Organism used	Type	Name of toxic metals removed
Anaeromyxobacter sp.	Bacteria	Uranium
Clostridium sphenoides	Bacteria	Uranium
Halomonas sp.	Bacteria	Uranium
Serratia sp.	Bacteria	Chromium
Fusarium oxysporum	Fungus	Cadmium
Rhizopus oryzae	Fungus	Cadmium

11.1.5 Biomineralization

Biomineralization describes the process in which toxic metal ions combine with an ions or ligands produced from the microbes to form precipitation (Achal *et al.* 2012). There are several bacterial and fungal species are used in the biosorption of cadmium, chromium, lead and uranium from e-waste (Table 11).

Table 11: Biomineralization of toxic metals from e-waste (Tabak *et al.* 2005, Benzerara *et al.* 2011; Achal *et al.* 2012)

Organism used	Type	Name of toxic metals removed
Bacillus fusiformis	Bacteria	Lead
Cupriavidus metallidurans	Bacteria	Cadmium
Desulfotomaculum auripigmentum	Bacteria	Arsenic
Sporosarcina ginsengisoli	Bacteria	Arsenic
Aspergillus flavus	Fungus	Lead

12. Microbial-Enhanced Chemisorption of Metals

Chemisorption is similar to adsorption except there is a chemical reaction between the surface and the adsorbate (Volesky and Holan 1995). In microbial-enhanced chemisorption of metals a series of chemical reactions in which microbes first precipitate a bio-mineral of a non-target metal known as priming deposits, the priming deposits, then act as a nucleation focus for the subsequent deposition of the target metal (Tabak *et al.* 2005).

Conclusion

India is placed among the other global nations which have generated more e-waste in quantity and especially urban India needs an urgent approach to tackle this issue. Technical and policy-level interventions, implementation and capacity building and increasing the public awareness can convert this challenge into an opportunity to show the world that India is ready to deal with future problems and can set global credible standards concerning environmental and occupational health. The present chapter summarizes that e-waste contains a number of hazardous substances. Heavy metals and halogenated compounds are of particular concern. Improper handling and management of e-waste during recycling and other end-of-life treatment options may develop potentially significant risks to both human health and the environment. Current simple recycling carried out in many developing countries is causing risks that could to a large extent, be avoided through the use of improved treatment methods. Biohydrometallurgical techniques allow metal cycling by processes similar to natural biogeochemical cycles. Using biological techniques, the recovery efficiency can be increased whereas thermal or physico-chemical methods alone are less successful, as shown in copper and gold mining where low-grade ores are biologically treated to obtain metal values, which are not accessible by conventional treatments. Micro-remediation (Microorganisms) methods can improve scenario of current treatment practices available for e-waste. Besides, management practices for e-waste there is a need for doing more research in the area of micro-remediation using microorganisms so that these techniques can be used for the treatment of e-waste.

References

Achal V, Pan X, Fu Q, Zhang D (2012) Biomineralization based remediation of As (III) contaminated soil by *Sporosarcina ginsengisoli*. Hazard. Mater. 201, 178-184.

Agate AD (1996) Recent advances in microbial mining. World J. Microbiol. Biotechnol. 254, 487–495.

Ahalya N, Ramachandra T, Kanamadi R (2003) Biosorption of heavy metals. Res. J. Chem. Environ. 7(4), 71-79.

Atkinson B, Bux F, Kasan HC (1998) Considerations for application of biosorption technology to remediate metal-contaminated industrial effluents. Water SA. 24 (2), 129-135.

Balde CP, Wang F, Wong J, Kuehr R, Huisman J (2015) The global e-waste monitor – 2014, United Nations University, IAS – SCYCLE, Bonn, Germany.

Barbosa FJ, Tanus-Santos JE, Gerlach RF, Parsons PJ (2005) A critical review of biomarkers used for monitoring human exposure to lead: advantages, limitations, and future needs. Environ. Health Perspect. 113, 1669–1674.

Bastiaan C, Zoeteman J, Krikke HR, Venselaar J (2010) Handling WEEE waste flows: on the effectiveness of producer responsibility in a globalizing world. Int. J. Adv. Manuf. Tech. 47, 415-453.

Benzerara K, Miot J, Morin G, Ona- Nguema G, Skouri-Panet F, Ferard C (2011) Significance, mechanisms and environmental implications of microbial biomineralization. CR Geoscl J. 343(2), 160-167.

Beolchini F, Fonti V, Ferella F, Veglio F (2010) Metal recovery from spent refinery catalysts by means of biotechnological strategies. J. Hazard. Mater. 178, 529-534.

Bernardes A, Bohlinger I, Rodriguez D, Milbrandt H, Wuth W (1997) Recycling of printed circuit boards by melting with oxidizing/reducing top blowing process. TMS Publishers Annual Meeting. Orlando 267, 363-375.

Bosshard P, Bachofen R, Brandl H (1996) Metal leaching of fly ash from municipal waste incineration by *Aspergillus niger*. Environ. Sci. Technol. 30 (10), 3066-3070.

Brandl H, Bosshard R, Wegmann M (2001) Computer-munching microbes: metal leaching from electronic scrap by bacteria and fungi. Hydrometallurgy 59, 319-326.

Brandl H, Lehman S, Faramazi MA, Martinelli D (2008) Biomobilization of silver, gold and platinum from solid waste materials by HCN-forming microorganisms. Hydrometallurgy 94, 14-17.

Chen XR, Zhang HY, Tian XY (2000) DMSO application development prospects. Chem. Ind. Eng. Prog. 1, 53-56.

Chi TD, Lee T, Pandey BD, Yoo K, Jeong J (2011) Bioleaching of gold and copper from waste mobile phone PCBs by using a cyanogenic bacterium. Mineral Eng. 24 (11), 1219–1222.

Chiang HL, KH Lin, MH Lai, TC Chen, S (2007) Pyrolysis characteristics of integrated circuit boards at various particle sizes and temperatures. J. Hazard. Mater. 49, 151–159.

Clark DA, Norris PR (1996) *Acidimicrobium ferrooxidans* Gen. Nov, Sp. Nov.: mixed-culture ferrous iron oxidation with *sulfobacillus species*. Microbiology. 142, 785–790.

Cui J, Forssberg E (2003) Mechanical recycling of waste electric and electronic equipment: A review. J. Hazard. Mater. 160, 243–263.

Cui S, Zhang J (2008) Metallurgical recovery of metals from electronic waste. J. Hazard. Mater. 158, 228–256.

Debaraj M, Dong, Ralph DE, Jong-Hwan AHN, Young HA Rhee (2008) Bioleaching of metals from spent lithium ion secondary batteries using *Acidithiobacillus ferrooxidans*. Waste Manag. 28 (2), 333.

Demirba, A (2001) Heavy metal bioaccumulation by mushrooms from artificially fortified soils. Food Chem. 74 (3), 293-301.

Dong X, Park S, Lin X, Copps K and White MF (2008) Irs1 and Irs2 signaling is essential for hepatic glucose homeostasis and systemic growth. Clinic. Invest. 116, 101–114.

Eswaraiah C, Kavitha T, Vidyasagar S, Narayanan SS (2008) Classification of metals and plastics from printed circuit boards (PCB) using air classifier. Chem. Eng. Process. 165, 565-576.

Fang D, Zhao L, Yang ZQ, Shan HX, Gao Y, Yang Q (2009) Effect of sulfur concentration on bioleaching of heavy metals from contaminated dredged sediments. Environ. Tech. 30, 1241-1248.

Faramarzi MA, Stagars M, Pensini E, Krebs W, Brandl H (2004) Metal solubilization from metal-containing solid materials by cyanogenic *Chromobacterium violaceum*. Biotechnol. J. 113, 321-326.

Govarthanan M, Lee KJ, Cho M, Kim JS, Kamala-Kannan S, Oh BT (2012) Significance of autochthonous *Bacillus* sp. KK1 on biomineralization of lead I mine tailings. Chemosphere 90(8), 2267- 2272.

Guo J, Cao B, Guo J, Xu Z (2008) Plate produced by non-metallic materials of pulverized waste printed circuit boards. Environ. Sci. Technol. 567, 5267-5271.

Guo J, Rao Q, Xu Z (2009) Application of glass non-metals of waste printed circuit boards to produce phenolic moulding compound. J. Hazard. Mater. 56, 728-734.

Hall WJ, Williams PT (2007) Separation and recovery of materials from scrap printed circuit board. Resour. Conserv. Recy. J. 431, 691-709.

Hanafi J, Jobiliong E, Christiani A, Soenarta DC, Kurniawan J, Irawan J (2012) Material recovery and characterization of PCB from electronic waste. Procedia Soc Behav. Sci. J. 176, 331-338.

Hu MZ, Norman JM, Faison BD and Reeves ME (1996) Biosorption of uranium by *Pseudomonas aeruginosa* strain CSU: characterization and comparison studies. Biotechnol Bioeng. 51(2), 237-247.

Huang Y, Takaoka M, Takeda N, Oshita K (2007) Partial removal of PCDD/Fs, coplanar PCBs, and PCBs from municipal solid waste incineration fly ash by a column flotation process. Environ. Sci. Technol. 65, 257–262.

Ilyas S, Ruan C, Bhatti HN, Ghauri MA, Anwar MA (2010) Column bioleaching of metals from electronic scrap. Hydrometallurgy 101, 135-140.

Jingwei Wang, Jianfeng Bai, Jinqiu Xu, Boliang (2009) Bioleaching of metals from printed wire boards by *Acidithiobacillus ferrooxidans* and *Acidithiobacillus thiooxidans* and their mixture. J. Hazard Mater. 172 (2, 3), 1100.

Johri N, Jacquillet G, Unwin R (2010) Heavy metal poisoning: the effects of cadmium on the kidney. BioMetals 23 (5), 783–792.

Jomova K, Jenisova Z, Feszterova M, Baros S, Liska J, Hudecova D, Rhodes CJ, Valko M (2011) Arsenic: toxicity, oxidative stress and human disease. Appl. Toxicol. 31(2), 95–107.

Juwarkar A and Yadav SK (2010) Bioaccumulation and Biotransformation of Heavy Metals. Bioremediat. Technol. 167, 264-284.

Kinoshita K (2003) Metal recovery from non-mounted printed wiring boards via hydrometallurgical processing. Hydrometallurgy 46, 73-79.

Kita Y, Nishikawa H, Takemoto T (2006) Effects of cyanide and dissolved oxygen concentration on biological Au recovery. Biotechnol. 124, 545-551.

Krebs W, Brombacher C, Bosshard PP, Bachofen R and Brandl H (1997) Microbial Recovery of Metals from Solids. FEMS Microbiol. Rev. 20, 605-617.

Li P, Shrivastava Z, Gao and HC Zhang (2004) Printed Circuit Board Recycling: A state-of-the-art survey. J. IEEE Trans. Electron. Packaging Manufacturing 27, 33-42.

Lovley DR and Coates JD (1997) Bioremediation of metal contamination. Curr. Opin. Biotechnol. 8(3), 285-289.

Luga A, Morar R, Samuila A, Dascalescu L (2001) Electrostatic separation of metals and plastics from granular industrial wastes. J. IEEE Trans. Electron. Packaging Manufacturing 8(4), 47-54.

Malik A (2004) Metal bioremediation through growing cells. Environ. Int. 30(2), 261-278.

Mankhand TR, Singh KK, Gupta SK, Das S (2012) Pyrolysis of printed circuit boards. Int. J. Med. Edu. 76, 102-107.

Mishra D, Ahn JG, Kim DJ, Roychaudhury G, Ralph DE (2009) Dissolution kinetics of spent petroleum catalyst using sulfur oxidizing acidophilic microorganisms. J. Hazard. Mater. 167, 1231-1236.

Mishra D, Kim DJ, Ralph DE, Ahn JG, Rhee YH (2008) Bioleaching of metals from spent lithium ion secondary batteries using *Acidithiobacillus ferrooxidans*. Waste Manag. 28, 333-338.

Mishra D, Kim DJ, Ralph DE, Ahn JG, Rhee YH (2008) Bioleaching of spent hydro-processing catalyst using acidophilic bacteria and its kinetics aspect. J. Hazard. Mater. 152, 1082-1091.

Mishra D, Rhee YH (2010) Current research trends of microbiological leaching for metal recovery from industrial wastes. Curr. Res. Technol. Edu. Topics Appl. Microbiol. Microbial Biotechnol. 2, 1289–1292.

Monika JK (2010) E-waste management: as a challenge to public health in India. Indian J Commun. Med. 35(3), 382–385.

Mulligan CN, R Galvez-Cloutier, N Renaud (2001) Biological leaching of copper mine residues by *Aspergillus niger*. In: Biohydrometallurgy & the Environment toward the Mining of the 21ˢᵗ Century (Ed) Amils R and Ballester A, Elsevier Science, Netherlands, pp 453-461.

Murugan T (2008) Milling and separation of the multi-component printed circuit board materials and the analysis of elutriation based on a single particle model. Powder Technol. 56, 169–176.

Nagpal S, Dahlstrom D, Oolman T (1993) Effect of carbon dioxide concentration on the bioleaching of a pyrite-arsenopyrite ore concentrate. Biotechnol. Bioeng. 41, 459–464.

Padiyar N (2011) Nickel allergy-is it a cause of concern in everyday dental practice. Int. J. Contemp. Dent. 12(1), 80–81.

Pan J, Plant JA, Voulvoulis N, and Oates CJ, Ihlenfeld C (2010) Cadmium levels in Europe: implications for human health. Environ Geochem Health. 32(1), 1–12.

Pant D, Joshi D, Upreti MK, Kotnala RK (2012) Chemical and biological extraction of metals present in E waste: a hybrid technology. Waste Manag. 32, 979–990.

Pathak A, Dastidar MG, Sreekrishnan TR (2009) Bioleaching of heavy metals from sewage sludge by indigenous iron-oxidizing microorganisms using ammonium ferrous sulfate and ferrous sulfate as energy sources: A comparative study. J. Hazard Mater. 171, 273-278.

Poon CS (2008) Management of CRT glass from discarded computer monitors and TV sets. Waste Manag. 28, 1499.

Robinson, (2009) E-waste: An assessment of global production and environmental impacts. Sci. Total Environ. 408, 183–191.

Rohwerder T, GehrkeT, Kinzler K, Sand W (2003) Progress in bioleaching: fundamentals and mechanisms of bacterial metal sulfide oxidation, Appl Microbiol Biotechnol. 134, 239–248.

Rotter S (2002) Schwermetalle in haushaltsabfallen heavy metals in household waste. J. Hazard Mater. 56, 243-264.

Saanthiya D, Ting YP (2005) Bioleaching of spent refinery processing catalyst using *Aspergillus niger* with high-yield oxalic acid. Biotechnology 116, 171-184.

Saanthiya D, Ting YP (2006) Use of adapted *Aspergillus niger* in the bioleaching of spent refinery processing catalyst. Biotechnology 121, 62-74.

Sand W, Gehrke T, Hallmann R, Schippers A (1999) Sulfur chemistry, biofilm, and the (in) direct attack mechanism-critical evaluation of bacterial leaching. Appl. Microbiol. Biotechnol. 43, 961-966.

Sasse F, Emig G, (1998) Chemical recycling of polymer materials. Chem. Eng. Technol. 196, 777-789.

Seidel A, Zimmels, Armon R (2001) Mechanism of bioleaching of coal fly ash by *Thiobacillus thiooxidans*. Chem. Eng. 83, 123-130.

Shapiro M, Galperin V (2005) Air classification of solid particles: A review, Chem Eng Process. 57, 279-285.

Sharma Pramila, Fulekar MH, Pathak Bhawana (2012) E-Waste- A Challenge for Tomorrow. Res. J. Recent Sci. 1(3), 86-93.

Sohaili J, Kumari S, Muniyandi S, Suhaila Mohd (2012) A review on printed circuit boards waste recycling technologies and reuse of recovered nonmetallic materials. IJSER J. 5, 1-7.

Srinath T, Verma T, Ramteke P, Garg SK (2002) Chromium (VI) biosorption and bioaccumulation by chromate resistant bacteria. Chemosphere 48(4), 427-435.

Stephen JR, Macnaughtont SJ (1999) Developments in terrestrial bacterial remediation of metals. Curr. Opin. Biotechnol.10 (3), 230-233.

Tabak H, Lens P, Van Hullebusch ED and Dejonghe W (2005) Developments in bioremediation of soils and sediments polluted with metals and radionuclides 1. Microbial processes and mechanisms affecting bioremediation of metal contamination and influencing metal toxicity and transport. Environ. Sci. Biotech. 4(3), 115-156.

Tichy R, Rulkens WH, Grotenhuis J, Nydl V, Cuypers C Fajtl (1998) Bioleach metals. 3, 254-267.

Tsydenova O, Bengtsson M (2011) Chemical hazards associated with treatment of waste electrical and electronic equipment. Waste Manag. 31(1), 45–58.

Vestola EA, Kuusenaho MK, Narhi HM, Tuovinen OH, Puhakka JA, Plumb J, Kaksonen AH (2010) Acid bioleaching of solid waste materials from copper, steel and recycling industries. Hydrometallurgy 103, 74-79.

Volesky B (1991) Biosorption by fungal biomass. In: Biosorption of Heavy Metals (Ed) Volesky B, CRC Press, Canada, pp 139-172

Wang Y-S, Pan Z-Y, Lang J-M, Xu J-M, Zheng Y-G (2007) Bioleaching of chromium from tannery sludge by indigenous *Acidithiobacillus thiooxidans*. J. Hazard. Mater. 147, 319-324.

Widmer R, Oswald-Krapf H, Sinha-Khetriwal D, Schenellmann M, Boni H (2005) Environmental impact assessment review. WEEE Handbook. 25, 436-458.

Xu TJ, Ting YP (2009) Fungal bioleaching of incineration fly ash: metal extraction and modeling growth kinetics. Enzyme Microbial Tech. 44, 323-328.

Yamawaki T (2003) The gasification recycling technology of plastics WEEE containing brominated flame retardants. Fire Mater. 87, 315-319.

Yang T, Xu Z, Wen J, Yang L (2009) Factors influencing bioleaching copper from waste printed circuit board by *Acidithiobacillus ferrooxidans*. Hydrometallurgy 97, 29-32.

Yokoyama S, Iji M (1995) Recycling of thermosetting plastics waste from electronic component production processed, Proceedings of the EEE International Symposium on Electronics and the Environment. Orlando Florida Technol. 67, 132–137.

Yuan-Shan Wang, Zhi-Yan Pan, Jian-Min Lang, Jian-Miao Xu, Yu-GuoZheng (2007) Bioleaching of chromium from tannery sludge by indigenous *Acidithiobacillus thiooxidans*, J. Hazard. Mater. 147, 319-324.

Zhang S, Forssberg E (1997) Mechanical separation-oriented characterization of electronic scrap. Resour. Conserv. Recy. 87, 247-269.

Zhang S, Forssberg E, (1998) Optimization of electrodynamics separation for metals recovery from electronic scrap. Resour. Conserv. Recy. 78, 143-162.

Zhang S, Forssberg E, (1999) Intelligent liberation and classification of electronic scarp. Powder Technol. 65, 295–301.

Zhao L, Yang D, Zhu N-W (2008) Bioleaching of spent Ni-Cd batteries by continuous flow system: Effect of hydraulic retention time and process load. J. Hazard Mater. 160, 648-654.

Zhu P, Chen Y, Wang LY, Zhou M, Zhou J (2013) The separation of waste printed circuit board by dissolving bromine epoxy resin using organic solvent. Waste Manag. 46, 484-488.

1. Introduction

Effluent from textile dyeing units contain large amount of dyes and create an environmental problem, which increase toxicity and decrease the aesthetic value of rivers and lakes. A variety of physico-chemical methods are in use worldwide. However, there is an increasing concern as to their impact in effectively treating textile effluents as they introduce secondary pollutants during the 'remediation' process which are quite costly to run and maintain. Research on biological treatment has offered simple and cost effective ways of bioremediation textile effluents. Increase in population and modernized civilization has led to flourishing of textile industries in India. Textile sector is a complicated industrial chain and high diversity in terms of raw materials, processes, productions and equipment. It is estimated that textile account for 14% of India's industrial production and around 27% of its export earnings. India is the second largest producer of cotton yarn and silk and third largest producer of cotton and cellulose fibre. There are about 10,000 garment manufacturers and 2100 bleaching and dyeing industries in India. Majority are concentrated at Tirupur and Karur in Tamil Nadu, Ludiyana in Punjab and Surat in Gujarat. Dyeing is a combined process of bleaching and colouring, which generates huge volumes of wastewaters which results in environmental degradation. More than 100,000 commercial dyes are available to textile industries worldwide with over 700,000 tons of commercial dyes a year being produced. The mechanism of microbial decolourization of azo dyes involves the reductive cleavage of azo linkages under anaerobic conditions resulting in the formation of colorless aromatic amines. The biodegradation of relatively simple sulfonated amino-benzene and amino-naphthalene compounds under aerobic conditions is simple and effective (Pinheiro et al. 2004). The microorganisms are efficient in degradation of recalcitrant and xenobiotic compounds. The degradation ability of the organisms also depends on the oxidation potential of the dye (Tauber et al. 2008). Biosorption of the dyes is also a promising technique which will enhance the colour removal of the effluent stream containing a mixture of dyes. Biosorption can be defined as a process in which solids of natural origin are employed for sequestration or separation of solids from the effluent streams. The interaction between the cell wall ligands and adsorbates can be explained by ion exchange, complexation, coordination and micro-precipitation which are responsible for color removal. For which process living cells or dead biomass can be used as biosorbent (Bhole et al. 2004). Living cells require constant nutrient supply and the toxicity of dyes which will harm them. On the other hand, dead biomass can avoid such problems and can be regenerated for use (Malarvizhi et al. 2010). The bioremediation potential of microbes and their enzymes acting on synthetic dyes has been demonstrated, with others needing to be explored in the future as alternatives to conventional physico-chemical approaches (Ali 2010). It is obvious that each process has its own constraints in terms of cost, feasibility, practicability, reliability, stability, environmental impact, sludge production, operational difficulty, pretreatment requirements, the extent of the organic removal and potential toxic byproducts. Also, the use of a single process may not completely decolourize the wastewater and degrade the dye molecules. Even when some processes are reported to be successful in decolourising a particular wastewater, the same may not be applicable to other types of coloured wastewaters. Certainly, the effective removal of dye from dye industrial wastewater is a challenge to the manufactures and researchers, as some of the processes are neither economical nor effective. The amount of water consumed in textile industries must also be considered because the traditional textile finishing industry consumes - 100 L of water in the processing of 1 kg of textile material. Consequently the potential of water re-use should be an objective when applying a particular wastewater treatment. In this chapter, all these issues will come under focus and discussion of microbial degradation of textile dye wastes – with an eco-friendly approach.

2. Application of Dyes

Approximately 40,000 different synthetic dyes and pigments are used industrially, and about 450,000 tons of dyestuffs are produced worldwide. Azo dyes are the largest and more versatile class of dyes, accounting for up to 50% of the annual production. They are extensively used in many fields of up-to-date technology, e.g., in various branches of the textile industry, the leather tanning industry, paper production, food, colour photography, pharmaceuticals and medicine, cosmetics, hair colourings, wood staining, agricultural, biological and chemical research, light-harvesting arrays and photo-electrochemical cells. Moreover, synthetic dyes have been employed for the efficacious control of sewage and wastewater treatment, for the determination of specific surface area of activated sludge for ground water tracing, etc. (Zollinger 1987). Different classes of dyes are used according to the fibres to which they can be applied. Reactive dyes are most commonly used as they can be applied to both in natural (wool, cotton, silk) and synthetic (modified actylics) fibres. Reactive dyes differ from other class of dyes in that their molecules contain one or more reactive groups capable of forming a covalent bond with a compatible fibre group. They have become very popular due to their high wet-fastness, brilliance and range of hues (Hao et al. 2000). Their use has increased as synthetic fibres became more abundant. Acid and basic dyes are used for dyeing all natural fibres and some synthetics (polyesters, actylic and rayon). Direct dyes are classified by this way because they are applied directly to cellulose fibres. Furthermore, they are used for colouring rayon, paper, leather and to a small extent nylon.

The application of mordant dyes is limited to the colouring of wool, leather, furs and anodised aluminium. Solvent dyes are used for colouring inks, plastics and wax, fat and mineral oil products.

3. Impact of Textile Dyes on the Environment

Colour is usually the first contaminant to be recognized in a wastewater because a very small amount of synthetic dyes in water (<1 ppm) are highly visible, affecting the aesthetic merit, transparency and gas solubility of water bodies. They adsorb and reflect the sunlight entering water, thereby interfering with the aquatic species growth and hindering photosynthesis. Additionally, they can have acute and/ or chronic effects on organisms depending on their concentration and length of exposure. Removal of color from dye-containing wastewater is the first and major concern, but the point of degrading dyes is not only to remove color, but to eliminate, or substantially decrease, the toxicity.

Table 1: Colour concentration limits and quantum of water generated from industries

Industry		Quantum of water (generated standards)	Colour concentration (hazen units)	Colour limits (hazen units)	
				USPHS	BIS
Textile		120m³/Ton	1100- 1300	0-25	20
Pulp and paper	Large	130m³/Ton	100-600	0-10	5-101
	Small	150m³/Ton			
Tannery		28 m³/Ton	2100-2300	10-40	20
Kraft mill		40 m³/Ton	150-200	5-10	20
Sugar		0.4 m³/Ton	400-500	10-50	25

The textile industry accounts for 2/3 of the total dyestuff market and consume large volumes of water and chemicals for wet processing of textiles. The discharges of wastewater are the main cause of the harmful environmental impact of the textile industry. Robinson *et al.* (2001) estimated about 10-15% of textile dyes are discharged into waterways as effluent and such industries consist of high sodium, chloride, sulphate, hardness and carcinogenic dye ingredients. Effluents from textile industry are characterized by their high visible colour (3000-4500 units), chemical oxygen demand (COD) (800-1600 mg/L), and alkaline pH range of 9-11. Industry wise quantum of water used, concentration of colours and allowed colour limits is given in Table 1. They also possess large amount of organic chemicals, low biodegradability and total solids in the range of 6000-7000 mg/L. The chemical used in the textile processing are varied in chemical composition, ranging from inorganic compounds to polymers and organic products and depend on the nature of the raw material and product. Major pollution by textile wastewater comes from dyeing and wastewater is characterized by high suspended solids, COD, heat, colour, acidity and other soluble substances. The presence of dyes in aqueous ecosystem reduces sunlight penetration into deeper layers diminishing photosynthetic activity, declines the water quality, lowering the gas solubility which causes sensitive toxic effects on aquatic flora and fauna. Therefore, the release of harmful dyes in the environment can be an eco-toxic risk and can affect human through the food chain. Among different textile dyes used, azo dyes and nitrated polycyclic aromatic hydrocarbons are two groups of chemicals that are abundant in the environment. They cause rigorous contamination in river and ground water in the surrounding area of dyeing industries (Riu *et al.* 1998). The impact of azo dyes in food industry and their degraded products on human health has caused concern over a number of years. Moreover azo dyes have been linked to human bladder cancer, splenic sarcomas, hepatocarcinomas and nuclear anomalies in experimental animals and chromosomal aberrations. Some azo dyes induce liver nodules in experimental animals and there are a higher numbers of bladder cancers in dyeing industries workers. Benzidine based azo dyes are widely used in dye manufacturing, textile dyeing, colour paper printing and leather industries. Benzidine has long been recognized as a human urinary bladder carcinogenic and tumorigenic agent in a variety of laboratory animals (Haley 1975).

4. Dyes

A dye is a colored substance that has an affinity to the substrate to which it is being applied. The dye is generally applied in an aqueous solution, and may require a mordant to improve the fastness of the dye on the fibre (Booth and Gerald 2000). Both dyes and pigments are colored because they absorb some wavelengths of light more than others. In contrast to dyes, pigments are insoluble and have no affinity for the substrate. Some dyes can be precipitated with an inert salt to produce a lake pigment, and based on the salt used they could be aluminum lake, calcium lake or barium lake pigments.

5. Classification of Dyes

5.1. Basic or Cationic Dyes

This group was the first of the synthetic dyes to be taken out of coaltar derivatives (Fig. 1). As textile dyes, they have been largely replaced by later developments. However, they are still used in discharge printing, and for preparing leather, paper, wood and straw. More recently, they have been successfully used with some readymade fibres, especially the acrylics. Basic dyes were originally used

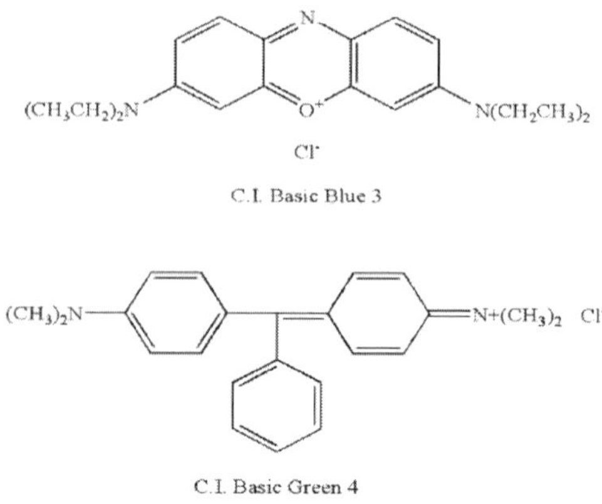

C.I. Basic Blue 3

C.I. Basic Green 4

Fig. 1: Structure of basic dye

to coloring the wool, silk, linen, hemp, etc., without the use of a mordant, or using agent. With a mordant like tannic acid they were used on cotton and rayon. Basic dyes give brilliant colors with exceptional fastness to acrylic fibres. They can be used on basic dyeable variants of nylon and polyester. Basic dyes represent 5% of all dyes listed in the color index. They are cationic compounds and used for the dyeing acid-group containing synthetic fibres like modified polyacryl.

They bind to the acid groups of the fibres. Most of the basic dyes are diarylmethane, triarylmethane, anthraquinone or azo compounds (Almansa *et al.* 2004). Nowadays basic dyes are no longer used to great extent on cotton or linen and seldom used on wool. Since they are cheap, however, they are used for hemp, jute and similar fibres.

5.2 Direct Dyes

The direct dyes are the second largest dye class in the color index with respect to the amount of the dyes. Direct dyes are relatively large molecules with high affinity for cellulose fibres. Van der Waals forces make them bind to the fibre. They are mostly azo dyes with more than one azo bond or phthalocyanine, stilbene or oxazine compounds (Fig. 2). About 1600 direct dyes are listed but only 30% of them are in current production.

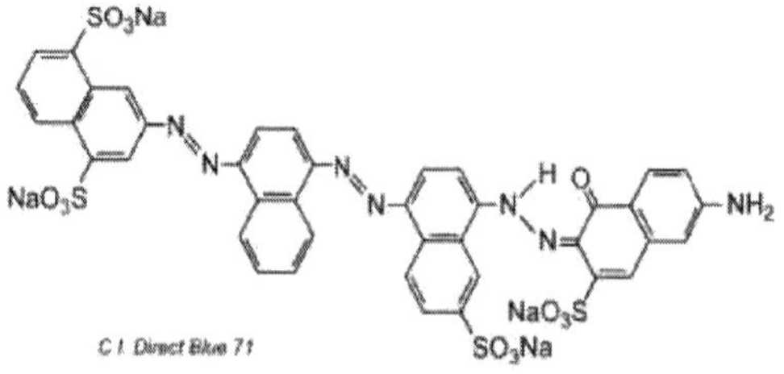

C.I. Direct Blue 71

Fig. 2: Structure of direct dye

Historically, the direct dyes followed the basic dyes and were widely hailed because they made it unnecessary to use a mordant or binder in dyeing cotton (Almansa *et al.* 2004). The colors are not as brilliant as those in the basic dyes but they have better fastness to light and washing, and such fastness can be measurably improved by after treatments (diazotized and developed.) Direct dyes can be used on cotton, linen, rayon, wool, silk and nylon. These dyes usually have azo linkage –N=N– and high molecular weight. They are water soluble because of sulfonic acid groups.

5.3 Acidic or Anionic dyes

This is a very large and important group of dyestuffs. While an acidic dye is a salt the color comes from the acidic component, while in the basic dye it is from the organic base. The first acid dyes are combinations of basic dyes with sulphuric or nitric acid. Acid dyes are anionic compounds that are mainly used for the dyeing nitrogen-containing fabrics like wool, polyamide, and silk and modified acryl. They bind to the cationic NH_4^+ ions of those fibres (Fig. 3). Most of the acid dyes are azo, anthraquinone or triarylmethane compounds. Adding metallic salts especially chrome to the dyed fabric in an after-treatment generally has increased colorfastness of acid dyes. Acid dyes cannot be used for wool tops but are used in dyeing wool piece goods, silk, nylon, and some of the other manmade fibres (Ali 2010). If a mordant is used they will successfully dye cotton and linen, though this is seldom done today. The ordinary type of acid dye is reserved largely for apparel fabrics and for knitting and rug yarns. A great deal of it is used on nylon carpeting.

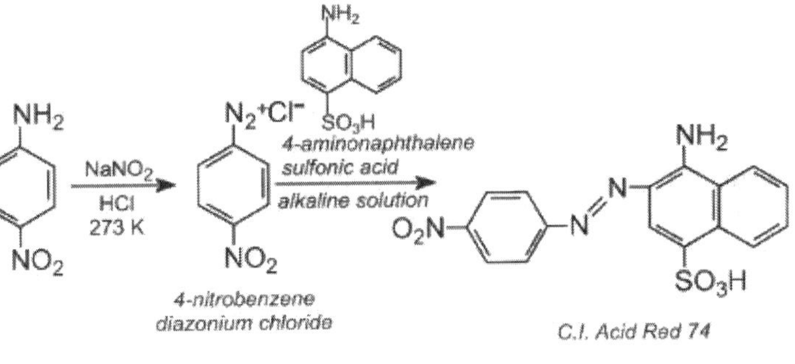

Fig. 3: Structure of an acidic dye

5.4 Metal Complex Dyes

This is an important group of acid dyes, which have been complexed with metallic ions to improve light fastness on wool and nylon. Many metal complex dyes are reported among the acid and reactive dyes (Arslan and Balciolul

1999). Metal complex dyes are the strong complexes of one metal atom (usually chromium, copper, cobalt or nickel) and 1 or 2 dye molecules (1:1 and 1:2 metal complex dyes respectively) (Fig. 4). About 16% of the azo dyes listed in the color index are metal complexes. The phthalocyanine metal complex dyes are also being used.

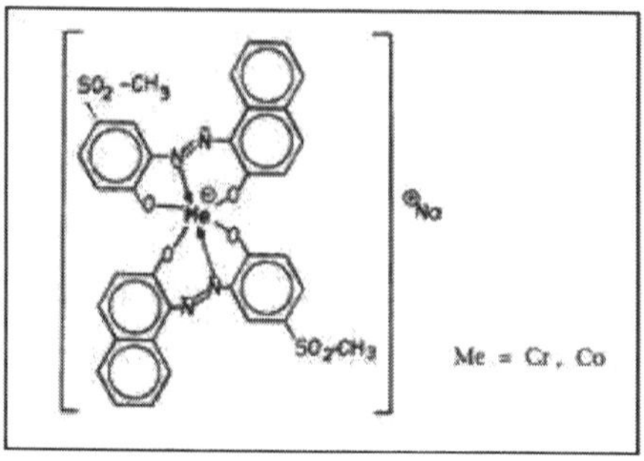

Fig. 4: Structure of metal complex dye

5.5 Sulphur Dyes

Sulphur dyes are complex polymeric aromatics with heterocyclic S-containing rings (Fig. 5). This dye group represents about 15% of the global dye production. Dyeing with sulphur dyes involves reduction and oxidation comparable to vat dyeing.

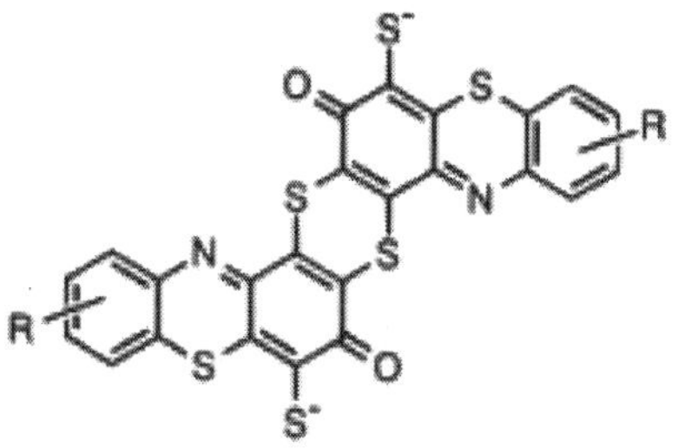

Fig. 5: Structure of sulphur dye

They are mainly used for dyeing cellulose fibres. The sulphur dyes provide very deep shades, which have excellent resistance to washing but poor resistance to sunlight. They will dye cotton, linen and rayon, but not brightly. A problem with sulphur dyes especially the black colors is that they make the fabric tender, or weaken its structure due to the formation of acids. Sulphur dyed fabrics are therefore treated with alkalis to neutralize the acids (Arslan and Balciolul 1999).

5.6 Azoic Dyes

Azoic dyes and ingrain dyes (naphthol dyes) are the insoluble products of a reaction between a coupling component, including naphthols, phenols or acetoacetylamides and a diazotised aromatic amine (Fig. 6). This reaction is carried out

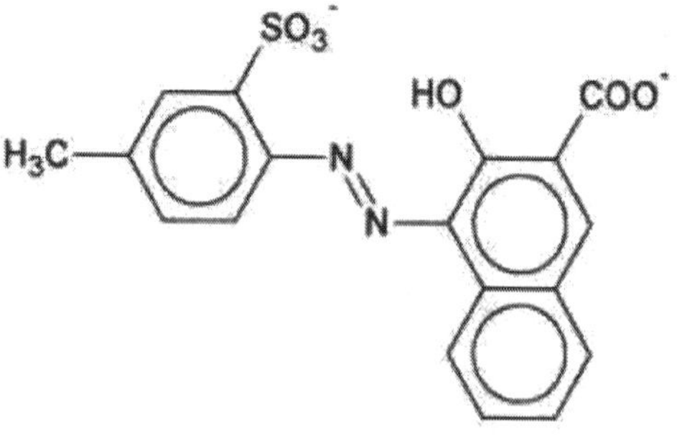

Fig. 6: Structure of an azoic dye

on the fibre. All naphthol dyes are azo compounds. These dyes are used primarily for bright red shades in dyeing and printing since most other classes of fast dyes are lacking in good red dyes. Azoic dyes, called naphthols in the industry, are actually manufactured in the fabric by applying one half of the dye. The other half is then put on and they combine to form the finished color. Unless they are carefully applied and well washed, they have poor fastness to rubbing or crocking.

5.7 Vat Dyes

The vat dyes are best known group of dyes in use today because of their all round fastness to washing and sunlight on cotton and rayon. The term vat comes from the old indigo method dyeing in a vat: indigo had to be reduced to light form (Fig. 7). Vat dyes are made from indigo, anthraquinone and carbazole. They are successfully used on cotton, linen, rayon, wool, silk and nylon. Vat dyes are also used in the continuous piece of dying process sometimes called the pigment application process. In this method the dyes are not reduced before application, but after they have been introduced into the fabric. Alternate reduction and oxidation process was used for dying these dyes to fibres. Sodium dithionite is used as reducing agent for this process. This ensures superior and economical dyeing. There are no light red vat dyes. Vat dyes are water insoluble dyes that are widely used for dyeing cellulose fibres. Vat refers to the vats that are used for the reduction of indigo plants through fermentation.

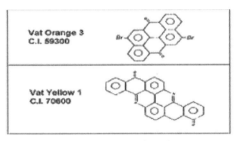

Fig. 7: Structure of vat dye

5.8 Collective Dyes

Collectives are the latest dyestuff and because they react chemically with cotton, viscose, linen, wool and silk they are very fast to washing treatments. They can be dyed and printed by many methods and for the first time, the whole spectrum of color can be put onto cloth using just one class of dyes. Substituting a reactive group on a direct dye produces these dyes (Fig. 8).

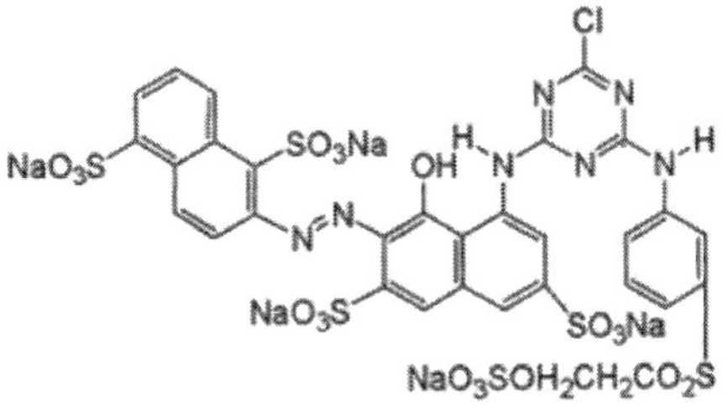

Fig. 8: Structure of collective dye

5.9 Reactive Dyes

The dyes, which containing the reactive groups are called as reactive dyes. In the color index, this is the second largest dye class, introduced in 1956. The reactive dyes form covalent bonds with -OH, -NH, or -SH groups of fibres. The reactive group is often a heterocyclic aromatic ring

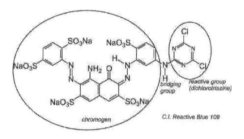

Fig. 9: Structure of reactive dye

substituted with chloride or fluoride or vinyl sulphone (Fig. 9). The hydrolysis of reactive group during dyeing lowers the degree of fixation of reactive dyes. These create the problem of colored effluent.

5.10 Mordant dyes

The dyes which required the mordant for dying fibres is called as mordant dyes. They are used for dyeing wool, leather, silk, paper and modified cellulose fibres. Most mordant dyes are azo, oxazine or triarylmethane compounds. The mordants are usually dichromates or chromium complexes (Fig. 10).

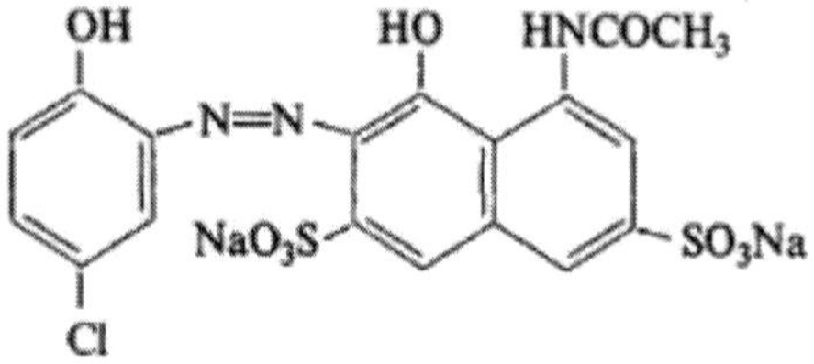

Fig. 10: Structure of mordant dye

5.11 Disperse Dyes

Disperse dyes form the third largest group of dyes in the color index. Disperse dyes are scarcely soluble dyes, thus it required high temperature or chemical softeners for dying to synthetic fibres *viz.* cellulose acetate, polyester, polyamide, acryl, etc. Dying takes place in dye baths with fine disperse solutions. They are usually small azo or nitro compounds, anthraquinones or metal complex azo compounds (Fig. 11).

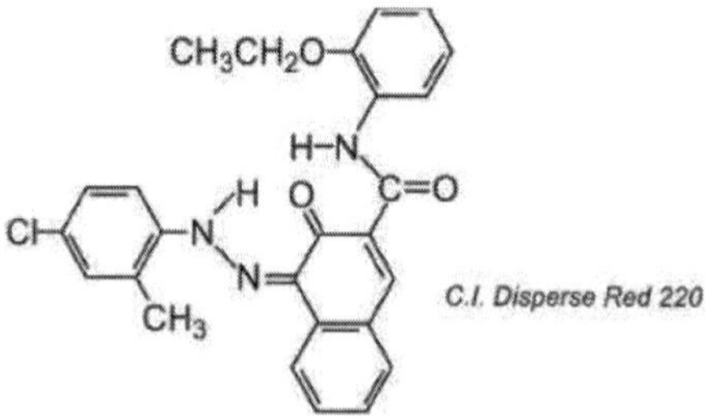

Fig. 11: Structure of disperse dye

5.12. Pigment Dyes

These are insoluble non-ionic compounds or insoluble salts retain their crystalline or particulate structure throughout their application. Pigment dyes (i.e. organic pigments) represent a small but increasing fraction of the pigments, the most widely applied group of colorants(Fig. 12). About 25% of all commercial dye names listed in the color index are pigment dyes.

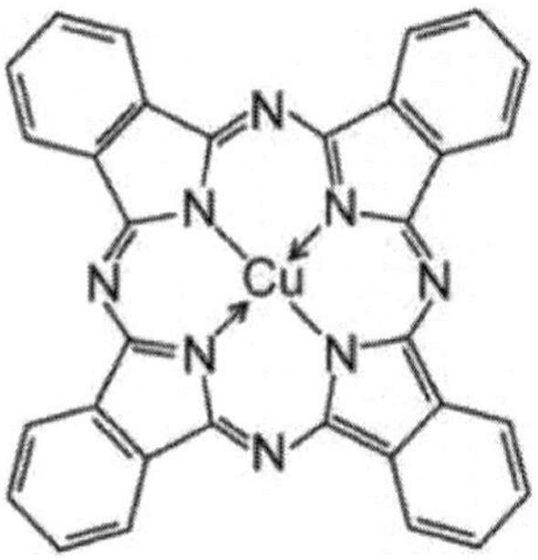

C.I. Pigment Blue 15

Fig. 12: Structure of pigment dye

Pigment dyeing required the use of dispersing agents. Pigments are usually used together with thickeners in print pastes for printing diverse fabrics. Most pigment dyes are azo compounds or metal complex phthalocyanines or anthraquinone or quinacridone.

5.13 Solvent Dyes

Solvent dyes are non-ionic dyes that are used for dyeing substrates in which they can dissolve, e.g. plastics, varnish, ink, waxes and fats. They are not often used for textile processing but their use is increasing. Most solvent dyes are diazo compounds that underwent some molecular rearrangement (Fig. 13). Also triarylmethane, anthraquinone and phthalocyanine solvent dyes are applied.

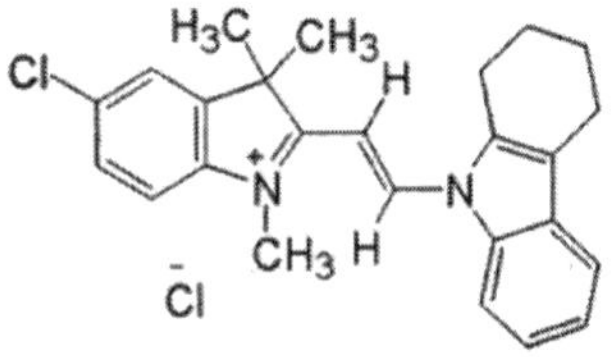

Fig. 13: Structure of solvent dye

5.14 Fluorescent Brighteners

Fluorescent brighteners mask the yellowish tint of natural fibres by absorbing ultraviolet (UV) light and weakly emitting visible blue. They are not dyes in the usual sense because they lack intense color.

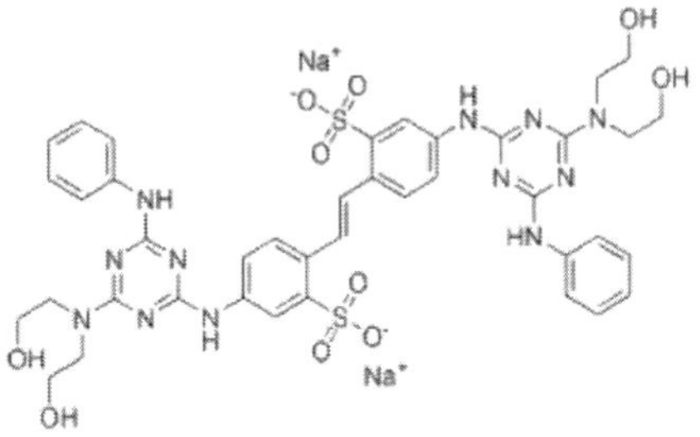

Fig. 14: Structure of fluorescent brighteners

Based on chemical structure, several different classes of fluorescent brighteners are distinguished such as stilbene derivatives, coumarin derivatives, pyrazolines, 1,2-ethene derivatives, naphthalimides and aromatic or heterocyclic ring structures. Many fluorescent brighteners contain triazinyl units and water-solubilising groups.

5.15 Other Dye Classes

Apart from the dye classes mentioned above, the color index also lists food dyes and natural dyes. Food dyes are not used as textile dyes and the use of natural dyes *viz.* anthraquinone, indigoid, flavenol, flavone or chroman compounds that can be used as mordant, vat, direct, acid or solvent dyes in textile processing operations is very limited.

6. Methods Used in Removal of Dyes

Effluent discharge from textile and dyestuff industries into water bodies and wastewater treatment systems is currently causing significant health concerns to environmental regulatory agencies. So, it is essential to treat the textile dye effluent for environmental safety. Government legislation is increasingly becoming stricter especially in the developed countries, regarding the removal of dyes from industrial effluents. The dye removal technologies can be divided into three categories like biological, chemical and physical (Fig. 15). Generally, in physical method dyes removed by adsorption, in chemical method chromophore has been modified through chemical reaction, biological method occurs through sorption and enzymatic degradation.

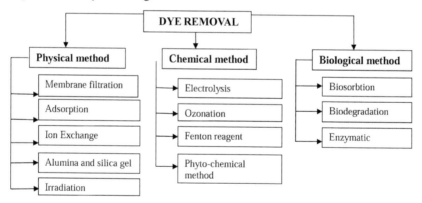

Fig. 15: Methods of dye removal

6.1 Physical Method

Adsorption refers to a process where a substance or material is concentrated at a solid surface from its liquid of gaseous surrounding. There are two types of adsorption based on the type of attraction between the solid surface and the adsorbed molecules. If this attraction forces are due to chemical bonding, the process is called chemical adsorption (Chemisorption). Adsorption techniques have gained favour recently due to their efficiency in the removal of pollutants too stable for conventional methods (Bansal and Goyal 2005). Membrane filtration technique has been widely used for drinking and wastewater treatment. The process of filtration consists of microfiltration, ultra-filtration, nano-filtration and reverse osmosis. Ultra-filtration, nano-filtration and reverse osmosis can be applied as main or post treatment processes for separation, purification and reuse of salts and large molecules including dyes from dye bath effluents and bulk textile processing wastewater (Slokar et al. 1998). Ion exchange techniques are effective in decolourizing cationic and anionic dyes and have not been used in extensively for dye wastewater treatment due to the opinion that ion exchangers cannot accommodate a wide range of dyes (Kim and Shoda 1999). Alumina is a synthetic porous crystalline gel available in the form of granules whereas, silica gel is a porous and non crystalline granule of different size prepared by the coagulation of colloidal silicic or salicylic acid. Alumina and silica gel have been studied by various workers for the removal of dyes (Kumar et al. 1998). The irradiation process to treat dye containing effluent in a dual tube bubbling reactor requires large volumes of dissolved oxygen for organic substances to be broken down effectively by radiation (Table 2) (Gupta et al. 2007).

6.2 Chemical Method

The electrochemical technique is very efficient to remove color from dye wastewater (Gupta et al. 2007). Electrochemical method is a relatively new process exhibiting efficient colour removal and degradation of recalcitrant pollutants. This process is very simple and is based on applying an electric current to wastewater by using sacrificial iron electrodes to produce ferrous hydroxide. These ferrous hydroxides remove soluble and insoluble acid dyes from the effluent. Ozonation is a technology initially used in 1970's and it was carried out by ozone generated from oxygen. Oxidation by ozone is capable of degrading chlorinated hydrocarbons, phenols, pesticides and aromatic hydrocarbons (Xu and Leburn 1999). Ozone (O_3) rapidly decolourize water soluble dyes but with non soluble dyes react much slower. Moreover, textile processing wastewater usually contains other refractory constituents that will react with ozone (Muthukumar et al. 2005). A solution of hydrogen peroxide (H_2O_2) and an iron catalyst, known as Fenton's reagent, is a suitable chemical means of treating wastewaters that are either resistant to biological treatment or poisonous to live biomass (Table 3). Chemical separation used the action of sorption or bonding to remove dissolved dyes from wastewater and has been shown to be effective in decolourization of both soluble and insoluble dyes (Pak and Changm, 1999). Fenton oxidation process can decolourize a wide range of dyes and in comparison to ozonation; the process is relatively cheap and results generally in a larger COD reduction. Photochemical or photocatalytic method

Table 2: Different technologies for colour removal

Process	Technology	Advantages	Disadvantages
Conventional treatment processes	Coagulation Flocculation Biodegradation	Simple, economically feasible Economically attractive publicly acceptable treatment	High sludge production, handling and disposal problems Slow process, necessary to create an optimal favourable environment, maintenance and nutrition requirements
	Adsorption on activated carbons	The most effective adsorbent, great, capacity, produce a high-quality treated effluent	Ineffective against disperse and vat dyes, the regeneration is expensive and results in loss of the adsorbent, non-destructive process
Established recovery	Membrane separations processes	Removes all dye types, produce a high-quality treated effluent	High pressures, expensive, incapable of treating large volumes
	Ion-exchange	No loss of sorbent on regeneration, effective	Economic constraints, not effective for disperse dyes
	Oxidation	Rapid and efficient process	High energy cost, chemicals required
Emerging removal processes	Advanced oxidation process	No sludge production, little or no consumption of chemicals, efficiency for recalcitrant dyes	Economically unfeasible, formation of by-products, technical constraints
	Selective bio-adsorbents	Economically attractive, regeneration is not necessary, high selectivity	Requires chemical modification, non-destructive process
	Biomass	Low operating cost, good efficiency and selectivity, no toxic effect on micro-organisms	Slow process, performance depends on some external factors (pH and salts)

Source: Robinson *et al.* (2001)

Table 3: Merits and demerits of physical/chemical methods

Physical/chemical methods	Merits	Demerits
Fentons reagent	Effective decolourization of both soluble and insoluble dyes	Sludge generation
Ozonation	Applied in gaseous state: no alteration of volume	Short half-life (20 min)
Photochemical	No sludge production	Formation of by-products
NaOCl	Initiates and accelerate azo bond cleavage	Release of aromatic amine
Cucurbituril	Good sorption capacity for various dyes	High cost
Electrochemical destruction	Breakdown compounds are non-hazardous	High cost of electricity
Activated carbon	Good removal of wide variety of dyes	Very expensive
Peat	Good adsorbent due to cellular structure	Specific surface areas for adsorption are lower than activated carbon
Wood chips	Good sorption capacity for acid dyes	Requires long retention times
Silica gel	Effective for basic dye removal	
Membrane filtration	Removes all dye types	Concentrated sludge production
Ion exchange	Regeneration: no adsorbent loss	Not effective for all dyes
Irradiation	Effective oxidation at lab scale	Requires a lot of dissolved O_2
Electrokinetic coagulation	Economically feasible	High sludge production

Source: Chacko et al. (2011)

degrades dye molecules to CO_2 and H_2O by UV treatment in the presence of H_2O_2. Degradation is caused by the production of high concentrations of hydroxyl radicals. UV light may be used to activate chemicals such as H_2O_2, and the rate of dye removal is influenced by the intensity of the UV radiation, pH, dye structure and the dye bath composition. UV light has been tested in combination with H_2O_2, TiO_2, Fenton reagents, O_3 and other solid catalysts for the decolourization of dye solution.

6.3 Biological Method

Biological method is generally considered to be the most effective and less energy intensive to removing the bulk of pollutants from wastewater. Different microorganisms have been used for the treatment of various dye effluents. The most important advantage of this method is the low running costs. Biosorption, biodegradation and ligninolytic enzymes have been explored as methods of biological treatments for removal of dye from effluents. Removal of dye by low cost adsorbents has been extensively reviewed. Fungal and bacterial biomass which is a byproduct of fermentations can be used as a cheap source of biosorbent (Gazso, 2001). Knapp et al. (2001) reported both bacterial and fungal cells are capable of partial or complete removal of industrial dyes by using adsorption process. However, with some fungi, adsorption is the only decolourization mechanism, but with white rot fungi both adsorption and degradation can occur simultaneously or sequentially. There are many reports on decolourization of dye wastewater by live or dead fungal biomass. However, only limited information is available on interactions between biomass and molecular structure of dyes. The major mechanism of removal of dye by dead cells is biosorption, which involves physico-chemical interactions (adsorption, deposition, and ion exchange) (Huang et al. 2007). Zhou and Zimmerman (1993) used actinobacteria as an adsorbent for decolourization of effluents containing anthroquinone, phalocyanine and azo dyes. The decolourization activity was significantly affected by dye concentration, amount of pellet, temperature and agitation of the media. Removal of dye effluents by biosorption is still in the initial stage. It was not practically approached for treating large volumes of dye effluents due to the disposal of the large volumes of biomass after biosorption. Biodegradation of the dye are broadly demonstrated by pure and mixed cultures of bacteria and fungi under aerobic and anaerobic conditions. Aerobic biodegradation process influence by several environmental and nutritional factors such as pH, temperature, amount of oxygen and co-metabolic carbon sources. Bacteria and fungi are the two major groups of microorganisms that have been extensively studied in the treatment of dye wastewater. The enzymes secreted by aerobic bacteria can breakdown the organic compounds.

Thus, the isolation of aerobic bacterial strains capable of degrading different dyes has been carried out for more than two decades (Rai *et al.* 2005). Biodegradation of the dye by certain groups of actinobacteria during dye removal has been extensively demonstrated. The degradation and mineralization of dyes is successful by certain white rot fungi. Ligninolytic enzymes secreted by white rot fungi bind non-specifically to the substrate and therefore can degrade a wide variety of recalcitrant compounds and even complex mixtures of pollutants including dyes (Ge *et al.* 2004). Various enzymes involved in dye decolourization are laccase, manganese peroxidase (MnP), manganese independent peroxidase (MIP), lignin peroxidase (LiP), tyrosinase, etc. Production of these enzymes and their activity in biodegradation of dyes is often judged from the appearance of the mycelial mat, which ultimately appears colourless (Kanapp and Newby 1995). The potential advantages of using enzymes instead of fungal cultures are mainly associated to the following factors: shorter treatment period, operation of high and low concentrations of substrates, absence of delays associated with the lag phase of biomass, reduction in sludge volume and no difficulty of process control (Michniewicz *et al.* 2008). However there are several practical limitations in the use of free enzymes such as the high cost associated with production, isolation and purification of enzymes and the short life times of enzymes. To overcome this limitations enzyme immobilization has shown to improve enzyme stability (Akhtar and Husain 2006). In addition, enzyme immobilization allows enzyme re-utilization and continuous operation in bioreactor which is very important for an industrial application of the enzyme.

7. Anaerobic Conditions for Decolourization of Dye

Anaerobic processes convert the organic contaminants principally into methane and CO_2. They usually occupy less space, can treat wastes containing up to 30,000 mg L^{-1} of COD, have lower running costs, and produce less sludge. Azo dye degradation occurs preferentially under anaerobic or O_2 limited concentrations, acting as final electron acceptors during microbial respiration. Oxygen, when it is present, may compete with the dyes. In many cases the decolourization of reactive azo dyes under anaerobic conditions is a co-metabolic reaction. Several mechanisms have been proposed for the decolourisation of azo dyes under anaerobic conditions (Rau and Stolz 2003). One of these is the reductive cleavage of the azo bond by unspecific cytoplasmic azo reductases using flavoproteins as cofactors. A second proposed mechanism is an intracellular, non-enzymatic reaction consisting of a simple chemical reduction of the azo bond by reduced flavin nucleotides. These reductive cleavages, with the transfer of 4 electrons and the respective aromatic amine formation, usually occur with low specific activities but are extremely nonspecific with regard to the organism involved and the dyes converted. Transport of the reduction equivalents from the cellular system to the azo compounds is also important, because the most relevant azo compounds are either too polar and/or too large to pass through the cell membrane (Rau and Stolz 2003). Mediators generally enable or accelerate the electron transfer of reducing equivalents from a cell membrane of a bacterium to the terminal electron acceptor, the azo dye (Van der zee and Villaverde 2005). Such compounds can either result from the aerobic metabolism of certain substances by bacteria themselves or be added to the medium. They are enzymatically reduced by the cells and these reduced mediator compounds reduce the azo group in a purely chemical reaction (Stolz 2001). It has been suggested that quinoide redox mediators with standard redox potentials (EO') between -320 and -50 mV could function as effective redox mediators in the microbial reduction of azo dyes. The first example of an anaerobic cleavage of azo dyes by redox mediators which are naturally formed during the aerobic metabolism of xenobiotic compound was reported by Keck et al. (2002). The need in some cases for external addition of chemical mediators may increase the cost of the process. In crude extracts and crude enzyme preparations, low molecular weight compounds may be naturally present and act as natural enhancing compounds. The physiology of the possible reactions that result in a reductive cleavage of azo compounds under anaerobic conditions differs significantly from the situation in the presence of O_2 because the redox active compounds rapidly react either with O_2 or azo dyes (Stolz 2001). Therefore, under aerobic conditions, O_2 and dyes compete for the reduced electron carriers. In bioremediation, an anaerobic treatment or pretreatment step can be a cheap alternative compared with an aerobic system because expensive aeration is omitted and the problem with bulking sludge is avoided.

8. Aerobic Conditions for Decolourization of Dye

In aerobic pathways, azo dyes are oxidized without the cleavage of the azo bond through a highly non-specific free radical mechanism, forming phenolic type compounds. This mechanism avoids the formation of the toxic aromatic amines arising under reductive conditions (Pereira et al. 2009). The main organisms involved in the oxidative degradation of dyes are the fungi, mainly wood rot fungi by the so called lignolytic enzymes. The application of these organisms or their enzymes in dye wastewater bioremediation have attracted increasing scientific attention because they are able to degrade a wide range of organic pollutants, including various azo, heterocyclic and polymerese dye (Wesenberg et al. 2003a). It is also noteworthy that they require mild conditions, with better activities and stability in acidic media and at temperatures from 20 to 35°C. Oxidases can also be found in some microorganisms, plants and animals. Peroxidases, including LiP and MnP, use H_2O_2 to promote the one electron oxidation mechanism of chemicals to free radicals. These enzymes have an important role on the cellular detoxification of organisms by eliminating H_2O_2. They are hemoproteins belonging to the oxidoreductase enzymes that catalyze the oxidation of phenols, biphenols, anilines, benzidines and related hetereoaromatic compounds, for which H_2O_2 is the final electron acceptor (Duran and Esposito 2000). The dye does not need to bind to the enzyme; instead oxidation occurs through simple electron transfer, either directly or through the action of low molecular weight redox mediators (Eggert et al. 1996). Lignin peroxidase was discovered earlier than MnP and exhibits the common peroxidase catalytic cycle. It interacts with its substrates via a ping-pong mechanism, i.e. firstly it is oxidized by H_2O_2 through the removal of two electrons that give compound I, and further oxidized through the removal of one electron to give compound II, which oxidizes its substrate, returning to the resting enzyme. The active intermediates of LiP (i.e. compounds I and II) have considerably higher reduction potentials than the intermediates of other peroxidases, extending the number of chemicals that can be oxidized. MnP mechanism differs from that of LiP in using Mn^{+2} as a mediator. Once Mn^{+2} has been oxidized by the enzyme, Mn^{+1} can oxidize organic substrate molecules.

9. Combination of Anaerobic/Aerobic Processes for Dye Treatment

Many dyes used in textile industry cannot be degraded aerobically as the enzymes involved are dye-specific. Generally, anaerobic reduction of azo dyes is more satisfactory than aerobic degradation, but the intermediate products (carcinogenic aromatic amines) must be further degraded. These colourless amines are very resistant to further degradation under anaerobic conditions, and therefore aerobic conditions are required for complete mineralization (Pandey et al. 2007). For the most effective wastewater treatment, two stage biological wastewater treatment systems are then necessary in which an aerobic treatment is introduced after the initial anaerobic reduction of the azo bond (Sponza and Ieik 2005). The balance between the anaerobic and aerobic stages in this treatment system must carefully be controlled because it may become darker during re-aeration of a reduced dye solution. This is to be expected, since aromatic amines are spontaneously unstable in the presence of O_2. Oxidation of the hydroxyl and amino groups to quinines and quinine imines can occur and these products can also undergo dimerization or polymerization, leading to the development of new darkly coloured chromophores (Pereira et al. 2010). However, with the establishment of correct operating conditions, many strains of bacteria are capable of achieving high levels of decolourization when used in a sequential anaerobic/aerobic treatment process (Steffan et al. 2005). Aerobic biodegradation of many aromatic amines has been extensively studied (Pinherio et al. 2004), but these findings may not apply to all aromatic amines. Specially sulfonated aromatic amines are often difficult to degrade (Tan et al. 2005). Aromatic amines are commonly not degraded under anaerobic conditions. Melgoza *et al.* (2004) studied the fate of disperse blue 79 in a two-stage anaerobic/aerobic process; the azo dye was bio-transformed to amines in the anaerobic stage and an increase of toxicity was obtained; the toxic amines were subsequently mineralized in the aerobic phase, resulting in the detoxification of the effluent.

10. Microbial Decolourization Mechanisms

Microbial communities are of primary importance in degradation of dye contaminated soils and water as microorganisms alter the chemical nature of dye and mobility through reduction, accumulation, mobilization and immobilization (Robinson *et al.* 2001). In recent years, biodegradation has become a viable alternative and proven to be a promising technology. Microorganisms have been successfully employed as sources for bioremediation (Khan and Husain 2007). Bioremediation is gaining its significance in utilizing the biological activity of microorganisms to degrade toxic chemicals in the environment. Microbial decolourization can occur *via* two principal mechanisms: biosorption and enzymatic degradation, or a combination of both have been used to remove dyes by biosorption.

11. Different Microorganisms are Used for Decolourization of Dyes

11.1 Fungi

Microorganisms which are as known nature's recyclers, convert toxic compounds to harmless products such as carbon dioxide and water (Joshi *et al.* 2010). The various organisms which degrade dyes are fungi, bacteria and actinobacteria (Asgher *et al.* 2006). Dye decolourizing fungi namely lignolytic and nonlignolytic fungi were used for the decolourization of dye wastewater. The lignolytic wood rot fungi are known to be the most efficient microorganisms for degradation of dye. Lignolytic fungi, including *Phanerochaete chrysosporium, Trichophyton rubrum* LSK-27, *Ganoderma* sp. WR-1, *Trametes versicolor, Funalia trogii, Irpex lacteus*, etc. are widely used for the decolourization of textile dyes (Singh *et al.* 2012). The *Myrothecium* sp. IMER1, a non-lignolytic fungi and yeasts *viz. S. cerevisiae, C. tropicalis, K. marxianus, C. zeylanoides*, and *I. occidentalis* were also used for the decolourization of textile dyes (Kumarpraveen *et al.* 2011).

Microbial degradation of Congo red by *Gliocladium virens* (Singh 2008) various hazardous dyes likes, Congo red, acid red, basic blue and bromophenol blue, direct green by the fungus *Trichoderma harzianum* (Singh and Singh 2010) and biodegradation of plant waste materials (Singh 2008) by using different fungal strains has been investigated . The results were similar to biodegradation of Congo red and bromophenol blue by the fungus *T. harzianam* in semi-solid medium (Singh and Singh 2010) and biodegradation of methylene blue, Gentian violet, crystal violet, cotton blue, Sudan black, malachite green and methyl red by few species of *Aspergillus* (Muthezhilan *et al.* 2008) in liquid medium. Cripps *et al.* (1990) also reported the biodegradation of three azo dyes (Congo red, orange II and tropaeolin O) by the fungus *P. chrysosporium* (Singh and Singh 2010). Rathnan *et al.* (2013) found that the isolated fungus *A. niger* and *A. oryzae* and mixed consortium is as an important source for bioremediation of toxic dye. *A. niger* showed greater decolourization production during 16 days incubation (Manikandan 2012). The mechanism of the fungal decolourization was biodegradation or bioadsorption of dyes. The biodegradation of dyes using white rot fungi was associated with an involvement of various lignolytic enzymes, such as lignin peroxidase, manganese peroxidase and laccase. Biodegradation of dyes using non-lignolytic fungi was associated with an involvement of billirubin oxidase, laccase and azoreductase (Wesenberg *et al.* 2003a).

11.2 Yeasts

More recently, some studies have shown that yeast species acted as a promising dye adsorbent capable to uptake higher dye concentration, such as *Galactomyces geotrichum*, *Saccharomyces cerevisiae* and *Trichosporon beigelii*, etc. (Joshi *et al.* 2010). The first two reports use the ascomycete yeast *Candida zeylanoides* isolated from contaminated soil to reduce model azo dyes (Mathur *et al.* 2005). The characterisation of an enzymatic activity is described in further studies with the yeast *Issatchenkia occidentalis* (Razia Khan *et al.* 2012) and the enzymatic system involved is biodegradation in *Saccharomyces cerevisiae* (Wesenberg *et al.* 2003b).

11.3 Bacteria

Halophiles have been reported to be involved in the dye decolourization. The moderately halotolerant *Bacillus* sp. were isolated for decolourization of azo dye red 2G to an extent of 64.89%. This rate of decolourization may be due to the high metabolic diversity being seen in the halophiles due to their extremophilic nature (Pearce *et al.* 2003). The degradation of Navitan Fast Blue S5R, a very important commercial diazo dye in the tannery and textile industries was investigated. *Pseudomonas aeruginosa* decolourized this dye at concentrations up to 1200 mgl[-1] and the organism was also able to decolourize various other tannery dyes at different levels (Valli and Suseela 2003). The organisms required ammonium salts and glucose to co-metabolize the dye. Organic nitrogen sources did not support appreciable decolourization, whereas, inorganic nitrogen showed an increasing effect on both growth and decolourization. An oxygen intensive azo reductase was also involved in the decolourization mechanism. Gurulakshmi *et al.* (2008) reported that an aerobic bacterial consortium consisting of two isolated strains and a strain of *Pseudomonas putida* was also developed for the aerobic degradation of a mixture of textile azo dyes and individual azo dyes at alkaline pH (9-10) and salinity (0.9 - 3.8 g/L) at ambient temperature (28 ± 2°C). The degradation efficiency of the strains in different media and at different dye concentrations was studied. The enzyme present in the crude supernatant was found to be reusable for the dye degradation. Extent of decolourization recorded by *Bacillus cereus* under ideal conditions was 95% and that by *B. megaterium* was 98% (Sharma *et al.* 2010). An investigation confirmed the decolourization of azo dye red by *B. cereus* and *B. megaterium* under *in vitro* conditions. Extent of decolourization recorded by *B. cereus* under ideal conditions was 95% and that by *B. megaterium* was 98% (Ventosa *et al.* 1998). Ayyasamy *et al.* (2015) reported the textile dye remazol golden yellow decolourization potentiality by native bacterium *Micrococcus endophyticus* (ES37).

11.4 Actinobacteria

Actinobacteria play vital role in the biodegradation processes during biological treatment. Basic characterization of actinobacteria is that they are aerobic, Gram positive, filamentous bacteria which form a 'crusty' colony on complex media and gave rise to fragmented mycelia in liquid broth. The textile dyes were completely decolourized by *Streptomyces* spp. after 8 to 10 days of incubation (Senan and Abraham 2004). The organisms are adapted to different concentrations of dyes from 0-250mg/100ml of dye concentration (Pinherio *et al.* 2004). The mechanism of microbial decolourization of azo dyes involves the reductive cleavage of azo linkages under anaerobic conditions resulting in the formation of colorless aromatic amines. The biodegradation of relatively simple sulfonated amino-benzene and amino-naphthalene compounds under aerobic conditions is simple and effective (Junnarkar *et al.* 2006). The time required for degradation increased when increasing concentrations of the direct red 5B from 50-1000 mg/L by *Sphingobacterium* sp. (Rai *et al.* 2005). But the decolourization was not observed at 250 mg/100 mL; instead the high dye concentration was toxic for the cell growth. The depression of dye decolourization at higher dye concentration is due to higher inhibition at high dyestuff concentration (Junnarkar *et al.* 2006).

Streptomyces spp. were able to readily decolourize Congo red, a structurally complex dye which contains two azo bonds and polyaromatic and sulfonated groups. Yet, the structurally simpler dye, orange II which has a single azo bond and sulfonated group was not cleaved at all. It is obvious from the literature that azoreductase action is variable among strains. For instance, *Kerstersia* spp. decolourized dyes carrying sulfonated groups on either side of the azo bond to a higher extent than dyes which carried a single sulfonated group. This is in contrast with decolourization reported by *Sphingomonas* species. Evidently, greater structural complexity of a dye does not signify greater difficulty in its breakdown. Instead it has been found to be dependent on several factors namely, the type, position and number of functional groups. Additionally, the presence of phenolic, amino, acetoamido and 2-methyoxyphenol groups on the ring were found to lead to an increase in biodegradation (Swamy and Ramsay 1999).

Amycolatopsis japonica and *A. orientalis* these strains are easily distinguishable based on the aerial spore and diffusible pigment color. *A. japonica* has white aerial spores, releasing a dark olive brown pigment with a deep yellow-brown reverse color, while *A. orientalis* has light blue aerial spores, releases a light olive brown pigment with an orange-yellow reverse color (Tan *et al.* 2006). Both the strains are identified as great degraders of dyes in textile industry. The majority of strains tested were capable of decolourizing dyes within a single class. *A. orientalis* was unique in the fact that it crossed this boundary, capable of degrading diazo, azo and triphenylmethane dyes. The strains of *Streptomyces* namely *S. lividans* and *S. coelicolor* A3, *S. griseus*, *S. ambofaciens* and *S. rimosus* were used in textile industry for the purpose of dye degradation and waste water treatments (Kieser *et al.* 2000).

The intra-cellular enzymes were extracted from the isolates of actinobacteria at 0.05% (mg/100 mL) dye concentration with enrichments was subjected to enzyme assay. LiP, tyrosinase, laccase assay showed that the degradation of dye was mainly due to mixture of these enzymes. The enzyme activity decreased when decreasing the dye concentration and hence the total protein content also decreased. The highest activity of laccase, tyrosinase, LiP was found to be 0.223, 0.104, 0.309 (U/mg/min) on day 9, day 6 and day 3 respectively which was also stated in saying that after complete decolourization of dye enzymes like tyrosinase, laccase and manganese peroxidase activity were decreased in 96 h in the batch culture (Kalme *et al.* 2006) he biosorption technique showed 90% dye adsorption by dead biomass of actinobacteria in 1 h for 0.02% and 0.05% of reactive yellow dye concentration.

It is also known that a mixture of organisms (consortium) degrade a dye better as compared to individual organisms as their complexity enables them to act on a variety of pollutants (Pandey *et al.* 2007; Vijayakumar *et al.* 2015). The decolourization takes place in anoxic conditions the organisms grow well and in anaerobic condition they degrade the dye (Dafale *et al.* 2008). But the intermediates formed in aerobic conditions are more toxic and difficult to degrade. The degradation rate also depends on the different conditions such as high salinity of the medium, temperature, pH and also static or shaking conditions (Mona *et al.* 2008). The degradation is found to be most effective under static conditions. The best media for degradation was Vogel's mineral media as compared to mineral salt media and basal mineral media (Asgher *et al.* 2006). The media is also supplemented with different carbon and nitrogen sources for better degradation of dye. The carbon source generally used is glucose, sucrose, starch and nitrogen sources used are yeast extract, beef extract and tryptone (Gou *et al.* 2009).

11.5. Algae

It has been reported that more than thirty azo compounds can be biodegraded and decolourized by *Chlorella pyrenoidosa, C. vulgaris* and *Oscillatoria tenius*, with azo dyes decomposed into simpler aromatic amines (Yan and Pan 2004). Algae can play an important role in the removal of azo dyes and aromatic amines in stabilization ponds (Banat *et al.* 1996). Both living and non-viable algae have been used in the removal of color from wastewater (Lim *et al.* 2010). *C. vulgaris* and *Scenedesmus quadricauda* immobilized on alginate can remove a higher percentage of color from textile dyes than suspended algae (Ergene *et al.* 2009). Decolourization ability of algae is given in Table 4.

12. Enzymatic Degradation

Enzymes from certain fungi have shown high ability of degradation of synthetic dyes. The microbial decolourization was associated with the involvement of various reductive enzymes *viz.* laccase, lignin peroxidases and manganese peroxidases.

12.1 Laccases

Laccases are the most numerous members of multi-copper oxidase protein family. It catalyzes the oxidation of substituted phenolic and non-phenolic compounds in the presence of oxygen as an electron acceptor (Lepez *et al.* 2004). Phylogenetically, these enzymes have developed from small sized prokaryotic azurins to eukaryotic plasma proteins ceruloplasmin. They contain four histidine rich copper binding domains, which coordinate the type 1, type 2, and type 3 copper atoms that differ in their environment and spectroscopic properties. The molecular weight of laccases varies from 60-390 kDa. They are classified into two category *viz.* the blue laccases (presence of type 1 copper site) and laccases that lack the type 1 copper site. The blue laccases showed the absorbance maxima at 610 nm. The laccases that lack type 1

Table 4: Decolourization of some azo dyes using algae

Microalgae and source	Decolouration process & conditions	(Initial concentration) Dyes, % colour removal	References
Chlorella vulgaris from a collection	Adsorption; pH 7, 150 rpm, 25°C	Supranol Red 3BW, 35.62 mg dye/g biomass	Lim *et al.* 2010
Scenedesmus quadricauda from a lake	Adsorption; pH 2-8, 150 rpm, 30°C, 300 min.	Remazol brilliant blue R, 48.3 mg dye/g biomass	Ergene *et al.* 2009
Spirogyra species from a forestry waste	Adsorption; pH 2-10, 180 rpm, 20-50°C, 5 h.	15ppm direct brown, 70% color	Sivarajasekar *et al.* 2009
Chlorella vulgaris from polluted water	Degradation; pH 7, 25°C, 7 days; azo reductase	20ppm each, methyl red, 82% colour; orange II, 47% colour; G-red (FN-3G), 59% color	El-Sheekh *et al.* 2009
Oscillatoria curviceps from collection	Degradation; pH 7, 100 rpm, 4°C, 8 days; azoreductase, polyphenol oxidase	100 ppm acid black, 84% colour	Priya *et al.* 2011

copper site did not show the absorbance maxima at 610 nm. The first report of prokaryotic laccase is from the rhizospheric bacterium *Azospirillum lipoferum*. Blanquez *et al.* (2004) used *Trametes versicolour* in the form of pellets to treat a black liquors discharge for detoxifying and reducing the colour, aromatic compounds, and chemical oxygen demand (COD). They found that colour and aromatic compounds were reduced up to 70–80% and COD was reduced up to 60%. They concluded that *T. versicolour* is able to produce laccase. *T. versicolour* completely decolourizes the Amaranth, Tropaeolin O, Reactive Blue 15, Congo Red, and Reactive Black 5 with no dye sorption while it partially decolourizes Brilliant Red 3G-P, Brilliant Yellow 3B-A and Remazol Brilliant Blue R with some dye sorption. They found that after decolourization, toxicity of few dyes remained the same while some became nontoxic (Ramsay and Nguyen 2002). Romero *et al.* (2006) found that bacteria *Stenotrophomonas maltophilia* decolourizes some synthetic dyes (methylene blue, methyl green, toluidine blue, Congo red, methyl orange, and pink) as well as the industrial effluent (Almansa *et al.* 2004).

12.2 Peroxidases

Peroxidases are assisted in the degradation of lignin moieties by supportive enzymes and mediators: low-molecular weight compounds that improve lignin biotransformation by readily diffusing into the lignocellulosic matrix and by providing high redox potentials that enhance the variety of substrates like laccases and peroxidases were able to degrade (Wesenberg *et al.* 2003b). Congo red was a substrate for the lignolytic

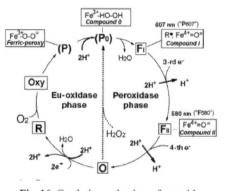

Fig. 16: Catalytic mechanism of peroxidase

enzyme lignin peroxidase (Schroeder *et al.* 2007). The excellent performance of *T. lignorum* and *F. oxysporum* in the biodegradation of textile dyes of different chemical structures reinforces the potential of these fungi for environmental decontamination similar to white rot fungi (Zhang *et al.* 1999).

The capacity of fungi to reduce azo dyes is related to the formation of exo enzymes such as peroxidases and phenol oxidases. Peroxidases are hemoproteins that catalyze reactions in the presence of hydrogen peroxide (Fig. 16) (Levin *et al.* 2004).

12.3 Lignin Peroxidase (LiP)

This enzyme belongs to the family of oxidoreductases, specifically acting on peroxide as an acceptor (peroxidases) and can be included in the broad category of ligninases. The systematic name of this enzyme class is 1,2-bis (3,4- dimethoxyphenyl) propane-1,3-diol: H_2O_2 oxidoreductase. LiP is N-glycosylated protein with molecular weight 38-47 kDa. It contains heme in the active site and shows a classical peroxidase mechanism (Fig. 17) (Tien 1961).

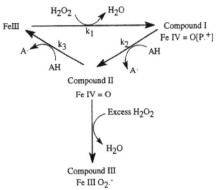

Fig. 17: Catalytic cycle of a lignin peroxidase

LiP catalyzes several oxidations in the side chains of lignin and related compounds by one-electron abstraction to form reactive radicals.

12.4 Azoreductase

The presence of extra-cellular oxygen sensitive azoreductase in anaerobic bacteria *viz. Clostridium* and *Eubacterium* that decolourized sulfonated azo dyes during growth on solid or complex media has been reported (Schroeder *et al.* 2007). Azoreductases are flavoprotiens (NAD(P)H: flavin oxidoreductase). It is localized intra-cellular or extra-cellular site of the bacterial cell membrane. These azoreductases required the NADH/NADPH/FADH as an electron donor for the reduction of an azo bond (Fig. 18) (Rus *et al.* 2000). The substrate specificity of azoreductases depends on the functional group present near azo bond. The oxygen sensitive orange II azoreductase from

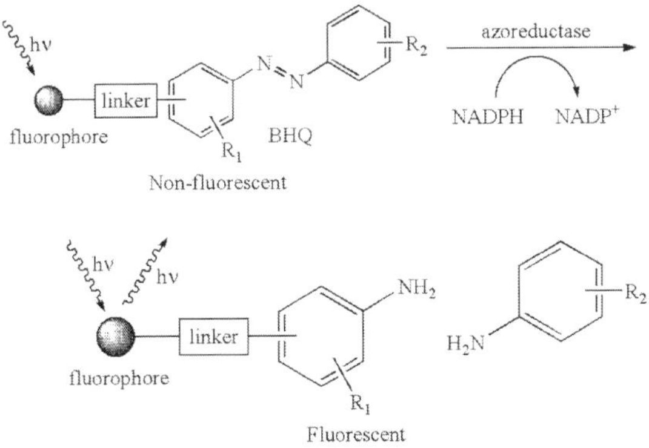

Fig. 18: Catalytic cycle of azoreductase

Pseudomonas sp. KF46 showed highest specificity towards the carboxyl group substituted sulfophenyl azo dyes. Orange II azoreductase showed lowest Km (0.8 µM) for 1-(4-carboxyphenylazo)-2-naphthol and highest Km (14.8 µM) for 1-(4'-sulfoaminophenylazo)-2-naphthol as substrate. Orange II azoreductase was unable to decolourize the azo dyes which contain charged groups in the proximity to the azo group (Zimmermann *et al.* 1982). The induction of azoreductase during decolourization of azo dyes under static condition was reported earlier (Duran and Esposito 2000).

Azoreductase generated the toxic amines after reduction of an azo bond. *Bacillus natto* and *B. sphaericus* (Chivukula and Renganathan 1995). The purified polyphenol oxidase was used for the oxidation of the colored and phenolic substances.

12.5 Polyphenol Oxidase

Polyphenol oxidase (PPO) enzymes catalyse the o-hydroxylation of monophenols (phenol molecules in which the benzene ring contains a single hydroxyl substituent) to o-diphenols (phenol molecules containing two hydroxyl substitutes).

Further, they catalyse the oxidation of o-diphenols to produce o-quinones (Fig. 19). The amino acid tyrosine contains a single phenolic ring that may be oxidized by the action of PPOs to form o-quinone (Solomon *et al.* 1996). Hence, PPOs may also be referred as tyrosinases. Polyphenol oxidases are enzymes that catalyze the oxidation of certain phenolic substrates to quinones in the presence of molecular oxygen. Polyphenol oxidases have been reported in the bacteria viz. *Streptomyces glaucescens*, *S. antibioticus* and *Bacillus licheniformis*.

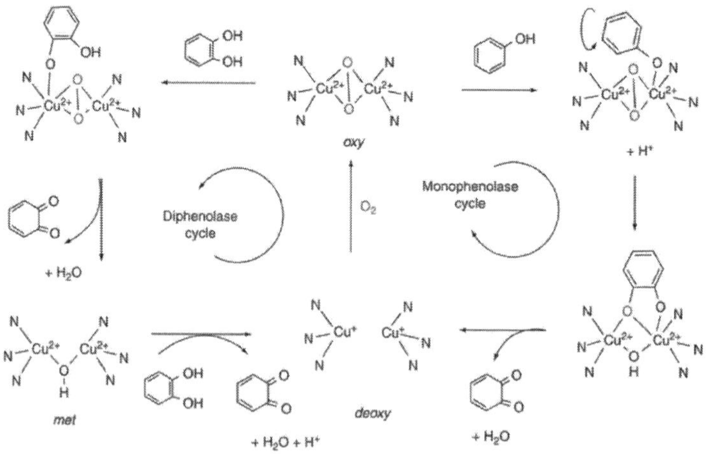

Fig. 19: Catalytic cycle of polyphenol oxidase

12.6 Tyrosinase

Tyrosinases are copper-containing dioxygen activating enzymes found in many species of bacteria and are usually associated with melanin production (Razia Khan *et al.* 2012). These proteins have a strong preference for phenolic and diphenolic substrates and are somewhat limited in their reaction scope, always producing an activated quinone as product (Fig. 20).

Despite this fact they have potential in several biotechnological applications, including the production of novel mixed melanins, protein cross-linking, phenolic biosensors, and production of L-DOPA, phenol and dye removal and biocatalysis. Although most studies have used *Streptomyces* sp. enzymes for dye removing process.

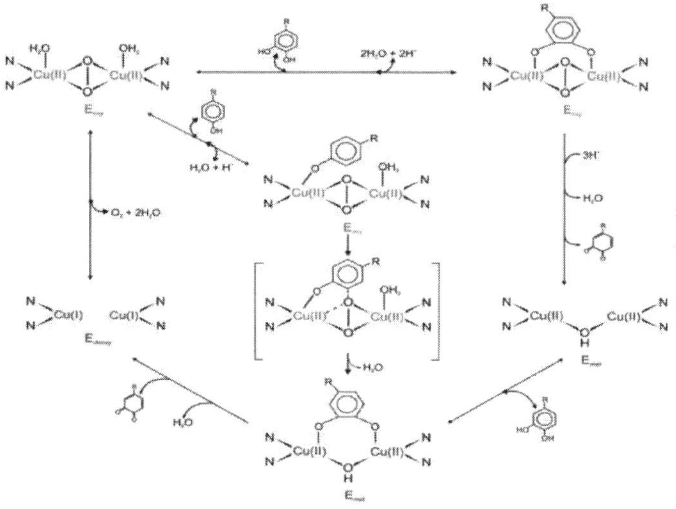

Fig. 20: Catalytic cycle on tyrosinase

13. Factors Affecting Biodegradation of Dyes

Ecosystems are dynamic environments with variable abiotic conditions, e.g. pH, temperature, dissolved oxygen, nitrate concentration, metals, salts, etc. Microorganisms are affected by changes in these parameters, and therefore their decomposing activities are also affected. Other non-dye related parameters are the type and source of reduction equivalents, bacterial consortium and cell permeability. Enzymes synthesized by microorganisms are also sensitive to the culture conditions in which they are applied. Usually, they proceed better in environments more suitable for the original microorganism. Textile wastewaters result from different classes of dyes, and consequently vary in their composition and pH. Therefore it is important that evaluating the potential of different microorganisms for dye degradation, to consider the effects of other present compounds. Optimization of abiotic conditions will greatly help in the development of industrial-scale bioreactors for bioremediation. Optimal pH and temperatures are always related to the environment where the organisms were collected, but usually fungi work better, either for growth or compounds degradation, in acidic or neutral media, whereas bacteria prefer alkaline conditions. Dye decolourization proceeds better under acidic conditions for fungi and fungal enzymes (Abadulla et al. 2000; Zille et al. 2005) but under alkaline conditions for bacteria and bacterial enzymes (Zimmermann et al. 1982). The optimal temperatures of microorganisms usually range from 20 to 35°C, but there are also others that tolerate higher values, although the stability may be compromised. Beyond the optimum temperature, the degradation activities of the microorganisms decrease because of slower growth, 1 reproduction rate and the deactivation of enzymes responsible for degradation. Biodegradation of azo dyes and textile effluents can be affected by dye related parameters, such as class and type of azo dye, reduction metabolites, dye concentration, dye side-groups and organic dye additives. Microbial activity can decrease with increasing dye concentration, which can be attributed to the toxicity of the dyes to the growing microbial cells at higher concentrations (in the biodegradation) and/or cell saturation (biosorption). In the most enzymatic decolourisation studies, the kinetics are described by Michaelis-Menten model and an increase of the rate with an increase in the dye concentration is observed up to a certain concentration (saturation). At dye amounts higher than the optimal, the rates usually remain constant due to saturation, but there are also some cases of inhibition at concentrations higher than the optimal. The inhibition concentration of dye is not the same for all the organisms, and for the same organism, the inhibition concentration will depend on the dye. Adaptation of a microbial community to the compound is very useful in improving the rate of decolourisation process, due to the natural expression of genes encoding for enzymes responsible for its degradation, when previously exposed (Ramalho et al. 2005). The fact that some dyes are biodegraded and others not, even under the same conditions, is explained by the role of the chemical structure of the dye on the process. Even when belonging to the same class and type, dyes differ in their structure and present different pKa and potential redox. Zille et al. (2004) studied the biodegradation under aerobic conditions of azo dyes by yeasts with reducing activity and by an oxidative enzyme, laccase, with or without mediator; they compared these 2 approaches on the basis of the electrochemical properties of dyes and bioagents. A linear increase of dye decolourisation with decreasing redox potential of dye was obtained with laccase and laccase/mediator systems; in the reductive approach, they observed that the less negative the redox potential of the azo dye, the more favourable (and faster) is its reduction. The redox potential should reportedly be below -450 to -500 mV for azo dye reduction to occur (Delee *et al.* 1998). It is worth mentioning that the redox potential is influenced by other external factors, such as the pH of the solution Table 5.

Table 5: Factors affecting decolourization and degradation of synthetic dyes

Factors	Descriptions
pH	The pH has a major effect on the efficiency of dye decolourization, the optimal pH for color removal in bacteria is often between 6.0 and 10.0. The tolerance to high pH is important in particular for industrial processes using reactive azo dyes, which are usually performed under alkaline conditions. The pH has a major effect on the efficiency of dye decolourization, the optimal pH for color removal in bacteria is often between 6.0 and 10.0 (Chen *et al.* 2003; Guo *et al.* 2007).
Temperature	Temperature is also again a very important factor for all processes associated with microbial vitality, including the remediation of water and soil. It was also observed that the decolourization rate of azo dyes increases up to the optimal temperature, and afterwards there is a marginal reduction in the decolourization activity.
Dye concentration	Earlier reports show that increasing the dye concentration gradually decreases the decolourization rate, probably due to the toxic effect of dyes with regard to the individual bacteria and/or inadequate biomass concentration, as well as blockage of active sites of azo reductase by dye molecules with different structures. Dyes are deficient in carbon and nitrogen sources, and the biodegradation of dyes without any supplement of these sources is very difcult. Microbial cultures generally require complex organic sources, such as yeast extract, peptone, or a combination of complex organic sources and carbohydrates for dye decolourization and degradation.
Oxygen and agitation	Environmental conditions can affect the azo dyes degradation and decolourization process directly, depending on the reductive or oxidative status of the environment, and indirectly, influencing the microbial metabolism. It is assumed that under anaerobic conditions reductive enzyme activities are higher; however a small amount of oxygen is also required for the oxidative enzymes which are involved in the degradation of azo dyes.
Dye structure	Dyes with simpler structures and low molecular weights exhibit higher rates of color removal, whereas the removal rate is lower in the case of dyes with substitution of electron withdrawing groups such as SO_3H, SO_2NH_2 in the para position of phenyl ring, relative to the azo bond and high molecular weight dyes.
Electron donor	It has been observed that the addition of electron donors, such as glucose or acetate ions, apparently induces the reductive cleavage of azo bonds. The type and availability of electron donors are important in achieving good colour removal in bioreactors operated under anaerobic conditions.
Redox mediator	Redox mediators (RM) can enhance many reductive processes under anaerobic conditions, including azo dye reduction.

Source: (Razia Khan *et al.* 2013)

14. Toxicity of Dyes

It is very important to know whether biodegradation of a dye leads to detoxification of the dye or not. This can be done by performing phytotoxicity and microbial toxicity tests of the original dye and its biodegradation products. In phytotoxicity studies, the seeds of model plants can be treated with a particular concentration of the original dye and also with its biodegradation products. Kalyani *et al.* (2009) conducted phytotoxicity study of Reactive Red 2 and its degradation products using *Sorghum vulgare* and *Phaseolus mungo* as model plants. They treated the plant seeds with water, reactive red 2 (5,000 ppm), and its extracted metabolite (5,000 ppm) separately and compared percent germination and the lengths of plumule and radicle. From the results, it was found that the metabolites produced after the biodegradation of Reactive Red 2 were less toxic as compared to the original dye.

Dyes may significantly affect photosynthetic activity in aquatic life and microbial toxicity on *S. paucimobilis* showed growth inhibitory zone (0.8 cm) surrounding the well containing dye, while degradation product did not show inhibitory zone which also confirmed the nontoxic nature of the extracted metabolite. These findings suggest non-toxic nature of the product formed. Previous reports showed malachite green G degradation into Leuco-malachite green that is equally toxic to microbes (Wesenberg *et al.* 2003b) and reduce light penetration.

Conclusion

The management of textile industrial effluents is a complicated task, taking into consideration the complexity of the waste compounds that may be present, in addition to the dyes, and the numerous established options for treatment and reuse of water. Wide ranges of water, pH, temperature, salt concentration and in the chemical structure of numerous dyes in use today add to the complication. Economical removal of colour from effluents remains an important problem, although a number of successful systems employing various physicochemical and biological processes have been successfully implemented. Regulatory agencies are increasingly interested in new, efficient, and improved decolourisation technologies. Solid and evolving scientific knowledge and research is of the utmost relevance for the effective response to current needs. In view of the requirement for a technically and economically satisfactory treatment, a flurry of emerging technologies are being proposed and are at different stages of being tested for commercialization. A broader validation of these new technologies and the integration of different methods in the current treatment schemes will be most likely in the near future, rendering them both efficient and economically viable. Conventional physicochemical treatments are not always efficient. The high cost, the generation of sludge and of other pollutants, and the need for sophisticated technologies are limiting factors as well. Bioremediation of textile effluents is still seen as an attractive solution due to its reputation as a low-cost, sustainable and publicly acceptable technology. Microorganisms are easy to grow and the use of their isolated enzymes for textile dyes degradation is not expensive in relative terms because there is no need for high purity levels in treating effluents. Many microorganisms and enzymes have been isolated and explored for their ability and capacity to degrade dyes. Others have been modified by the genetic engineering tools to obtain "super and faster degraders". Several low cost and efficient sorbents including natural wastes are very promising, not only due to the lower cost and high availability, but also a new utility is granted to those wastes. The increasing manufacture and application of synthetic dyes, taking into account their impact in the environment, needs an effective response in terms of modern and viable treatment processes of coloured effluents, prior to their discharge as waste into waterways: Biodegradation of synthetic dyes using different microorganisms and isolated enzymes offer a promising approach by themselves or in combination with conventional treatments. The complexity of dyes degradation and the existence of an immense variety of structurally different dyes, indicates the need for more research.

References

Abadulla E, Tzanov T, Costa S, Robra KH, Cavaco-Paulo A, Gubitz G (2000) Decolorization and detoxification of textile dyes with a laccase from *Trametes hirsute*. Appl. Environ. Microbiol. 66, 3357-3362.

Akhtar S, Husain Q (2006) Potential applications of immobilized bitter gourd (*Momordica charantia*) peroxidase in the removal of phenols from polluted water. Chemosphere 65, 1228–1235.

Ali H (2010) Biodegradation of synthetic dyes-a review. Water Air Soil Pollut. DOI:l007/s 11270-010-0382-4.

Almansa E, Kandelbauer A, Pereira L, Cavaco-Paulo A, Gubtiz G (2004) Influence of structure on dye degradation with laccase mediator systems. Biocat. Biotransform. 22, 315-324,

Arslan I, Balciolul A (1999) Degradation of commercial reactive dyestuffs by heterogenous and homogenous advanced oxidation processes: a comparative study. Dyes Pigments 43, 95-108.

Asgher M, Shah SAH, Ali M, et al (2006) Decolourization of some reactive textile dyes by white rot fungus isolated in Pakistan. World J. Microbiol. Biotechnol. 22, 89-93.

Ayyasamy PM, Suresh Raja SS, Subashni B, Palanivelan R (2015) Bio-statistical evaluation of cultural conditions on industrial textile dye decolourisation using a native bacterium *Micrococcus endophyticus* (ES37). J. Water Reuse Desalination 5(4), 557-568.

Banat IM, Nigam P, Singh D, Marchant R (1996) Microbial decolorisation of textile dye-containing effluents: a review. Bioresour. Technol. 58, 217-227.

Bansal RC, Goyal M (2005) Activated Carbon Adsorption, Taylor and Francis Group Publisher London.

Bhole BD, Ganguly B, Madhuram A, Deshpande D, Joshi J (2004). Biosorption of methyl violet, basic fuchsin et their mixture using dead fungal biomass, Curr. Sci. 86(12), 1641-1645.

Blanquez P, Casas N, Font X, Gabarrella, X, Sarràa M, Caminalb G and Vicenta T (2004) Mechanism of textile metal dye biotransformation by *Trametes versicolor*. Water Res. 38(8), 2166–2172.

Booth, Gerald (2000) Dyes, General Survey. Wiley-VCH. DOI:10.1002/14356007.a09-073.

Chen CH, Chang CF, Ho CH, Tsai TL, Liu SM (2003). Biodegradation of crystal violet by a *Shewanella* sp. NTOU1. Chemosphere 72, 1712 1720.

Chivukula M, Renganathan V (1995) Phenolic azo dye oxidation by laccase from *Pyricularia oriyzae*. Appl. Environ. Microbiol. 61, 4374-4377

Cripps C, Bumpus J.A (1990). Biodegradation of azo and heterocyclic dyes by *Phanerochaete chrysosporium*. Appl. Environ. Microbiol. 56, 1114-1118.

Dafale N, Nageswara Rao N, Sudhir U, Meshram, Satish RW (2008) Decolorization of azo dyes and simulated dye bath wastewater using acclimatized microbial consortium – Biostimulation and halo tolerance Bioresour. Technol. 99:2552-2558.

Delee W, O'Neill C, Hawkes FR, Pinheiro HM (1998) Anaerobic treatment of textile effluents: a review. J. Chem. Technol. Biotechnol. 73, 323-335.

Duran N, Esposito E (2000) Potential applications of oxidative enzymes and phenoloxidase-like compounds in wastewater and soil treatment: a review. Appl. Catal. B: Environ. 28, 83-99.

Eggert C, Temp U, Eriksson KEL (1996) Laccase-producing white-rot-fungus lacking lignin peroxidase and manganese peroxidase. In: Enzymes for Pulp and Paper Processing (Ed) Jefries TW and Viikari L, ACS Symposium Series 655, Washington, DC, pp 130- 150.

El-Sheekh MM, Gharieb MM, Abou-El-Souod GW (2009). Biodegradation of dyes by some green algae and cyanobacteria. Int. Biodeter. Biodegr. 63, 699-704.

Ergene A, Ada K, Tan S, Katrcou H (2009). Removal of remazol brilliant blue R dye from aqueous solutions by adsorption onto immobilized *Scenedesmus quadricauda*; Equilibrium and kinetic modeling studies. Desalination 249, 1308-1314.

Gazso LG (2001) The key microbial processes in the removal of toxic metals and radionuclides from the environment. Central Eur. J. Occup. Environ. Med. 7, 178-185.

Ge Y, Yan L, Qinge K (2004) Effect of environment factors on dye decolourization by *P. sordid* ATCC90872 in an aerated reactor. Process Biochem. 39, 1401-1405.

Gou M, Qu Y, Zhou J, Ma F, Tan L (2009) Azo dye decolorization by a new fungal isolate, *Penicillium* sp. QQ and fungal-bacterial co-cultures J. Hazard. Mater. 170, 314-319.

Gupta VK, Jain R, Varshney S (2007) Electrochemical removal of the hazardous dye reactofix red 3 BFN from industrial effluents. J. Colloidal Interface Sci. 312, 292–296.

Gurulakshmi M, Sudarmani DNP and Venba R (2008). Biodegradation of leather acid dye by *Bacillus subtilis*. Advanced BioTech. 7, 12-18.

Haley TJ (1975) Benzidine revisited; a review of the literature and its congeners. Clinic. Toxicol. 8, 13-42

Hao OJ, Kim H, Chiang PC (2000) Decolorization of waste water. Crit. Rev. Environ. Sci. Technol. 30, 449- 505.

Huang YH, Hsueh CL, Huang CP, Su LC, Chen CY (2007). Adsorption thermodynamic and kinetic studies of Pb(II) removal from water onto a versatile Al2O3-supported iron oxide. Sep. Purif. Technol. 55, 23-29.

Joshi SM, Inamdar SA, Telke AA, Tamboli, DP, Govindwar SP (2010) Exploring the potential of natural bacterial consortium to degrade mixture of dyes and textile effluent. Int. J. Biodeter. Biodegrad. 64, 622-628.

Chacko JT, Subramaniam K (2011) Enzymatic degradation of azo dyes: a review. Int. J. Environ. Sci. 1(6), 1250-1260.

Junnarkar N, Srinivas Murty D, Bhatt NS, Madamwar D (2006) Decolourization of diazo dye direct red 81 by a novel bacterium consortium. World J. Microbiol. Biotechnol. 22, 163-168.

Kalyani DC, Telke AA, Dhanve RS, Jadhav JP (2009). Eco-friendly biodegradation and detoxification of reactive red 2 textile dye by newly isolated *Pseudomonas* sp. SUK1. J. Hazard Mater. 163, 735-742.

Kalme SD, Parshetti GD, Jadhav SU, Govindwar SP (2006) Biodegradation of benzidine based dye Direct Blue-6 by *Pseudomonas desmolyticum* NCIM 2112. Bioresour. Technol. 7, 1405-1410.

Keck A, Rau 1, Reemtsma T, Mattes R, Stolz A, Klein 1 (2002) Identification of quinoide redox mediators that are formed during the degradation of napbthalene-2-sulfonate by *Sphingomonas xenophaga* BN6. Appl. Environ. Microbiol. 68, 4341-4349.

Khan AA, Husain Q (2007). Potential of plant polyphenol oxidases in the decolorization and removal of textile and non-textile dyes. J. Environ. Sci. 19, 396-402.

Kieser T, Bibb MJ, Buttner MJ, Chater KF, Hopwood DA (2000). Practical *Streptomyces* Genetics, The John Innes Foundation, Norwich, UK.

Kim SJ, Shoda P (1999) Purification and characterization of a novel peroxidase from *Geotrichum candidum* dec1 involved in decolourization of dyes. Appl. Environ. Microbiol. 65, 1029-1035.

Knapp J, Newby P (1995) The microbiological decolorization of an industrial effluent containing a diazo-linked chromophore. Water Res. 29(7), 1807–1809.

Knappa JS, Vantoch-Wood EJ, Zhang F (2001). Use of wood-rotting fungi for the decolourization of dyes and industrial effluent. In: Fungi in Bioremediation (Ed) Gadd GM, Cambridge University Press, pp 242–304.

Kumar MNVR, Sridhar TR, Bhavani KD, Dutta PK (1998) Trends in color removal from textile mill effluents. Colorage. 40, 25-34.

Kumarpraveen G, Sumangala N, Bhat K (2011) Fungal degradation of azo dye red 3BN and optimization of physico-chemical parameters. Int. J. Environ. Sci. 1(6), 2278-3202.

Lepez C, Valade AG, Combourieu B, Mielgo I, Bouchon B, Lema M (2004) Mechanism of enzymatic degradation of the azo dye orange II determined by *ex situ* 1H nuclear magnetic resonance and electrospray ionization ion trap mass spectrometry. Ann. Biochem. 335, 135-149.

Levin L, Papinutti L, Forchiassin F (2004) Evaluation of argentinean white rot fungi for their ability to produce lignin-modifying enzymes and decolourize industrial dyes. Bioresour. Technol. 94(2), 169-176.

Lim SL, Chu WL, Phang SM (2010). Use of *Chlorella vulgaris* for bioremediation of textile wastewater. Bioresour. Technol. 101, 7314-7322.

Malarvizhi R, Wang MH, Ho YS (2010) Research trends in adsorption technologies for dye containing wastewaters. World Appl. Sci. J. 8, 930-942.

Manikandan N, Surumbar Kuzhali S, Kumuthakalavalli, R (2012) Decolorisation of textile dye effluent using fungal microflora isolated from spent mushroom substrate (SMS). J. Microbiol. Biotech. Res, 2(1), 57-62.

Mathur N, Bhatnagar P, Bakre P (2005). Assessing mutagenicity of textile dyes from Pali (Rajasthan) using Ames bioassay. Appl. Ecol. Environ. Res. 4(1), 111-118.

Melgoza RM, Cruz A, Buitron G (2004) Anaerobic/aerobic treatment of colorants present in textile effluents. Water Sci. Technol. 50, 149-155.

Michniewicz A, Ledakowicz S, Ullrich R, Hofrichter M (2008) Kinetics of the enzymatic decolourization of textile dyes by laccase from *Cerrena unicolor*. Dyes Pigments. 77(2), 295-302.

Mona HM, Mabrouk EM, Yusef HH (2008) J. Appl. Sci. Res. 4, 262-269.

Muthezhilan R, Yoganath N, Vidhya S, Jayalakshmi S (2008) Dye degrading mycoflora from industrial effluents. Res. J. Microbiol. 3(3), 204-208.

Muthukumar M, Sargunamani D, Senthilkumar M, Selvakumar N (2005) Studies on decolourisation, toxicity and the possibility for recycling of acid dye effluents using ozone treatment. Dyes Pigments 64, 39- 44.

Pak D, Changm W (1999) Decolourizing dye wastewater with low temperature catalytic oxidation. Water Sci. Technol. 40, 115-121.

Pandey A, Singh P, Iyengar L (2007) Bacterial decolorization and degradation of azo dyes. Int. Biodet. Biodeg. 59, 73-84.

Pearce C I, Lioyd JR, Guthrie JT (2003). The removal of colour from textile wastewater using whole bacterial cells: a review. Dyes pigments 58, 179-186.

Pereira L, Coelho AV, Viegas CA, Santos MMC, Robalo MP, Martins LO (2009) Enzymatic biotransformation of the azo dye Sudan orange G with bacterial CotA-laccase. J. Biotechnol. 139, 68-77.

Pereira L, Pereira R, Pereira MFR, VanderZee FP, Cervantes FJ, Alves MM (2010) Thermal modification of activated carbon surface chemistry improves its capacity as redox mediator for azo dye reduction. J. Hazard. Mater. 183:931-939.

Pinheiro HM, Touraud E, Thomas O (2004) Aromatic amines from azo dye reduction: status review with emphasis on direct UV spectrophotometric detection in textile industry wastewaters. Dyes Pigments 61, 121-139.

Priya B, Sivaprasanth KR, Jensi VD, Uma L, Subramanian G, Prabaharan D (2010). Characterization of manganese superoxide dismutase from a marine cyanobacterium *Leptolyngbya valderiana* BDU 20041. Saline Syst. 6, 6-15.

Rai HS, Bhattacharyya MS, Singh J, Bansal TK, Vats P, Banerjee UC (2005). Removal of dyes from the effluent of textile and dyestuff manufacturing industry: a review of emerging techniques with reference to biological treatment. Crit. Rev. Environ. Sci. Technol. 35, 219-238.

Ramalho PA, Saraiva S, Cavaco-Paulo A, Casal M, Cardoso MH, Ramalho MT (2005) Azo reductase activity of intact *Saccharomyces cerevisiae* cells is dependent on the fre 1p component of plasma membrane ferric reductase. Appl. Environ. Microbiol. 71, 3882-3888.

Ramsay J A and Nguyen T (2002) Decolourization of textile dyes by *Trametes versicolour* and its effect on dye toxicity. Biotechnol. Lett. 24, 1757-1761.

Rathnan RK, Anto SM, Rajan L, Sreedevi ES Ambili M, Balasaravanan T (2013). Comparative studies of decolorization of toxic dyes with laccase enzymes producing mono and mixed cultures of fungi. Nehru E-J. 1(1), 21-24.

Rau J, Stolz A (2003) Oxygen-insensitive nitroreductases NfsA and NfsB of *Escherichia coli* function under anaerobic conditions as laws one dependent azo reductase. Appl. Environ. Microbiol. 69, 3448-3455, DOI: 10.1128/AEM.69.6.3448-3455.2003.

Razia Khan P, Bhawana M, Fulekar H (2012) Microbial decolorization and degradation of synthetic dyes: a review. Rev. Environ. Sci. Biotechnol. 12(1), 75-97.

Riu J, Schonsee I, Barcelo D (1998) Determination of sulfonated azo dyes in ground water and industrial effluent by automated solid-phase extraction followed by capillary electrophoresis/mass spectrometry. J. Mass Spect. 33, 653.

Robinson T, McMullan G, Marchant R, Nigam P (2001) Remediation of dyes in textile effluent: a critical review on current treatment technologies with a proposed alternative. Bioresour. Technol. 77, 247-255.

Romero S, Blanquez P, Caminal G Font X, Sarra M, Gabarrell X, Vincent T (2006) Different approaches to improving the textile dye degradation capacity of *Trametes versicolor*. Biochem. Eng. J. 31(1), 42-47.

Rus R, Rau J, Stolz A (2000) The function of cytoplasmatic reductases in the reduction of azo dyes by bacteria. Appl. Environ. Microbiol. 66, 1429-1434.

Schroeder M, Pereira L, Couto SR, Erlacher A, Schoening K-U, Cavaco-Paulo A, Guebitz GM (2007) Enzymatic synthesis of Tinuvin. Enzyme Microbial Technol. 40, 1748-1752.

Senan RC, Abraham TE (2004) Bioremediation of textile azo dyes by aerobic bacterial consortium. Biodegradation 15(4), 275-280.

Sharma P, Singh L, Mehta J (2010) COD reduction and color removal of simulated textile mill wastewater by mixed bacterial consortium. Rasayan J. Chem. 3(4), 731-735.

Singh AK, Singh R, Soam A, Shahi SK (2012) Degradation of textile dye orange 3R by *Aspergillus* strain (MMF3) and their culture optimization. Curr. Discovery 1(1), 7-12.

Singh L (2008) Exploration of fungus *Gliocladium virens* for textile dye (Congo red) accumulation/ degradation in semi-solid medium: A microbial approach for hazardous degradation, "ISME 12", Cairns, Australia.

Singh L, Singh VP (2010) Microbial degradation and decolourization of dyes in semisolid medium by the fungus – *Trichoderma harzianum*. Int. J. Sci. Technol. 5(3), 147-153.

Sivarajasekar N, Baskar R, Balakrishnan V (2009) Biosorption of an azo dye from aqueous solutions onto *Spirogyra*. J. Univ. Chem. Technol. Metal. 44, 157-164.

Slokar Y, Majcen Le Marechal A (1998) Methods of decolouration of textile wastewaters. Dyes Pigments 37(4), 335-356.

Solomon EI, Sundaram UM, Machonkin TE (1996) Multicopper oxidases and oxygenases. Chem. Rev. 96, 2563-2605.

Sponza DT, Ieik M (2005) Reactor performance and fate of aromatic amines through decolorization of direct black 38 dye under anaerobic/aerobic sequentilas. Process Biochem. 40, 35-44.

Steffan S, Bardi L, Marzona M (2005) Azo dye biodegradation by microbial cultures immobilized in alginate beads. Environ. Int. 31, 201-205.

Stolz A (2001) Basic and applied aspects in the microbial degradation of azo dyes. Appl. Microbiol. Biotechnol. 56, 69-80, DOI: 10.1007/s002530100686.

Swamy J, Ramsay JA (1999). The evaluation of white rot fungi in the decoloration of textile dyes. Enzyme Microbial Technol. 24, 130-137.

Tan GYA, Ward AC, Goodfellow M (2006). Exploration of *Amycolatopsis* diversity in soil using genus-specific primers and novel selective media. Syst. Appl. Microbiol. 29, 557-569.

Tan NCO, van Leeuwen A, van Voorthuinzen EM, Slenders P, Prenafeta-Boldu FX, Temmink H, Lettinga G, Field JA (2005) Fate and biodegradability of sulfonated aromatic amines. Biodegradation 16, 527-537.

Tauber MM, Gubitz GM, Rehorek A (2008) Degradation of azo dyes by oxidative processes laccase and ultrasound treatment. Bioresour. Technol. 99, 4213-4220.

Tien C (1961) Adsorption kinetics of a non-flow system with nonlinear equilibrium relationship. AIChE J. 7, 410-419.

Valli Nachiyar C, Suseela Rajkumar G (2003) Degradation of a tannery and textile dye, Navitan fast blue S5R by *Pseudomonas aeruginosa*. World J. Microbiol Biotechnol. 19(6), 609-614.

Van der Zee FP, Villaverde S (2005) Combined anaerobic-aerobic treatment of azo dyes- a short review of bioreactors studies. Water Res. 39, 1425-1440.

Ventosa, Antonia, Nieto, Joaquín, Joren Aharon (1998). Biology of moderately halophilic aerobic bacteria. Microbiol. Mol. Biol. Rev. 62(2), 504-544.

Vijayakumar R, Vaijayanthi G, Sandhiya K (2015) Decolourization potentiality of actinobacteria on synthetic textile dyes and effluents a comparative study. Sci. Trans. Environ. Technol. 8(4), 178-184.

Wesenberg D, Kyriakides I, Agathos SN (2003a) White-rot fungi and their enzymes for the treatment of industrial dye effluents. Int. J. Chemtech Applications 2(2), 126-136.

Wesenberg D, Kyriakides I, Agathos SN (2003b) White-rot fungi and their enzymes for the treatment of industrial dye effluents. Biotechnol. Adv. 22, 161-187.

Xu Y, Leburn RE (1999) Treatment of textile dye plant effluent by nanofiltration membrane. Separ. Sci. Technol. 34, 2501-2519.

Yan H and Pan G (2004) Increase in biodegradation of dimethyl phthalate by *Closterium lunula* using inorganic carbon. Chemosphere 55, 1281-1285.

Zhang F, Knapp J, Tapley K (1999) Development of bioreactor systems for decolourization of orange II using white rot fungus. Enzyme Microbial Technol. 24(1-2), 48-53.

Zhou W, Zimmerman W (1993) Decolourization of industrial effluents containing reactive dyes by actinomycetes. FEMS Microbiol. Lett. 107, 157-162.

Zille A, Gemacka B, Rehorek A, Cavaco-Paulo A (2005) Degradation of azo dyes by *Trametes villosa* laccase under long time oxidative conditions. Appl. Environ. Microbiol. 71: 6711-718.

Zille A, Ramalho P, Tzanov T, Millward R, Aires V, Cardoso MH, Ramalho MT, Gtibitz GM, Cavaco-Paulo A (2004) Predicting dye biodegradation from redox potentials. Biotechnol. Prog. 20:1588-1592.

Zimmermann T, Kulla HG, Leisinger T (1982) Properties of purified Orange II azoreductase, the enzyme initiating azo dye degradation by *Pseudomonas* KF46. Eur. J. Biochem. 129, 197-203.

Zollinger H (1987) Colour Chemistry-Synthesis, Properties and Applications of Organic Dyes and Pigments, VCH Pub., New York.

9

Engineered Microbes for Remediation of Toxic Pollutants and Restoration of Contaminated Environments

Jeyabalan Sangeetha, Shrinivas Jadhav, Devarajan Thangadurai
Muniswamy David, Abhishek Mundaragi and Purushotam Prathima

Abstract

With recent advances in molecular biology and rapid developments in new analytical techniques, engineering of microbes for the bioremediation processes has been extensively studied as recombinant microbes can degrade various toxic pollutants effectively. The prospect of microbial engineering has greater potentialities for agriculture and environmental applications and this may provide us deeper insight in understanding the microbial interactions and biotransformation processes. Though much work has been carried out on bioremediation using genetically engineered microorganisms, its application is restricted to environment due to the concerns of health and environment over ecosystem. Nevertheless, evaluation of risk assessment has also been carried out in order to ascertain the potential benefits of transgenic microbes.

Keywords: Bioremediation, Ecorestoration, Genetically engineered microorganisms, Polyaromatic hydrocarbons, Recalcitrant materials, Xenobiotics

1. Introduction

Xenobiotics (Greek, *xenos* - strange; *bios* - life) refers to compounds of chemical origin that do not occur in nature, thus are considered foreign to ecosystem. Xenobiotics have uncommon chemical formulation to which microorganisms have never been exposed to earlier. During the course of evolution, this fact has led to genetic susceptibility of microorganisms to xenobiotics. Thus, xenobiotics may resist complete biodegradation by microbes. Xenobiotic compounds may be synthetic organochlorine or naturally derived chemicals like polyaromatic hydrocarbons (PAHs) and some traces of crude oil and coal. By definition, xenobiotics are the compounds alien to life; this does not necessarily imply all xenobiotics are harmful, but some are certainly toxic to life (Rozgaj 1994).

2. Xenobiotics: Benefits and Effects

Xenobiotics, the chemical compounds, involve substances which are present at higher concentration than that of usual one. Molecules such as antibiotics in broader term are considered xenobiotics since they are not naturally formed inside one's system, nor are they part of one's diet. Natural human hormones entering into fishes found downstream of sewage outfall, or pheromones produced by one organism in defence to another can cause notable physiological changes. This implies that natural compounds can also be xenobiotics, if they find their path into another organism (Mansuy 2013). Xenobiotic compounds are generally considered recalcitrant (structure immanent stability), because of the fact that they possess unphysiological chemical bonds, which are harder to cleave by microbial catabolic enzymes. Some natural compounds show same recalcitrant features as xenobiotics; those are halogen substituents or nitro groups found in some antibiotics and chemicals.

The reason why xenobiotics resist complete biodegradation is because these are not always recognised by naturally occurring microflora due to their complex chemical structure and hence these rarely enter common metabolic pathways (Rozgaj 1994). Whereas some compounds are transformed only in assistance of another compound, which happens to be additional nutrient source. Furthermore, few compounds are degraded from sequential manner of action by series of different microbes. Rapid rate of multiplication and fast metabolic potentials allow bacteria and fungi to readily adapt to new substances in nature, making them chief degraders of xenobiotic substances.

3. Bioremediation of Xenobiotics: An Overview

Bioremediation, a biological degradation method where sites contaminated with xenobiotic compounds are made orderly clean by microbial actions resulting in reduced pollution and toxicity of the pollutants. This is an economic and ecofriendly approach which is rapidly developing in the field of environmental restoration. Bioremediation involves harvesting microbial ability to degrade and detoxify recalcitrant materials like hydrocarbons, aliphatic and aromatic compounds, industrial effluents, pesticides and their derivatives. Presence of vast majority of microbial diversity in nature expands the possibility of chemical compounds that can be degraded and restored. Notable example for this is biological cleanup attempt of a huge accidental oil spill by tanker Exxon Valdez which affected Bligh reef in the Gulf of Alaska in March 1989, spilling almost 41,000 m^3 of crude oil and contaminating about 2,000 km of coastline (Margesin *et al.* 1999). Thus, use of microorganisms in addressing environmental contamination has become a choice of increasingly relevant technology and for restoration of contaminated sites economically.

General interest appears in studying diversity of indigenous microorganisms capable of degrading different compounds because of their generalised tendency to act upon pollutants. Many efforts have been put forward in an attempt to study the response of microbial community toward pollutants, and to isolate and identify potent degraders of pollutants with genes involved in degradation (Greene *et al.* 2000; Watanabe *et al.* 2002). The detailed analysis of microbial diversity is however studied under two different terms: culture dependent and culture independent (Juck *et al.* 2000). It has been known for long time that polluted sites port a wide range of unanimous microbial population that play a vital role in degradation process (Margesin *et al.* 2003) which can only be elucidated with culture independent techniques. Whereas conventional characterisation of microbial strains have been put into shadows, as it determined by the ability of the strain to grow under defined environmental conditions (Bakonyi *et al.* 2003). During recent decades, molecular tools like16S rRNA analysis has facilitated the study of natural microbial population without cultivation (Kubicek *et al.* 2003). However, there is no single tool which allows defined analysis of soil microbial community. Therefore, the use of polyphasic approach has made scientists to bring molecular, microbial and geochemical techniques under one definition to avail better understanding of interaction between microorganisms and their natural habitats (Ramsing *et al.* 1996; Teske *et al.* 1996).

Microorganisms, however has the ability to develop selective resistance under environmental pollution stress against number of recalcitrant materials which were previously thought non-degradable (Table 1). Whereas, the fact that many pollutants are still persistent in nature implies microbial inadequacy in addressing such pollutants. This is where engineered microbial population with better efficiency to degrade pollutants comes into picture. Recent advances in molecular biology and genetic engineering has led to more rational steps in addressing xenobiotic problems (Nielsen 2001). It has been made possible by rapid development in new analytical tools and several cloning methods for inducing the directed genetic changes of microbes and to analyse the consequence of the induced microbial changes at cellular levels. Additionally, many sophisticated state of the art analytical instrumentation like gas chromatography, gas chromatography–mass spectrometry (GC-MS), nuclear magnetic resonance

Table 1: List of microorganisms capable of degrading xenobiotics and potential candidates for genetic engineering purposes.

Microorganism	Target xenobiotics	References
Achromobacter spp.	Lower and higher chlorinated biphenyls	Floodgate (1984); Grishchenkov *et al.* (2000); Seeger *et al.* (2001); Taguchi *et al.* (2001); Petric *et al.* (2007); Kafilzadeh *et al.* (2011)
Acidithiobacillus ferrooxidans	Hg(II), As, Cd, Cu, Co, Zn	White *et al.* (1998); Takeuchi and Sugio (2006); Jaysankar *et al.* (2008)
Acinetobacter spp.	Aromatic hydrocarbons	Mrozik *et al.* (2003)
Aeromonas spp.	Aliphatic and aromatic hydrocarbons	Mrozik *et al.* (2003)
Alcaligenes faecalis	Hg(II), As, Cd, Cu, Co, Zn	White *et al.* (1998); Takeuchi and Sugio (2006); Jaysankar *et al.* (2008)
Alcaligenes spp.	Aromatic hydrocarbons	Mrozik *et al.* (2003)
Alcanivorax borkumensis	Heterocyclic compounds (pyridine)	Sims and O'Loughlin (1989); Yakimov *et al.* (2007); dos Santos *et al.* (2008)
Alternaria alternate	Dye degradation, gelatine emulsion	Bumpus (2004); Abruscia *et al.* (2007); Chaplain *et al.* (2011)
Amorphoteca spp.	Petroleum oil hydrocarbons	Russell *et al.* (2011)
Aromatoleum aromaticum strain EbN1	Anaerobic hydrocarbon (toluene and ethylbenzene)	Russell *et al.* (2011)
Arthrobacter sp. strain R1	2-picoline	O'Loughlin *et al.* (2000)
Aspergillus spp.	Aliphatic hydrocarbons, starches, hemicelluloses, celluloses, pectins, sugar polymers, fats, oils, chitin, keratin, aromatic hydrocarbons, crude oil hydrocarbons, pesticides	Bennett (2010); Machida and Gomi (2010); Chaplain *et al.* (2011)
Aspergillus nidulans var. nidulans	Dye degradation, gelatine emulsion	Bumpus (2004); Abruscia *et al.* (2007); Chaplain *et al.* (2011)
Aspergillus niger	Polychlorinated biphenyls	Dmochewitz and Ballschmiter (1988); Donnelly and Fletcher (1995)
Aspergillus niger AB10	Cadmium	Pal *et al.* (2010)
Aspergillus ustus	Dye degradation, gelatine emulsion	Bumpus (2004); Abruscia *et al.* (2007); Chaplain *et al.* (2011)
Aspergillus versicolor	Dye degradation, gelatine emulsion	Bumpus (2004); Abruscia *et al.* (2007); Chaplain *et al.* (2011)

(Contd.)

Microorganism	Target xenobiotics	References
Azospirillum lipoferum	Organophosphorus insecticide (malathion)	Kanade *et al.* (2012)
Bacillus pumilus	Hg(II), As, Cd, Cu, Co, Zn	White *et al.* (1998); Takeuchi and Sugio (2006); Jaysankar *et al.* (2008)
Bacillus spp.	Aromatic hydrocarbons, Hg(II), As, Cd, Cu, Co, Zn, lower and higher chlorinated biphenyls, pesticides (atrazine, chlorpyrifos, dichlorodiphenyltrichloroethane)	Floodgate (1984); Struthers *et al.* (1998); White *et al.* (1998); Grishchenkov *et al.* (2000); Seeger *et al.* (2001); Taguchi *et al.* (2001); Mrozik *et al.* (2003); Takeuchi and Sugio (2006); Petric *et al.* (2007); Jaysankar *et al.* (2008); Surekha Rani *et al.* (2008); Kafilzadeh *et al.* (2011); Kanade *et al.* (2012)
Brevibacillus spp.	Aromatic hydrocarbons	Mrozik *et al.* (2003)
Brevibacterium iodinium	Hg(II), As, Cd, Cu, Co, Zn	White *et al.* (1998); Takeuchi and Sugio (2006); Jaysankar *et al.* (2008)
Burkholderia spp.	Lower and higher chlorinated biphenyls	Floodgate (1984); Grishchenkov *et al.* (2000); Seeger *et al.* (2001); Taguchi *et al.* (2001); Petric *et al.* (2007); Kafilzadeh *et al.* (2011)
Candida boidinii	Aromatic and aliphatic hydrocarbons, aromatic organopollutants, polycyclic aromatic hydrocarbons, biphenyls, dibenzofurans, nitroaromatics, pesticides, and plasticizers	Eaton (1985); Cabras *et al.* (1988); Sasek *et al.* (1993); Fritsche and Hofrichter (2000)
Candida ernobii	Aliphatic hydrocarbons, crude oil, petroleum products, diesel oil	Bartha (1986); Ijah (1998); Fritsche and Hofrichter (2005); De Cássia Miranda *et al.* (2007); Mucha *et al.* (2010)
Candida lipolytica	Aliphatic hydrocarbons, crude oil, petroleum products, diesel oil	Bartha (1986); Ijah (1998); Fritsche and Hofrichter (2005); De Cássia Miranda *et al.* (2007); Mucha *et al.* (2010)
Candida lipolytica	Aromatic and aliphatic hydrocarbons, aromatic organopollutants, polycyclic aromatic hydrocarbons, biphenyls, dibenzofurans, nitroaromatics, pesticides, and plasticizers	Eaton (1985); Cabras *et al.* (1988); Sasek *et al.* (1993); Fritsche and Hofrichter (2000)

(Contd.)

Microorganism	Target xenobiotics	References
Candida methanosorbosa BP-6	Aniline	Bartha (1986); Ijah (1998); Fritsche and Hofrichter (2005); De Cássia Miranda (2007); Mucha et al. (2010)
Candida tropicalis	Aliphatic hydrocarbons, crude oil, petroleum products, diesel oil, heavy metals, Cu(II), Ni(II), Co(II), Cd(II), Mg(II), Cr(VI)	Bartha (1986); Ijah (1998); Fritsche and Hofrichter (2005); Ksheminska et al. (2006); Wang and Chen (2006); De Cássia Miranda et al. (2007); Ksheminska et al. (2008); Mucha et al. (2010); Saisubhashini et al. (2011); Bahafid et al. (2011); Bahafid et al. (2013)
Cephalosporium spp.	Crude oil hydrocarbons	Russell et al. (2011)
Cladophialophora spp.	Toluene	Prenafeta-Boldu et al. (2001)
Cladosporium spp.	Aliphatic hydrocarbon	Chaillan et al. (2004); Singh (2006); Steliga (2012)
Cladosporium cladosporioides	Dye degradation, gelatine emulsion	Bumpus (2004); Abruscia et al. (2007); Chaplain et al. (2011)
Comamonas spp.	Lower and higher chlorinated biphenyls	Floodgate (1984); Grishchenkov et al. (2000); Seeger et al. (2001); Taguchi et al. (2001); Petric et al. (2007); Kafilzadeh et al. (2011)
Coniophora puteana	Aromatic pollutants (petroleum), chlorinated compounds (pesticides)	Russell et al. (2011)
Corynebacterium spp.	Aliphatic and aromatic hydrocarbons	Mrozik et al. (2003)
Cunninghamella spp.	Aromatic hydrocarbons	Russell et al. (2011)
Cyberlindnera fabianii	Heavy metals, Cu(II), Ni(II), Co(II), Cd(II), Mg(II), Cr(VI)	Ksheminska et al. (2006); Wang and Chen (2006); Ksheminska et al. (2008); Bahafid et al. (2011); Saisubhashini et al. (2011); Bahafid et al. (2013)
Dehalococcoides ethenogenes strain 195	Halogenated hydrocarbons	Russell et al. (2011)
Dehalococcoides sp. strain CBDB1	Halogenated hydrocarbons	Russell et al. (2011)
Deinococcus radiodurans	Polycyclic aromatic hydrocarbons, toluene, ionic mercury	Lovley (2003)
Desulfitobacterium chlororespirans	Chlorophenols, brominated compounds, bromoxynil, iodinated compounds	Cupples et al. (2005)
Desulfitobacterium hafniense strain Y51	Halogenated hydrocarbons	Russell et al. (2011)

(Contd.)

Microorganism	Target xenobiotics	References
Enterobacter spp.	Aromatic hydrocarbons	Mrozik *et al.* (2003)
Enterococcus spp.	Azo dyes	dos Santos *et al.* (2007); Hong *et al.* (2007); Lodato *et al.* (2007); Chaube *et al.* (2010)
Escherichia spp.	Aromatic hydrocarbons	Mrozik *et al.* (2003)
Exophiala spp.	Toluene	Prenafeta-Boldu *et al.* (2001)
Fibroporia vaillantii	Aromatic pollutants (petroleum), chlorinated compounds (pesticides)	Russell *et al.* (2011)
Fomitopsis pinicola	Aromatic pollutants (petroleum), chlorinated compounds (pesticides)	Russell *et al.* (2011)
Fusarium spp.	Aromatic hydrocarbon	Russell *et al.* (2011)
Geobacter metallireducens	Uranium	
Hansenula polymorpha	Heavy metals, Cu(II), Ni(II), Co(II), Cd(II), Mg(II), Cr(VI)	Ksheminska *et al.* (2006); Wang and Chen (2006); Ksheminska et al. (2008); Bahafid *et al.* (2011); Saisubhashini *et al.* (2011) Bahafid *et al.* (2013)
Janibacter spp.	Lower and higher chlorinated biphenyls	Floodgate (1984); Grishchenkov *et al.* (2000); Seeger *et al.* (2001); Taguchi *et al.* (2001); Petric *et al.* (2007); Kafilzadeh *et al.* (2011)
Klebsiella spp.	Aromatic hydrocarbons	Mrozik *et al.* (2003)
Leptodontium spp.	Toluene	Francesc *et al.* (2001)
Luteibacter spp.	Polycyclic aromatic hydrocarbons, polychlorinated biphenyls	Hontzeas *et al.* (2004); Leigh *et al.* (2006); Ma *et al.* (2010)
Microbacterium spp.	Lower and higher chlorinated biphenyls	Floodgate (1984); Grishchenkov *et al.* (2000); Seeger *et al.* (2001); Taguchi *et al.* (2001); Petric *et al.* (2007); Kafilzadeh *et al.* (2011)
Mucor racemosus	Dye degradation, gelatine emulsion	Bumpus (2004); Abruscia *et al.* (2007); Chaplain *et al.* (2011)
Mycobacterium spp.	Aliphatic and aromatic hydrocarbons	Mrozik *et al.* (2003)
Neosartorya spp.	Petroleum oil hydrocarbons	Russell *et al.* (2011)
Paenibacillus spp.	Lower and higher chlorinated biphenyls	Floodgate (1984); Grishchenkov *et al.* (2000); Seeger *et al.* (2001); Taguchi *et al.* (2001); Petric *et al.* (2007); Kafilzadeh *et al.* (2011)

(Contd.)

Microorganism	Target xenobiotics	References
Penicillium spp.	Crude oil hydrocarbons	Russell *et al.* (2011)
Penicillinum spp.	Aromatic hydrocarbons	Russell *et al.* (2011)
Penicillium chrysogenum	Dye degradation, gelatine emulsion	Bumpus (2004); Abruscia *et al.* (2007); Chaplain *et al.* (2011)
Penicillium spp.	Pesticides	Chaplain *et al.* (2011)
Pestalotiopsis spp.	Aromatic pollutants, chlorinated compounds	Russell *et al.* (2011)
Phaeolus schweinitzii	Aromatic pollutants (petroleum), chlorinated compounds (pesticides)	Russell *et al.* (2011)
Phoma glomerata	Dye degradation, gelatine emulsion	Bumpus (2004); Abruscia *et al.* (2007); Chaplain *et al.* (2011)
Pichia anomala	Heavy metals, Cu(II), Ni(II), Co(II), Cd(II), Mg(II), Cr(VI)	Ksheminska *et al.* (2006); Wang and Chen (2006); Ksheminska et al. (2008); Bahafid *et al.* (2011); Saisubhashini *et al.* (2011); Bahafid *et al.* (2013)
Pichia guilliermondii	Heavy metals, Cu(II), Ni(II), Co(II), Cd(II), Mg(II), Cr(VI)	Ksheminska *et al.* (2006); Wang and Chen (2006); Ksheminska et al. (2008); Bahafid *et al.* (2011); Saisubhashini *et al.* (2011); Bahafid *et al.* (2013)
Pleurotus ostreatus	Toxic components of aromatic pollutants (petroleum and diesel), chlorinated compounds (pesticides), Polyurethane	Thomas (2000); Russell *et al.* (2011)
Proteus spp.	Azo dyes	Dos Santos *et al.* (2007); Hong *et al.* (2007); Lodato *et al.* (2007); Chaube *et al.* (2010)
Providencia stuartii	Pesticides (atrazine, chlorpyrifos, dichlorodiphenyltrichloroethane)	Struthers *et al.* (1998); Surekha Rani *et al.* (2008); Kanade *et al.* (2012)
Pseudeurotium zonatum	Toluene	Francesc *et al.* (2001)

(Contd.)

Microorganism	Target xenobiotics	References
Pseudomonas spp.	Aromatic hydrocarbons, azo dyes, heavy metals, lower and higher chlorinated biphenyls, radionuclides, uranium (VI), thorium (IV), recalcitrant compounds (chloro-benzoates and alkyl-benzoates)	Floodgate (1984); Ramos *et al.* (1987); Rojo *et al.* (1987); Grishchenkov *et al.* (2000); Seeger *et al.* (2001); Taguchi *et al.* (2001); Mrozik *et al.* (2003); Pinaki *et al.* (2004); dos Santos *et al.* (2007); Hong *et al.* (2007); Lodato *et al.* (2007); Petric *et al.* (2007); Chaube *et al.* (2010); Glick (2010); Kafilzadeh *et al.* (2011)
Pseudomonas aeruginosa	Hg(II), As, Cd, Cu, Co, Zn	White *et al.* (1998); Takeuchi and Sugio (2006); Jaysankar *et al.* (2008)
Pseudomonas putida	Organic solvents, toluene, naphthalene, styrene, polystyrene foam	Marqués and Ramos (1993); Anzai *et al.* (2000); Gomes *et al.* (2005); Ward *et al.* (2006)
Ralstonia spp.	Lower and higher chlorinated biphenyls	Floodgate (1984); Grishchenkov *et al.* (2000); Seeger *et al.* (2001); Taguchi *et al.* (2001); Petric *et al.* (2007); Kafilzadeh *et al.* (2011)
Rhizopus arrhizus	Uranium, thorium	Treen-Sears *et al.* (1984); Pal *et al.* (2010)
Rhizopus arrhizus M1	Lead	Pal *et al.* (2010)
Rhodococcus spp.	Polycyclic aromatic hydrocarbons, polychlorinated biphenyls	Floodgate (1984); Grishchenkov *et al.* (2000); Seeger *et al.* (2001); Taguchi *et al.* (2001); Mrozik *et al.* (2003); Hontzeas *et al.* (2004); Leigh *et al.* (2006); Petric *et al.* (2007); Ma *et al.* (2010); Kafilzadeh *et al.* (2011)
Rhodotorula aurantiaca	Aliphatic hydrocarbons, crude oil, petroleum products, diesel oil	Bartha (1986); Ijah (1998); Fritsche and Hofrichter (2005); De Cássia Miranda *et al.* (2007); Mucha *et al.* (2010)
Rhodotorula pilimanae	Heavy metals, Cu(II), Ni(II), Co(II), Cd(II), Mg(II), Cr(VI)	Ksheminska *et al.* (2006); Wang and Chen (2006); Ksheminska et al. (2008); Bahafid *et al.* (2011); Saisubhashini *et al.* (2011); Bahafid *et al.* (2013)
Rhodoturula rubra	Aliphatic hydrocarbons, crude oil, petroleum products, diesel oil	Bartha (1986); Ijah (1998); Fritsche and Hofrichter (2005); De Cássia Miranda *et al.* (2007); Mucha *et al.* (2010)
Saccharomyces cerevisiae	Polycyclic aromatic hydrocarbons, biphenyls, dibenzofurans, nitroaromatics, pesticides, plasticizers, heavy metals, Cu(II), Ni(II), Co(II), Cd(II), Mg(II), Cr(VI)	Eaton (1985); Cabras *et al.* (1988); Sasek *et al.* (1993); Fritsche and Hofrichter (2000); Ksheminska *et al.* (2006); Wang and Chen (2006); Ksheminska *et al.* (2008); Bahafid *et al.*(2011); Saisubhashini *et al.* (2011); Bahafid *et al.* (2013)

(Contd.)

Microorganism	Target xenobiotics	References
Schizosaccharomyces pombe	Copper	Saisubhashini et al. (2011)
Serpula lacrymans	Aromatic pollutants (petroleum), chlorinated compounds (pesticides)	Russell et al. (2011)
Shewanella decolorationis	Azo dyes	dos Santos et al. (2007); Hong et al. (2007); Lodato et al. (2007); Chaube et al. (2010)
Shigella spp.	Aromatic hydrocarbons	Mrozik et al. (2003)
Sphingomonas spp.	Lower and higher chlorinated biphenyls	Floodgate (1984); Grishchenkov et al. (2000); Seeger et al. (2001); Taguchi et al. (2001); Petric et al. (2007); Kafilzadeh et al. (2011)
Staphylococcus spp.	Pesticides (atrazine, chlorpyrifos, dichlorodiphenyltrichloroethane); aromatic hydrocarbons	Struthers et al. (1998); Mrozik et al. (2003); Surekha Rani et al. (2008); Kanade et al. (2012)
Stenotrophomonas spp.	Pesticides (atrazine, chlorpyrifos, dichlorodiphenyltrichloroethane)	Struthers et al. (1998); Surekha Rani et al. (2008); Kanade et al. (2012)
Streptococcus spp.	Aromatic hydrocarbons	Mrozik et al. (2003)
Talaromyces spp.	Petroleum oil hydrocarbons	Russell et al. (2011)
Trichoderma longibrachiatum	Dye degradation, gelatine emulsion	Bumpus (2004); Abruscia et al. (2007); Chaplain et al. (2011)
Trichosporon cutaneum	Phenol	Mörtberg and Neujahr (1985)
Trichosporon pullulans	Aliphatic hydrocarbons, crude oil, petroleum products, diesel oil, pesticides	Bartha (1986); Ijah (1998); Fritsche and Hofrichter (2005); De Cássia Miranda et al. (2007); Mucha et al. (2010); Chaplain et al. (2011)
Wickerhamomyces anomalus	Heavy metals, Cu(II), Ni(II), Co(II), Cd(II), Mg(II), Cr(VI)	Ksheminska et al. (2006); Wang and Chen (2006); Ksheminska et al. (2008); Bahafid et al. (2011); Saisubhashini et al. (2011); Bahafid et al. (2013)
Williamsia spp.	Polycyclic aromatic hydrocarbons, polychlorinated biphenyls	Hontzeas et al. (2004); Leigh et al. (2006); Ma et al. (2010)
Yarrowiali polytica	Heavy metals, Cu(II), Ni(II), Co(II), Cd(II), Mg(II), Cr(VI)	Ksheminska et al. (2006); Wang and Chen (2006); Ksheminska et al. (2008); Bahafid et al. (2011); Saisubhashini et al. (2011); Bahafid et al. (2013)

(NMR), two-dimensional gel electrophoresis, matrix-assisted laser desorption ionization-time of flight (MALDI-TOF), liquid chromatography-mass spectrometry (LC-MS) and DNA chips have been deployed in analysing metabolic pathways at cellular level.

4. Microbial Engineering: Applications and Risks

For the purpose of improving degradation processes, intensive work has been made during last few decades in recombinant DNA technology. However, this work has been confined only to laboratory conditions. Only one field trial of microbial remediation of hazardous material has been successively implemented (Sayler *et al.* 1999). Recombinant DNA technology is involving genetic engineering strategies (such as gene cloning) in bringing about two genomic sequences together in one organism that would not otherwise be found in normal genome. Recombinant microbes can be obtained by genetic engineering procedures or by natural genetic exchange between two organisms. Applications of these genetically engineered microorganisms (GEMs) in cleansing environmental wastes have found tremendous scope, but unfortunately have mainly been confined to laboratory experiments. This is because of regulatory bodies, risk assessment concerns and uncertainty of their potential practical application in field conditions. There are at least four principle steps in application of genetically engineered microbes for bioremediation processes. These include: (1) alteration in enzyme specificity of an organism, (2) biodegradation pathway studies and regulation of pathways, (3) bioprocess development for larger applications, and, (4) analytical method validation for residue analysis and toxicity measurements.

Since inception, genetic engineering in remediation studies has gone long way and seen developments in past decades. Although many large scale field applications have been successfully completed, yet future of engineered microorganisms remains unclear. This is due to US Environmental Protection Agency's risk based regulatory approach towards hindering both research and applications of engineered products in bioremediation (Miller 1997). This may appear true, but the problem is further associated with the risk of other issues linked with engineered products and cost competitiveness with that of other technical solutions in bioremediation. These aspects are true not only with engineered microorganisms, but even with the application of natural organisms as well (Kato and Davis 1996). Most of the companies operating remediation processes with narrow margin of profit making it undesirable to commercially involve engineered microbes in bioremediation process. Nonetheless, there happens to be continuous research in understanding molecular and genetic basis of degradative mechanism in biodegradation. There is a general perception that

US is gaining leadership in this area since many of the Organisation for Economic Co-operation and Development (OECD) nations have greater public sensitivity to genetically engineered organisms than US.

In spite of several impressions, interesting genetic systems have been and can be exploited for the developments of useful application in bioremediation (Sayler *et al*. 1988; Menn *et al*. 2000). By using genetic engineering strategies recently, Timmis and Pieper (1999) defined a number of opportunities to enhance the degradation performance. For example, rate-limiting steps in known metabolic pathways can be genetically manipulated to yield increased degradation rates, or completely new metabolic pathways can be incorporated into bacterial strains for the degradation of previously recalcitrant compounds. In addition, other strategies using engineered microorganisms for process monitoring and control, toxicity and stress response assessment, and endpoint analysis in bioremediation have also been summarized (Menn *et al*. 2000).

Many reports were recorded regarding the degradation of environmental pollutants by implying different bacterial strains. The use of GEMs has been applied to bioremediation process monitoring of PCB by *A. eutrophus* H850Lr (Van Dyke *et al*. 1996); naphthalene and anthracene by *P. fluorescens* HK44 (Sayler *et al*. 1999). For stress response purpose *P. fluorescens* 10586s/pUCD607 have been used in biodegradation study of BTEX by Sousa *et al*. (1998). Development of efficient and cost-effective bioremediation processes is the goal for environmental biotechnology. For *in situ* bioremediation GEMs were utilized, in which the association of microbiological, ecological, biochemical mechanisms, and field engineering were designed to be essential components. Extensive information is available on the biochemical pathway analysis, operon structure, and molecular biology of biodegradative pathways important in bioremediation. Much of this information was found to be critical; hence the development of appropriate GEMs was restricted to aerobic catabolic and co-metabolic pathways (Sayler *et al*. 1998).

An impediment to actualizing field release is securing required governmental clearances, which is often a difficult and lengthy endeavour. Although necessary to ensure environmental and public health safety, the process often leads to an overall aversion to GEM implementation in environmental systems, with researchers concentrating rather on the optimization and commercial development of naturally occurring (intrinsic) microbial degradation (Cha *et al*. 1999). This unfortunately prevents the integration of state-of-the-art engineered microbes into field release studies.

Toxic Substances Control Act (TSCA) of Environmental Protection Agency in the United States regulates the production and application of GEMs (Day 1993; Drobnik 1999). The risk assessment process used by TSCA involves ten official forms reviewing potential health and environmental effects to be detailed within a pre-manufacture notice (PMN). The University of Tennessee submitted a PMN (#P95-1601) in July 1995 signifying its intent to conduct field research tests with *P. fluorescens* HK44 for applications in bioremediation (Sayre 1997). University received an approval in March 1996 for the release of guidelines for PMN research design. This was because of species classification ambiguities, possibility of transfer of introduced genes to other organisms (Davidson 1999), the co-release of an introduced antibiotic resistance marker (Poy 1997; Droge 1998), and uncertainties concerning the metabolism of PAHs to more toxic compounds (Leblond 2000). Further laboratory tests were conducted to fulfil those concerns; the field release was officially commenced on October 30, 1996 and terminated on December 31, 1999.

5. Engineering Microbes for Bioremediation Processes

In the early 1980s the development and innovations in genetic engineering techniques and intensive research on metabolic potential of microorganisms allowed to design GEMs. Currently, they are extensively used in several sectors such as medical, food and agriculture and in bioremediation. The construction of GEMs, that can degrade complex toxic compounds, is possible for the reason that many enzymes, metabolic pathways and their regulating genes are known and biochemical mechanisms are well understood. This provides a deeper insight and gives ample opportunity to design GEMs with new catabolic (degradative) pathways. There are several reports on the degradation of toxic pollutants by microorganisms. Development of effective and cost-efficient bioremediation process is a real challenge till date. During the past two decades, with the advent of recombinant DNA (rDNA) technology, biodegradation of toxic pollutants has grabbed a limelight. rDNA technology has been intensively studied and applied in *in situ* bioremediation. Although, bioremediation shows potential applicability, our inadequate understanding of biological involvement to the consequence of bioremediation and its impact on the ecosystem has been an obstruction to make the technology further consistent and safer. The combination of microbiological and ecological knowledge, biochemical pathways and optimization and validation of bioremediation process are crucial elements for successful *in situ* bioremediation using GEMs. GEMs can be a cheaper and more reliable remediation process. There is an extensive array of potentialities for genetic engineering of bacteria for bioremediation purposes. That includes four principal approaches: (1) modifying enzyme specificity, (2) construction of a new catabolic pathway and its regulation, (3) incorporation of marker gene for detection of recombinant in polluted site, and (4) bioaffinity/bioreporter sensor applications for chemical sensing, toxicity reduction and end point analysis (Menn *et al.* 2008). Moreover, the advancements in modern molecular techniques such as metagenomics and genetic engineering have been much useful in selection and identification of genetically engineered microbes. Those are simple, safe, rapid and reliable. Furthermore, techniques such as *in situ* PCR (*in situ* polymerase chain reaction), DGGE (denaturing gradient gel electrophoresis), T-RFLP (terminal restriction fragment length polymorphism), ARDRA (amplified rDNA restriction analysis) have provided deeper insights in understanding the microbial community and catabolic pathways (Sayler and Ripp 2000; Filonov *et al.* 2005; Jussila *et al.* 2007; Massa *et al.* 2009; Viebahn *et al.* 2009; Wasilkowski *et al.* 2012; Joutey *et al.* 2013).

The following indigenous microorganisms proved to be the potential source for bioremediation, bacterial genera, namely, *Acinetobacter*, *Aeromicrobium*, *Alcaligenes*, *Arthrobacter*, *Bacillus*, *Burkholderia*, *Brevibacterium*, *Dietzia*, *Flavobacterium*, *Gordonia*, *Mycobacterium*, *Pseudomonas*, *Rhodococcus* and *Sphingomonas*. Bacteria, especially from genus *Pseudomonas* sp., have been extensively studied in the bioremediation processes. Fungi such as *Amorphoteca*, *Aspergillus*, *Candida*, *Cephalosporium*, *Graphium*, *Neosartorya*, *Pencillium*, *Pichia*, *Rhodotorula*, *Yarrowia* and *Talaromyces*; algae such as *Agmenellum*, *Amphora*, *Anabaena*, *Aphanocapsa*, *Chlorella*, *Chlamydomonas*, *Coccochloris*, *Cylindrotheca*, *Dunaliella*, *Fucus*, *Microcoleus*, *Nostoc*, *Oscillatoria*, *Petalonia*, *Porphyridium* and *Ulva* have extensively been studies for their bioremediation potential. Recent developments and advancements in molecular biology and rDNA technology has seen an enormous growth in selecting the novel strains among the diverse microbial population with desired properties for the bioremediation. GEMs and intrinsic microorganisms have been extensively utilized for bioremediation of polluted environment. The survival and specific activity of the introduced microbes will aid to the success of GEMs. Moreover the success rate depends on the biotic and abiotic factors acting upon them (Jonathan *et al.* 2003; Dixit *et al.* 2015).

The success of bioremediation may be attributed to the incidents that happened in Exxon Valdez Oil Spill of 1989 in Prince William Sound and the Gulf of Alaska that accounted for the huge interest in bioremediation technology. Williams and Murray (1974) described the first catabolic plasmid TOL (117 bp) that was isolated from *Pseudomonas putida* mt-2. In 1974, for the first time Chakrabarty, an Indo-American microbiologist patented two GEMs, viz. *Pseudomonas aeruginosa* (NRRL B-5472) and *Pseudomonas putida* (NRRL B-5473) that contained catabolic genes (plasmids) for naphthalene, salicylate and camphor degradation. *Pseudomonas fluorescens* strain HK44 is a genetically engineered bacterium that has been successfully experimented and performance was evaluated in a field release (lysimeter) for the first time by Sayler *et al.* (1999). Their aim was to evaluate the feasibility of GEMs in degrading the naphthalene with continuous observation of the bioremediation process. This dual property of degradation of naphthalene and luminescent signal was useful for risk assessment (Menn *et al.* 2008).

Every bacterium carries a specific catabolic gene in the form of plasmid that in turn is responsible for degradation. These plasmids are categorized into four types: OCT plasmid that degrades octane, hexane and decane; XYL plasmid that degrades xylene and toluenes; CAM plasmid degrading camphor; and, NAH plasmid which degrades naphthalene (Joutey *et al.* 2013). In GEMs, multiple plasmids are incorporated for multitasking so that a variety of environmental pollutants can be degraded in a given time. For instance, Friello *et al.* (1976) designed *Pseudomonas* strain with multiple plasmids capable of oxidizing aliphatic, aromatic and other pollutants.

The following GEMs have higher degradative ability and have been designed and demonstrated successfully for the degradation of a range of pollutants including aromatic hydrocarbons and chlorinated compounds under controlled laboratory conditions. For branched aromatic hydrocarbons, *Pseudomonas* spp. (Ramos *et al.* 1987); for chlorobenzoates, *Pseudomonas* sp. B13, *Pseudomonas aeruginosa* AC869 (pAC31), *Pseudomonas* sp. US1 ex. (Chatterjee and Chakrabarty 1983; Rojo *et al.* 1987; Sahasrabudhe and Modi 1991); for polychlorinated biphenyls (PCB) and chlorobiphenyls, *Ralstonia eutropha* AE707/AE1216, *Pseudomonas* sp. – hybrid strains, *Pseudomonas cepacia* JHR22, *Pseudomonas acidovorans* M3GY, *Pseudomonas putida* (pDA261), *Comamonas testosteroni* VP44(pE43)/VP44(pPC3), *Escherichia coli* JM109 (pSHF1003)/(pSHF1007), *Pseudomonas putida* IPL5 (Shields *et al.* 1985; Mokross *et al.* 1990; Havel and Reineke 1991; Mccullar *et al.* 1994; Kumamaru *et al.* 1998; Hrywna *et al.* 1999; Strong *et al.* 2000); for trichloroethylene (TCE) degradation *Escherichia coli* HB101/pMY402 and FM5/pKY287, *Pseudomonas pseudoalcaligenes* KF707-D2 and *Pseudomonas putida* KF715-D5, *Pseudomonas* sp. strain JR1A::*ipb* – hybrid strains, *Escherichia coli* JM109(pDTG601), *Pseudomonas putida* G786(pHG-2), *Pseudomonas putida* F1/pSMMO20, *Burkholderia cepacia* G4 5223-PR1, *Ralstonia eutropha* AEK301/pYK3021 (Wackett and Gibson 1988; Shields *et al.* 1989; Winter *et al.* 1989; Shields and Reagin 1992; Hirose *et al.* 1994; Jahng and Wood 1994; Wackett *et al.* 1994; Kim *et al.* 1996; Pflugmacher *et al.* 1996); and, for degradation of 2,4-dichlorophenoxyacetic acid (2,4-D), *Pseudomonas putida* PPO300(pRO101) and PPO301(pRO103), *Pseudomonas cepacia* RHJ1 (Don and Pemberton 1985; Harker *et al.* 1989; Haugland *et al.* 1990; Menn *et al.* 2008) have been designed and used for the degradation of various environmental pollutants.

For bioremediation of heavy metals, *E. coli* strain metalloregulatory protein ArsR, *E. coli* strain SpPCS, *Methylococcus capsulatus* CrR, *P. putida* strain Chromate reductase (ChrR), *Ralstonia eutropha* CH34, *Deinococcus radiodurans* merA, *E. coli* strain organomercurial lyase, *E. coli* JM109 Hg2+ transporter, *Pseudomonas* K-62 Organomercurial lyase, *Achromobacter* sp. AO22 mer, *P. fluorescens* 4F39 Phytochelatin synthase (PCS) (Brim *et al.* 2000; Valls *et al.* 2000; Lopez *et al.* 2002; Murtaza *et al.* 2002; Ackerley *et al.* 2004; Kostal *et al.* 2004; Zhao *et al.* 2005; Kiyono *et al.* 2006; Kang *et al.* 2007; Ng *et al.* 2009; Hasin *et al.* 2010) GEMs have been used.

In general, GEMs are useful in bioremediation of pollutants as well as removal of other toxicants from the surrounding environment including harmful substances from air, soil and water. Nevertheless, GEMs provide a cheaper alternative to physical and chemical remediation processes their release in to the open environment has raised concerns over issues of human and environmental health. Suitable detection technologies are required to monitor and trace the survival of GEMs among large numbers of diverse microbial populations (Errampalli *et al.* 1999; Ripp *et al.* 2000; Menn *et al.* 2008). Though there has been extensive research carried out to control GEMs by biological containment by the way of genetic marker systems such as use of green fluorescent protein (GFP) and suicidal genes. Health and environmental concerns and environment polices have limited the *in situ* application of GEMs. In future if these concerns are ruled out then the GEMs can be a true boon to environmental biotechnology.

6. Microbial Degradation of Recalcitrant Materials

Nitroaromatic compounds (NACs) are important industrial feedstocks because of their versatile nature and presence of nitro group. NACs have their presence in many industrial processes as precursors and additives. These are extensively used in the synthesis of plasticizers, pharmaceuticals, pesticides (parathion, dinoseb, dinitrocresole, nitrofen) and explosives [2,4,6-trinitrotoluene (TNT), hexahydro-1,3,5-trinitro-triazine (RDX), octahydro-1,3,5,7-tetranitro-1,3,5,7-tetrazine (HMX)] (Spain 1995). Conversely, the very property which makes NACs valuable industrial agents is its stability; however, long-term persistency and toxic potentials pose severe concerns when these are emitted out in to the environment. During the recent past, microbes capable of handling these NACs, once thought impossible to degrade have been isolated and identified.

Researchers have been exploring microbial diversity with respect to different geographical regions with the hope of isolating bacteria worthy of degrading NACs (Jain *et al.* 2005). Bacterial strain *Arthrobacter protophormiae* resistant to NAC compound o-nitrobenzoate (ONB), r-nitrophenol (PNP) and 4-nitrocatechol (NC) have been isolated from agriculture field sprayed with different pesticides. Degradation pathway for ONB has been summarized in detail by stain *Arthrobacter protophormiae* (Chauhan and Jain 2000). Strain *Burkholderia cepacia* has been isolated from soil samples and identified as resistant to PNP and NC with varied range of physical parameters (Prakash *et al.* 1996; Chauhan *et al.* 2000). The oxidative degradation of 4-nitrocatechol was found to occur by a novel pathway in this strain involving a reductive dehydroxylation step which is a well known reaction in anaerobic metabolism of aromatic compounds, but was reported to be operative aerobically in *B. cepacia* for the first time. Involvement of plasmid in PNP and NC metabolism in this strain was also established (Chauhan *et al.* 2000). The kinetics of biodegradation of PNP was studied in *A. protophormiae* and *B. cepacia*. These kinetic studies provided useful information which could be used in the designing of biological treatment plants (Bhushan *et al.* 2000).

Hydrocarbons are organic compounds constituting entirely of hydrogen and carbon. Aliphatic fractions of hydrocarbons are the major constituent of crude oil. These are characterised by the presence of straight chain, branched chain and cyclic chain carbon moieties. Aliphatic hydrocarbons can be easily degraded because of these moieties (Atlas 1991) and reports have showed hydrocarbon degrading bacterial population grow relatively good on these compounds (Lal and Khanna 1996; Mishra *et al.* 2001). There are reports addressing short chain hydrocarbon degradation (C16) (Bhattacharya *et al.* 2003) and long chain hydrocarbons (C44) (Tsao *et al.* 1998). Apart from these, bacterial species such as *Acinetobacter*, *Pseudomonas*, *Alcaligenes*, *Burkholderia*, *Arthrobacter*, *Flavobacterium*, *Bacillus*, etc. have been identified as potent degraders of aliphatic fractions of petroleum hydrocarbons (Watanabe 2001). Inclusion of surfactant producing bacteria is another strategy reported by Herman *et al.* (1997) which can overcome mass transfer limitations and dissolution rates of hydrocarbons, thus enhancing remediation potentials of the strains. Cycloalkanes are toxic to cellular structures and are more resistant to bacterial degradation (Uribe *et al.* 1990). However studies show that bio-surfactant producing bacteria such as *Pseudomonas citronellolis*, *Brevibacterium erythrogenes* and *Saccharomyces cerevisiae* have been successfully utilized against cycloalkanes (Yakimov et al. 1998; Hernandez et al. 2003).

Aromatic hydrocarbons are organic molecules consisting of benzene derivatives. These have benzene ring on their structure and are ubiquitous in nature (Dagley 1990). Being benzene derivatives, aromatic hydrocarbons have great stability and inertness which make them persist in environment for longer duration of time (Kastner 1991). At the same time, many bacterial species have adapted to use aromatic hydrocarbon as their source of energy (Gibson and Subranian 1984). Rendering to their hydrophobicity higher fractions of aromatic hydrocarbons are immiscible than lower aromatic hydrocarbons making them less available to biological uptake (Manilal and Alexander 1991). Watanabe *et al.* (2000) reported the presence of phylotypes associated with a subclass of proteobacteria in sites contaminated with crude oil. The bacteria affiliated to e-proteobacterial subgroup were reported to be associated with petroleum contaminated groundwater (Watanabe *et al.* 2000). Since petroleum hydrocarbons remain persistent under anaerobic condition, groundwater purity is at greater risk. Stapleton *et al.* (2000) demonstrated significant bioremediation potential of degradative bacteria from hydrocarbon-contaminated aquifers which had phylogenetic resemblance with common soil bacteria. This included members from the group *Pseudomonas, Ralstonia, Burkholderia, Sphingomonas, Flavobacterium* and *Bacillus.* Kniemeyer *et al.* (2003) also demonstrated anaerobic degradation of aromatic hydrocarbons by marine sulfate reducing bacteria that use sulfate as the electron acceptor.

Another class of aromatic fraction of hydrocarbon are polycyclic aromatic hydrocarbons (PAHs). These classes of hydrocarbons, among any other are of serious concern to human health because of their persistency and recalcitrance. These happen to be a specific class of petroleum hydrocarbons because of their pyrogenic nature and the complexity of the assemblages in which they occur. There has been a comprehensive research on biodegradation of these PAHs by microbial community (Kanaly and Harayama 2000). Many bacterial, algal and fungal species are known to resist PAHs and are potential degraders of these compounds. The most commonly reported bacterial species include *Acinetobacter calcoaceticus, Alcaligens denitrificans, Mycobacterium* sp., *Pseudomonas putida, Pseudomonas fluorescens, Pseudomonas vesicularis, Pseudomonas cepacia, Rhodococcus* sp., *Corynebacterium renale, Moraxella* sp., *Bacillus cereus, Beijerinckia* sp., *Micrococcus* sp., *Pseudomonas paucimobilis* and *Sphingomonas* sp. (Yakimov *et al.* 1998). Furthermore, there are intense reports on bacterial degradations of PAHs and degradation of these compounds having three benzene rings (Kanaly and Harayama 2000). However these reports only address degradation of low molecular weight PAHs, there are limited reports on degradation of high molecular weight PAHs with more number of benzene rings. In general, high molecular weight PAHs appear to be recalcitrant for most of the microflora studied; and persistency of these PAHs in environment increases with direct proportion to molecular weight of the compound.

Conclusion

Environment microflora, a rich species in environment sites, provides a huge repository which can be exploited for ecorestoration purposes. However, very little of this huge repository has been explored till date. Despite known values of microbial potentials, our understanding of their interaction with diverse substrates is lacking. This is because majority of the microorganisms are non-culturable at defined laboratory conditions. Exploring new molecular and genomic approaches and recombination strategies for bioremediation studies may have appreciating outcome in addressing environmental contamination by xenobiotic compounds.

References

Abruscia C, Marquinaa D, Del Amob A, Catalina F (2007) Biodegradation of cinematographic gelatin emulsion by bacteria and filamentous fungi using indirect impedance technique. Int. Biodeter. Biodegr. 60, 137-143.

Ackerley DF, Gonzalez CF, Keyhan M, Blake R, Matin A (2004) Mechanism of chromate reduction by the *Escherichia coli* protein, NfsA, and the role of different chromate reductases in minimizing oxidative stress during chromate reduction. Environ. Microbiol. 6, 851-860.

Anzai KH, Park JY, Wakabayashi H, Oyaizu H (2000) Phylogenetic affiliation of the pseudomonads based on 16S rRNA sequence. Int. J. Syst. Evol. Microbiol. 50(4), 1563-1589.

Atlas RJ (1991) Microbial hydrocarbon degradation-bioremediation of oil spills. Chem. Tech. Biotechnol. 52, 149-156.

Bahafid W, Sayel H, Tahri-Joutey N, El Ghachtouli N (2011) Removal mechanism of hexavalent chromium by a novel strain of *Pichia anomala* isolated from industrial effluents of Fez (Morocco). J. Env. Sci. Eng. 5, 980-991.

Bahafid W, Tahri-Joutey N, Sayel H, Iraqui-Houssaini M, El Ghachtouli N (2013) Chromium adsorption by three yeast strains isolated from sediments in Morocco. Geophys. J. Roy. Astron. Soc. 5, 422-429.

Bakonyi T, Derakhshifar I, Grabensteiner L, Nowotny N (2003) Development and evaluation of PCR assays for the detection of *Paenibacillus* larvae in honey samples: comparison with isolation and biochemical characterization. Appl. Environ. Microbiol. 69, 1504-1510.

Bartha R (1986) Biotechnology of petroleum pollutant degradation. Microbial Ecol. 12, 155-172.

Bennett JW (2010) An overview of the genus *Aspergillus*. In: *Aspergillus*: Molecular Biology and Genomics (Eds) Machida M and Gomi K, Caister Academic Press, Portland, USA, pp 1-17.

Bhattacharya D, Sarma PM, Krishnan S, Mishra S, Lal B (2003) Evaluation of the genetic diversity among some strains of *Pseudomonas citronellolis* isolated from oily sludge contaminated sites. Appl. Environ. Microbiol. 69, 1435-1441.

Bhushan B, Chauhan A, Samanta SK, Jain RK (2000) Kinetics of biodegradation of ¶-nitrophenol by different bacteria. Biochem. Biophys. Res. Commun. 274, 626-630.

Brim H, McFarlan SC, Fredrickson JK, Minton KW, Zhai M, Wackett LP, Daly MJ (2000) Engineering *Deinococcus radiodurans* for metal remediation in radioactive mixed waste environments. Nat. Biotechnol. 18, 85-90.

Bumpus JA (2004) Biodegradation of azo dyes by fungi. In: Fungal Biotechnology in Agricultural, Food and Environmental Applications (Ed) Arora DK, Marcel Dekker, New York, pp 457-480.

Cabras P, Meloni M, Pirisi FM, Farris GA, Fatichenti F (1988) Yeast and pesticide interaction during aerobic fermentation. App. Microb. Biotechnol. 29(2-3), 298-301.

Cha DK, Chiu PC, Kim SD, Chang JS (1999) A complete review of hazardous waste treatment technologies includes biological as well as chemical and physical treatment methodologies. Water Environ Res. 71, 870-885.

Chaillan F, Le Flèche A, Bury E, Phantavong YH, Grimont P, Saliot A, Oudot J (2004) Identification and biodegradation potential of tropical aerobic hydrocarbon degrading microorganisms. Res. Microb. 55(7), 587-595.

Chaplain V, Mamy L, Vieublei-Gonod L, Mougin C, Benoit P, Barriuso E, Neilieu S (2011) Fate of pesticides in soils: toward an integrated approach of influential factors. In: Pesticides in the Modern World - Risks and Benefits (Ed) Stoytcheva M, Intech, Rijeka, Croatia, pp 535-560.

Chatterjee DK, Chakrabarty AM (1983) Genetic homology between independently isolated chlorobenzoate-degradative plasmids. J. Bact. 153(1), 532-534.

Chaube P, Indurkar H, Moghe S (2010) Biodegradation and decolorisation of dye by mix consortia of bacteria and study of toxicity on *Phaseolus mungo* and *Triticum aestivum*. Asiatic J. Biotech. Res. 1, 45-56.

Chauhan A, Jain RK (2000) Degradation of o-nitrobenzoate via anthranilic acid (o-aminobenzoate) by *Arthrobacter protophormiae*: a plasmid-encoded new pathway. Biochem. Biophys. Res. Commun. 267, 236-244.

Chauhan A, Samanta SK, Jain RK (2000) Degradation of 4-nitrocatechol by *Burkholderia cepacia*: a plasmid-encoded novel pathway. J. Appl. Microbiol. 88, 764-772.

Cupples AM, Sanford RA, Sims GK (2005) Dehalogenation of bromoxynil (3,5-dibromo-4-hydroxybenzonitrile) and ioxynil (3,5-diiodino-4-hydroxybenzonitrile) by *Desulfitobacterium chlororespirans*. Appl. Env. Microbiol. 71(7), 3741-3746.

Dagley S (1990) Microbial metabolism of aromatic compounds. In: Chemical and Biochemical Fundamentals. University of Minnesota, USA, pp 483–503.

Davison J (1999) Genetic exchange between bacteria in the environment. Plasmid 42, 73-91.

Day SM (1993) US environmental regulations and policies - their impact on the commercial development of bioremediation. Trends Biotechnol. 11, 324-328.

De Cássia Miranda R, de Souza CS, de Barros Gomes E, Lovaglio RB, Lopes CE, de Fátima Vieira de Queiroz Sousa M (2007) Biodegradation of diesel oil by yeasts isolated from the vicinity of Suape Port in the State of Pernambuco - Brazil. Braz. Arch. Biol. Technol. 50(1), 147-152.

Dixit R, Wassiullah, Malviya D, Pandiyan K, Singh UB, Asha S, Shukla R, Singh BP, Rai JP, Sharma PK, Lade H, Paul D (2015) Bioremediation of heavy metals from soil and aquatic environment: an overview of principles and criteria of fundamental processes. Sustainability 7, 2189-2212.

Dmochewitz S, Ballschmiter K (1988) Microbial transformation of technical mixtures of polychlorinated biphenyls (PCB) by the fungus *Aspergillus niger*. Chemosphere 17, 111-121.

Don RH, Pemberton JM (1985) Genetic and physical map of the 2,4-dichlorophenoxyacetic acid degradative plasmid pJP4. J. Bacteriol. 161, 466-468.

Donnelly PK, Fletcher JS (1995) PCB metabolism by ectomycorrhizal fungi. Bull. Environ. Contam. Toxicol. 54, 507-513.

Dos Santos AB, Cervantes JF, Van Lier BJ (2007) Review paper on current technologies for decolourisation of textile wastewaters: perspectives for anaerobic biotechnology. Bioresour. Technol. 98, 2369-2385.

dos Santos VA, Yakimov MM, Timmis KN, Golyshin PN (2008) Genomic insights into oil biodegradation in marine systems. In: Microbial Biodegradation: Genomics and Molecular Biology (Ed) Díaz E, Caister Academic Press, UK, pp 269-296.

Drobnik J (1999) Genetically modified organisms (GMO) in bioremediation and legislation. Intl Biodet Biodegrad. 44, 3-6.

Droge M, Puhler A, Selbitschka W (1998) Horizontal gene transfer as a biosafety issue: a natural phenomenon of public concern. J. Biotechnol. 64, 75-90.

Eaton DC (1985) Mineralization of polychlorinated biphenyls by *Phanerochaete chrysosporium*: a ligninolytic fungus. Enzyme Microbial Technol. 7, 194-196.

Errampalli D, Leung K, Cassidy MB, Kostrzynska M, Blears M, Lee H, Trevors JT (1999) Applications of the green fluorescent protein as a molecular marker in environmental microorganisms. J. Microbiol. Methods 35, 187-199.

Filonov AE, Akhmetov LI, Puntus IF, ESikova TZ, Gafarov AB, Izmalkova TY, Sokolov SL, Kosheleva IA, Boronin AM (2005) The construction and monitoring of genetically tagged, plasmid-containing, naphthalene-degrading strains in soil. Microbiology 74(4), 526-532.

Floodgate G (1984) The fate of petroleum in marine ecosystems. In: Petroleum Microbiology (Ed) Atlas RM, Macmillan, New York, USA, pp 355-398.

Friello DA, Mylroie JR, Chakrabarty AM (1976) Use of genetically engineered multi-plasmid microorganisms for rapid degradation of fuel hydrocarbons. In: Proc. 3rd Intl. Biodegradation Symposium (Ed) Sharpley JM and Kaplan AM, Applied Science Publishers Ltd, London, pp 205-214.

Fritsche W, Hofrichter M (2000) Aerobic degradation by microorganisms. In: Biotechnology: Environmental Processes (Ed) Rehm H-J and Reed G, Wiley-VCH, Weinheim, pp 145-167.

Fritsche W, Hofrichter M (2005) Aerobic degradation of recalcitrant organic compounds by microorganisms. In: Environmental Biotechnology: Concepts and Applications (Ed) Jördening H-J and Winter J, Wiley-VCH, Weinheim, DOI: 10.1002/3527604 286.ch7

Gibson DT, Subranian V (1984) Microbial degradation of aromatic hydrocarbons. In: Microbial Degradation of Organic Compounds (Ed) Gibson DT, Marcel Dekker, New York, pp 181-252.

Glick BR (2010) Using soil bacteria to facilitate phytoremediation. Biotech. Adv. 28, 367-374.

Gomes NC, Kosheleva IA, Abraham WR, Smalla K (2005) Effects of the inoculant strain *Pseudomonas putida* KT2442 (pNF142) and of naphthalene contamination on the soil bacterial community. FEMS Microbiol. Ecol. 54(1), 21-33.

Greene EA, Kay JG, Jaber K, Stehmeier LG, Voordouw G (2000) Composition of soil microbial communities enriched on a mixture of aromatic hydrocarbons. Appl. Environ. Microbiol. 66, 5282-5289.

Grishchenkov VG, Townsend RT, McDonald TJ, Autenrieth RL, Bonner JS, Boronin AM (2000) Degradation of petroleum hydrocarbons by facultative anaerobic bacteria under aerobic and anaerobic conditions. Process Biochem. 35(9), 889-896.

Harker AR, Olsen RH, Seidler RJ (1989) Phenoxyacetic acid degradation by the 2,4-dichlorophenoxyacetic acid (TFD) pathway of plasmid pJP4: mapping and characterization of the TFD regulatory gene, tfdR. J. Bacteriol. 171, 314-320.

Hasin AA, Gurman SJ, Murphy LM, Perry A, Smith TJ, Gardiner PE (2010) Remediation of chromium (VI) by a methane-oxidizing bacterium. Environ. Sci. Technol. 44, 400-405.

Haugland RA, Sangodkar UMX, Chakrabarty AM (1990) Repeated sequences including RS1100 from *Pseudomonas cepacia* AC1100 function as IS elements. Mol. Gen. Genet. 220, 222-228.

Havel J, Reineke W (1991) Total degradation of various chlorobiphenyls by cocultures and *in vivo* constructed hybrid pseudomonads. FEMS Microbiol. Lett. 78, 163-170.

Herman DC, Zhang Y, Miller RM (1997) Rhamnolipid (biosurfactant) effect on cell aggregation and biodegradation of residue hexadecane under saturated flow conditions. Appl. Environ. Microbiol. 63, 3622-3627.

Hernandez LAR, Gieg LM, Suflita JM (2003) Biodegradation of an alicyclic hydrocarbon by sulphate-reducing enrichment from a gas condensate-contaminated aquifer. Appl. Environ. Microbiol. 69, 434-443.

Hirose J, Suyama A, Hayashida S, Furukada K (1994) Construction of hybrid biphenyl (bph) and toluene (tod) genes for functional analysis of aromatic ring dioxygenases. Gene 138, 27-33.

Hong Y, Xu M, Guo J, Xu Z, Chen X, Sun G (2007) Respiration and growth of *Shewanella decolorations* S12 with an azo compound as the sole electron acceptor. Appl. Environ. Microbiol. 73, 64-72.

Hontzeas N, Zoidakis J, Glick BR (2004) Expression and characterization of 1-aminocyclopropane-1-carboxylate deaminase from the rhizobacterium *Pseudomonas putida* UW4: a key enzyme in bacterial plant growth promotion. Biochim. Biophys. Acta 1703, 11-19.

Hrywna Y, Tsoi TV, Maltseva OV, Quensen JF III, Tiedje JM (1999) Construction and characterization of two recombinant bacteria that grow on ortho- and para-substituted chlorobiphenyls. Appl. Environ. Microbiol. 65, 2163-2169.

Ijah UJJ (1998) Studies on relative capabilities of bacterial and yeast isolates from tropical soil in degradating crude oil. Waste Manage. 18, 293.

Jahng D, Wood TK (1994) Trichloroethylene and chloroform degradation by a recombinant pseudomonad expressing soluble methane monooxygenase from *Methylosinus trichosporium* OB3b. Appl. Environ. Microbiol. 60, 2473-2482.

Jain RK, Kapur M, Labana S, Lal B, Sarma PM, Bhattacharya D, Thakur IS (2005) Microbial diversity: application of microorganisms for the biodegradation of xenobiotics. Curr. Sci. 89(1), 101-112.

Jaysankar D, Ramaiah N, Vardanyan L (2008) Detoxification of toxic heavy metals by marine bacteria highly resistant to mercury. Marine Biotechnol. 10(4), 471-477.

Jonathan DV, Ajay S, Owen PW (2003) Recent advances in petroleum microbiology. Microbiol. Mol. Biol. Rev. 67, 503-549.

Joutey NT, Bahafid W, Sayel H, Ghachtouli NE (2013) Biodegradation: involved microorganisms and genetically engineered microorganisms. In: Biodegradation - Life of Science (Ed) Chamy R, InTech, Rijeka, Croatia, pp 289-319.

Juck D, Charles T, Whyte LG, Greer CW (2000) Polyphasic community analysis of petroleum hydrocarbon contaminated soils from two northern Canadian communities. FEMS Microbiol. Ecol. 33, 241-249.

Jussila MM, Zhao J, Suominen L, Lindström K (2007) TOL plasmid transfer during bacterial conjugation *in vitro* and rhizoremediation of oil compounds *in vivo*. Environ. Pollut. 146(2), 510-524.

Kafilzadeh F, Sahragard P, Jamali H, Tahery Y (2011) Isolation and identification of hydrocarbons degrading bacteria in soil around Shiraz Refinery. Afr. J. Microbiol. Res. 4(19), 3084-3089.

Kanade SN, Adel AB, Khilare VC (2012) Malathion degradation by *Azospirillum lipoferum* Beijerinck. Sci. Res. Rep. 2(1), 94-103.

Kanaly RA, Harayama S (2000) Biodegradation of high molecular weight polycyclic aromatic hydrocarbons by bacteria. J. Bacteriol. 182, 2059-2067.

Kang SH, Singh S, Kim JY, Lee W, Mulchandani A, Chen W (2007) Bacteria metabolically engineered for enhanced phtochelatin production and cadmium accumulation. App. Environ. Microbiol. 73, 6317-6320.

Kastner M (1991) Degradation of aromatic and polyaromatic compounds. In: Biotechnology: The Science and the Business (Ed) Moses V and Cape RE, Harwood Academic Press, Switzerland, pp 274-304.

Kato K, Davis KL (1996) Current use of bioremediation for TCE cleanup: results of a survey. Remediation 3, 1-13.

Kerr, RP, Capone DG (1988) The effect of salinity on the microbial mineralization of two polycyclic aromatic hydrocarbons in estuarine sediments. Mar. Environ. Res. 26, 181-198.

Kim Y, Ayoubi P, Harker AR (1996) Constitutive expression of the cloned phenol hydroxylase gene(s) from *Alcaligenes eutrophus* JMP134 and concomitant trichloroethylene oxidation. Appl. Environ. Microbiol. 62, 3227-3233.

Kiyono M, Pan-Hou H (2006) Genetic engineering of bacteria for environmental remediation of mercury. J. Health Sci. 52, 199-204.

Kniemeyer O, Fischer T, Wilkes H, Glöckner FO, Widdel F (2003) Anaerobic degradation of ethylbenzene by a new type of marine sulfate-reducing bacterium. Appl. Environ. Microbiol. 69, 760-768.

Kostal JRY, Wu CH, Mulchandani A, Chen W (2004) Enhanced arsenic accumulation in engineered bacterial cells expressing ArsR. Appl. Environ. Microbiol. 70, 4582-4587.

Ksheminska H, Fedorovych D, Honchar T, Ivash M, Gonchar M (2008) Yeast tolerance to chromium depends on extracellular chromate reduction and Cr(III) chelation. Food Technol. Biotech. 46(4), 419-426.

Ksheminska HP, Honchar TM, Gayda GZ, Gonchar MV (2006) Extracellular chromate-reducing activity of the yeast cultures. Central Eur. J. Biol. 1(1), 137-149.

Kubicek CP, Bissett J, Druzhinina I, Kullnig-Gradinger C, Szakacs G (2003) Genetic and metabolic diversity of *Trichoderma*: a case study on South-East Asian isolates. Fungal Genet. Biol. 38, 310-319.

Kumamaru T, Suenaga H, Mitsuoka M, Watanabe M, Furukawa K (1998) Enhanced degradation of polychlorinated biphenyls by directed evolution of biphenyl dioxygenase. Nat. Biotechnol. 16, 663-666.

Lal B, Khanna S (1996) Mineralization of [14C] octacosane by *Acinetobacter calcoaceticus* S30. Can. J. Microbiol. 42, 1225-1231.

Leblond JD, Applegate BM, Menn FM, Schultz TW, Sayler GS (2000) Structure-toxicity assessment of metabolites on the aerobic bacterial transformation of substituted naphthalenes. Environ. Toxicol. Chem. 19, 1235-1246.

Leigh BM, Prouzova P, Mackova M, Macek T, Nagle DP, Kinuthia F, Mwangi JS (2006) Polychlorinated biphenyl (PCB)-degrading bacteria associated with trees in a PCB-contaminated site. Appl. Environ. Microbiol. 72(4), 2331-2342.

Lodato A, Alfieri F, Olivieri G, Donato D, Marzocchella A, Salatino A (2007) Azo-dye conversion by means of *Pseudomonas* sp. OX1. Enzyme Microbial Technol. 41, 646-652.

Lopez A, Lazaro N, Morales S, Margues AM (2002) Nickel biosorption by free and immobilized cells of *Pseudomonas fluorescens* 4F39: a comparative study. Water Air Soil Pollut. 135, 157-172.

Lovley DR (2003) Cleaning up with genomics: applying molecular biology to bioremediation. Nat. Rev. Microbiol. 1(1), 35-44.

Ma B, Chen HH, He Y, Xu JM (2010) Isolations and consortia of PAH-degrading bacteria from the rhizosphere of four crops in PAH-contaminated field. 19th World Congress of Soil Science, Soil Solutions for a Changing World, Brisbane, Australia, pp 63-66.

Manilal VB, Alexander M (1991) Factors affecting the microbial degradation of phenanthrene in soil. Appl. Microbiol. Biotechnol. 35, 401-405.

Mansuy D (2013) Metabolism of xenobiotics: beneficial and adverse effects. Biologie Aujourd'hui 207(1), 33-37.

Margesin R, Labbe D, Schinner F, Greer CW, Whyte LG (2003) Characterization of hydrocarbon degrading microbial population in contaminated and pristine alpine soils. Appl. Environ. Microbiol. 69, 3985-3092.

Margesin R, Schinner F (1999) Biological decontamination of oil spills in cold environments. J. Chem. Technol. Biotechnol. 74, 1-9.

Marqués S, Ramos JL (1993) Transcriptional control of the *Pseudomonas putida* TOL plasmid catabolic pathways. Mol. Microbiol. 9(5), 923-929.

Masayuki M, Katsuya G (2010) *Aspergillus*: Molecular Biology and Genomics. Caister Academic Press, UK.

Massa V, Infantin OA, Radice F, Orlandi V, Tavecchio F, Giudici R, Conti F, Urbini G, Di Guardo A, Barbieri P (2009) Efficiency of natural and engineered bacterial strains in the degradation of 4-chlorobenzoic acid in soil slurry. Int. Biodeterior. Biodegrad. 63(1), 112-115.

McCullar MV, Brenner V, Adams RH, Focht DD (1994) Construction of a novel PCB-degrading bacterium: utilization of 3,4'-dichlorobiphenyl by *Pseudomonas acidovorans* M3GY. Appl. Environ. Microbiol. 60, 3833-3839.

Menn FM, Easter JP, Sayler GS (2000) Field application of genetically engineered microorganisms for bioremediation process. Curr. Opin. Biotech. 11(3), 286-289.

Menn FM, Easter JP, Sayler GS (2008) Genetically engineered microorganisms and bioremediation. In: Biotechnology Set, Second Edn (Ed) Rehm HJ and Reed G, Wiley-VCH, Weinheim, Germany, DOI: 10.1002/9783527620951.ch21

Miller H (1997) The EPA's war on bioremediation. Nat. Biotechnol. 15, 486.

Mishra S, Jyot J, Kuhad RC, Lal B (2001) Evaluation of inoculum addition to stimulate *in situ* bioremediation of oily sludge contaminated soil. Appl. Environ. Microbiol. 67, 1675-1681.

Mokross H, Schmidt E, Reineke W (1990) Degradation of 3-chlorobiphenyl by *in vivo* constructed hybrid pseudomonads. FEMS Microbiol. Lett. 71, 179-186.

Mörtberg M, Neujahr HY (1985) Uptake of phenol in *Trichosporon cutaneum*. J. Bacteriol. 161, 615-619.

Mrozik A, Piotrowska-Seget Z, Labuzek S (2003) Bacterial degradation and bioremediation of polycyclic aromatic hydrocarbons. Polish J. Environ. Stud. 12(1), 15-25.

Mucha K, Kwapisz E, Kucharska U, Okruszeki A (2010) Mechanism of aniline degradation by yeast strain *Candida methanosorbosa* BP-6. Polish J. Microbiol. 59(4), 311-315.

Murtaza I, Dutt A, Ali A (2002) Biomolecular engineering of *Escherichia coli* organomercurial lyase gene and its expression. Indian J. Biotech. 1, 117-120.

Ng SP, Davis B, Polombo EA, Bhave MA (2009) Tn5051-like mer-containing transposon identified in a heavy metal tolerant strain *Achromobacter* sp. AO22. BMC Res. Notes 7, 2-38.

Nielsen J (2001) Metabolic engineering. Appl. Microbiol. Biotechnol. 55, 263-283.

O'Loughlin EJ, Traina SJ, Sims GK (2000) Effects of sorption on the biodegradation of 2-methylpyridine in aqueous suspensions of reference clay minerals. Environ. Toxicol. Chem. 19, 2168-2174.

Pal TK, Bhattacharyya S, Basumajumdar A (2010) Cellular distribution of bioaccumulated toxic heavy metals in *Aspergillus niger* and *Rhizopus arrhizus*. Int. J. Pharm. Biosci. 1(2), 1-6.

Petric I, Hrak D, Fingler S, Vonina E, Cetkovic H, Kolar BA, Koli NU (2007) Enrichment and characterization of PCB-degrading bacteria as potential seed cultures for bioremediation of contaminated soil. Food Technol. Biotech. 45(1), 11-20.

Pflugmacher U, Averhoff B, Gottschalk G (1996) Cloning, sequencing, and expression of isopropylbenzene degradation genes from *Pseudomonas* sp. strain JR1: identification of isopropyl benzene dioxygenase that mediates trichloroethene oxidation. Appl. Environ. Microbiol. 62, 3967-3977.

Pinaki S, Kazy SK, D'Souza SF (2004) Radionuclide remediation using a bacterial biosorbent. Int. Biodeter. Biodegr. 54(2-3), 193-202.

Poy PH (1997) Dissemination of antibiotic resistance - genetic engineering at work among bacteria. Med. Sci. 13, 927-933.

Prakash D, Chauhan A, Jain RK (1996) Plasmid-encoded degradation of ¶-nitrophenol by *Pseudomonas cepacia*. Biochem. Biophys. Res. Commun. 224, 375-381.

Prenafeta-Boldu FZ, Kuhn A, Luykx DMAM, Anke H, van Groenestijn JW, de Bont JAM (2001) Isolation and characterisation of fungi growing on volatile aromatic hydrocarbons as their sole carbon and energy source. Mycol. Res. 105(4), 477-484.

Ramos JL, Wasserfallen A, Rose K, Timmis KN (1987) Redesigning metabolic routes: manipulation of TOL plasmid pathway for catabolism of alkylbenzoates. Science 235, 593-596.

Ramsing NB, Fossing H, Ferdelman TG, Anderson F, Thamdrup B (1996) Distribution of bacterial populations in a stratified fjord (Mariager Fjord, Denmark) quantified by *in situ* hybridization and related to chemical gradients in the water column. Appl. Environ. Microbiol. 62, 1391-1404.

Ripp S, Nivens DE, Ahn Y, Werner C, Jarrel J, Easter JP, Cox CD, Burlage RS, Sayler GS (2000) Controlled field release of a bioluminescent genetically engineered microorganism for bioremediation process monitoring and control. Environ. Sci. Technol. 34, 846-853.

Rojo F, Pieper DH, Engesser KH, Knackmuss H-J, Timmis KN (1987) Assemblage of ortho cleavage route for simultaneous degradation of chloro- and methylaromatics. Science 238, 1395-1398.

Rozgaj R (1994) Microbial degradation of xenobiotics in environment. Arh Hig Rada Toksikol. 45(2), 189-198.

Russell JR, Huang J, Anand P, Kucera K, Sandoval AG, Dantzler KW, Hickman D, Jee J, Kimovec FM, Koppstein D, Marks DH, Mittermiller PA, Núñez SJ, Santiago M, Townes MA, Vishnevetsky M, Williams NE, Vargas MPN, Boulanger L-N, Bascom-Slack C, Strobel SA (2011) Biodegradation of polyester polyurethane by endophytic fungi. Appl. Environ. Microbiol. 77(17), 6076-6084.

Sahasrabudhe AV, Modi VV (1991) Degradation of isomeric monochlorobenzoates and 2,4-dichlorophenoxyacetic acid by a constructed *Pseudomonas* sp. Appl. Microbiol. Biotechnol. 34(4), 556-557.

Saisubhashini S, Kaliappan S, Velan M (2011) Removal of heavy metal from aqueous solution using *Schizosaccharomyces pombe* in free and alginate immobilized cells. Int. Proc. Chem. Biol. Environ. Eng. 6, 108-111.

Sasek V, Volfova O, Erbanova P, Vyas BRM, Matucha M (1993) Degradation of PCBs by white rot fungi, methylotrophic and hydrocarbon utilizing yeasts and bacteria. Biotechnol Lett. 15, 521-526.

Sayler GS, Cox CD, Burlage R, Ripp S, Nivens DE, Werner C, Ahn Y, Matrubutham U (1999) Field application of a genetically engineered microorganism for polycyclic aromatic hydrocarbon bioremediation process monitoring and control. In: Novel Approaches for Bioremediation of Organic Pollution (Ed) Fass R, Flashner Y and Reuveny S, Kluwer Academic/Plenum Publishers, New York, pp 241-254.

Sayler GS, Matrubutham U, Menn FM, Johnston WH, Stapleton RD (1998) Molecular probes and biosensors in bioremediation and site assessment. In: Bioremediation: Principles and Practice, Vol 1 (Ed) Sikdar SK and Irvine RL, Technomic Publishing Company, Lancaster, PN, pp 385-434.

Sayler GS, Ripp S (2000) Field applications of genetically engineered microorganisms for bioremediation processes. Curr. Opin. Biotechnol. 11(3), 286-289.

Sayre P (1997) Risk assessment for a recombinant biosensor. In: Biotechnology in the Sustainable Environment (Ed) Sayler GS, Sanseverino J and Davis KL, Plenum Press, New York, pp 269-279.

Seeger M, Cámara B, Hofer B (2001) Dehalogenation, denitration, dehydroxylation, and angular attack on substituted biphenyls and related compounds by a biphenyl dioxygenase. J. Bacteriol. 183, 3548-3555.

Shields MS, Hooper SW, Sayler GS (1985) Plasmid-mediated mineralization of 4-chlorobiphenyl. J. Bacteriol. 163, 882-889.

Shields MS, Montgomery SO, Chapman PJ, Cuskey SM, Pritchard PH (1989) Appl. Environ. Microbiol. 55, 1624-1629.

Shields MS, Montgomery SO, Chapman PJ, Cuskey SM, Pritchard PH (1989) Appl. Environ.Microbiol. 55, 1624-1629.

Sims GK, O'Loughlin EJ (1989) Degradation of pyridines in the environment. CRC Crit. Rev. Environ. Control 19(4), 309-340.

Singh H (2006) Mycoremediation: Fungal Bioremediation. Wiley-Interscience, New York, USA.

Sousa S, Duffy C, Weitz H, Glover AL, Bar E, Henkler R, Killham K (1998) Use of a lux-modified bacterial biosensor to identify constraints to bioremediation of BTEX-contaminated sites. J. Environ. Toxicol. Chem. 17, 1039-1045.

Spain JC (1995) Biodegradation of nitroaromatic compounds. Annu. Rev. Microbiol. 49, 523-555.

Stapleton RD, Bright NG, Sayler GS (2000) Catabolic and genetic diversity of degradative bacteria from hydrocarbon contaminated aquifers. Microb. Ecol. 39, 211-221.

Steliga T (2012) Role of fungi in biodegradation of petroleum hydrocarbons in drill waste. Polish J. Environ. Stud. 21(2), 471-479.

Strong LC, McTavish H, Sadowsky MJ, Wackett LP (2000) Field-scale remediation of atrazine-contaminated soil using recombinant *Escherichia coli* expressing atrazine chlorohydrolase. Environ. Microbiol. 2(1), 91-98.

Struthers JK, Jayachandran K, Moorman TB (1998) Biodegradation of atrazine by *Agrobacterium radiobacter* J14a and use of this strain in bioremediation of contaminated soil. Appl. Environ. Microbiol. 64, 3368-3375.

Surekha Rani M, Lakshmi V, Suvarnalatha Devi KP, Jaya Madhuri R, Aruna S, Jyothi K, Narasimha G, Venkateswarlu K (2008) Isolation and characterization of a chlorpyrifos degrading bacterium from agricultural soil and its growth response. Afr. J. Microbiol. Res. 2, 26-31.

Taguchi K, Motoyama M, Kudo T (2001) PCB/biphenyl degradation gene cluster in *Rhodococcus rhodochrous* K37, is different from the well-known *bph* gene clusters in *Rhodococcus* sp. P6, RHA1, and TA42. RIKEN Rev. 42, 23-26.

Takeuchi F, Sugio T (2006) Volatilization and recovery of mercury from mercury-polluted soils and wastewaters using mercury-resistant *Acidithiobacillus ferrooxidans* strains SUG 2-2 and MON-1. Environ. Sci. 13(6), 305-316.

Teske A, Wawer C, Muyzer G, Namsing NB (1996) Distribution of sulfate-reducing bacteria in a stratified fjord (Mariager Fjord, Denmark) as evaluated by most-probable-number counts and denaturing gradient gel electrophoresis of PCR-amplified ribosomal DNA fragments. Appl. Environ. Microbiol. 62, 1405-1415.

Thomas SA (2000) Mushrooms: higher macrofungi to clean up the environment. In: Environmental Issues, Battelle, Ohio, USA.

Timmis KN, Pieper DH (1999) Bacteria designed for bioremediation. Trends Biotechnol. 17, 201-204.

Treen-Sears ME, Martin SM, Volesky B (1984) Propagation of *Rhizopus juvanicus* biosorbent. Appl Environ Microbiol. 448, 137-141.

Tsao CW, Song HG, Bartha R (1998) Metabolism of benzene, toluene, and xylene hydrocarbons in soil. Appl. Environ. Microbiol. 64, 4924-4929.

Uribe S, Rangel P, Espinola G, Aguirre G (1990) Effects of cyclohexane, an industrial solvent, on the yeast *Saccharomyces cerevisiae* and an isolated yeast mitochondria. Appl. Environ. Microbiol. 56, 2114-2119.

Valls M, Atrian S, de Lorenzo V, La F (2000) Engineering a mouse metallothionein on the cell surface of *Ralstonia eutropha* CH34 for immobilization of heavy metals in soil. Nat. Biotechnol. 18, 661-665.

Van Dyke MI, Lee H, Trevors JT (1996) Survival of lux-marked *Alcaligenes eutrophus* H850 in PCB-contaminated soil and sediment. J. Chem. Technol. Biotechnol. 65, 115-122.

Viebahn M, Smit E, Glandorf DCM, Wernars K, Bakker PAHM (2009) Effect of genetically modified bacteria on ecosystems and their potential benefits for bioremediation and their biocontrol of plant diseases - A review. Climate Change, Intercropping, Pest Control and Beneficial Microorganism (Ed) Lichtfouse E, Springer, New York, pp 45-69.

Wackett LP, Gibson DT (1998) Degradation of trichloroethylene by toluene dioxygenase in whole-cell studies with *Pseudomonas putida* F1. Appl. Environ. Microbiol. 54, 1703-1708.

Wackett LP, Sadowsky MJ, Newman LM, Hur H-G, Li S (1994) Metabolism of polyhalogenated compounds by a genetically engineered bacterium. Nature 368, 627-629.

Wang J, Chen C (2006) Biosorption of heavy metals by *Saccharomyces cerevisiae*: a review. Biotech. Adv. 24, 427-451.

Ward PG, Goff M, Donner M, Kaminsky W, O'Connor KE (2006) A two step chemo-biotechnological conversion of polystyrene to a biodegradable thermoplastic. Environ. Sci. Tech. 40(7), 2433-2437.

Wasilkowski D, Swêdzioe Z, Mrozik A (2012) The applicability of genetically modified microorganisms in bioremediation of contaminated environments. Chemik 66(8), 817-826.

Watanabe K (2001) Microorganisms relevant to bioremediation. Curr. Opin. Biotechnol. 12, 237-241.

Watanabe K, Futamata H, Harayama S (2002) Understanding the diversity in catabolic potential of microorganisms for the development of bioremediation strategies. Anton Leeuwenhoek 81, 655-663.

Watanabe K, Watanabe K, Komama Y, Syutsubu K, Harayama S (2000) Molecular characterization of bacterial population in petroleum contaminated groundwater discharged from underground crude oil storage cavities. Appl. Environ. Microbiol. 66, 4803-4809.

White C, Sharman AK, Gadd GM (1998) An integrated microbial process for the bioremediation of soil contaminated with toxic metals. Nat. Biotech. 16, 572-575.

Williams PA, Murray K (1974) Metabolism of benzoate and the methylbenzoates by *Pseudomonas putida* (arvilla) mt-2: evidence for the existence of a TOL plasmid. J. Bacteriol. 120(1), 416-423.

Winter RB, Yen M, Enley BD (1989) Efficient degradation of trichloroethylene by recombinant *E. coli*. Biotech. 7, 282-285.

Yakimov MM, Golyshin PN, Lang S, Moore ER, Abraham WR, Lunsdorf H, Timmis KN (1998) *Alcanivorax borkumensis* gen. nov., sp. nov., a new, hydrocarbon-degrading and surfactant-producing marine bacterium. Int. J. Syst. Bacteriol. 48, 339-348.

Yakimov MM, Timmis KN, Golyshin PN (2007) Obligate oil-degrading marine bacteria. Curr. Opin. Biotechnol. 18(3), 257-266.

Zhao XW, Zhou MH, Li QB, Lu YH, He N, Sun DH, Deng X (2005) Simultaneous mercury bioaccumulation and cell propagation by genetically engineered *Escherichia coli*. Process Biochem. 40, 1611-1616.

10

Ecological Restoration and Bioremediation of Canadian Mining Boreal Ecosystems

Nadeau, M.B., Quoreshi, A. and Khasa, D.P.

Abstract

Traditional practices of restoring and vegetating mine spoils, which include mechanical tillage, the addition of tons of organic and mineral inputs, seeding of herbaceous plants, and regular fertilization, are often expensive and not sustainable on the long term in the Canadian boreal forest biome. Phytobial restoration and remediation may be an excellent alternative to these traditional practices while accelerating ecosystem reconstruction and sustainable reforestation of mined sites. In this review, the role and importance of root symbionts – ectomycorrhizal fungi, PGPR, and *Frankia* – on boreal tree fitness are discussed. Furthermore, highlights of recent research studies done by our team in Canada on ecological restoration and bioremediation of boreal mined sites are presented.

Keywords: Boreal forest, Ecological restoration, Ectomycorrhizal fungi, *Frankia* spp., Mined sites, PGPR

1. Introduction

Natural resource exploitation is a very important industrial activity worldwide. It promotes economic growth and provides access to essential and fundamental resources needed for the wellbeing of human civilization. In Canada, mining and mineral processing industries employ more than 380,000 people and contributed 54 billion $ to the country's GDP in 2013 (Marshall 2014). Activities include exploitation of a wide range of metallic minerals (Au, Ag, Fe, Pb, Cu, Ni, Zn, Mo, Cd, etc.), non-metallic minerals (Potash, Diamonds, Sand, Gravel,

Cement, Stone, Salt), and others substances such as coal and oil. On the downside, these industrial activities often disturb and destroy natural ecosystems as observed on open mined sites where the soil becomes very rocky, poor, without organic matter and often contaminated by heavy metals (Renault *et al.* 2001). Today, mining companies have the responsibility to restore and revegetate mine tailings after exploitation, which is considered to be very challenging due to poor soil conditions in which not too many plant species can grow (Renault *et al.* 2001). In Canada, the mining acts of all provinces and territories require that all mining companies must perform, in accordance with the management and restoration plan approved by the provincial ministry, all work necessary for land reclamation as soon as the mining activities have been completed. Traditional practices currently used in Canada to restore mined sites and revegetate mine spoils come from the agricultural sector where the soils are prepared mechanically, amended with tons of organic inputs, hydro-seeded with grasses and/or legumes, and fertilized several years in order to improve soil conditions before outplanting trees (Warman 1988). The big problem with this technique is that grasses and legumes tend to compete strongly against tree seedlings. Furthermore, these practices are very expensive and often not sustainable in the boreal forest biome; therefore, it is very important for our society to find cheaper and ecological friendly methods for the rehabilitation of mined sites.

2. Principles of Ecological Restoration and Bioremediation

2.1 Ecological Restoration

Ecological restoration is the practice of renewing and restoring, within a short period of time, degraded, damaged, or destroyed natural ecosystems and habitats that were altered by industrial human intervention (Jackson *et al.* 1995). This practice is an intentional activity used to promote and accelerate the recovery of altered ecosystems with respect to its health, integrity, and sustainability (Jackson *et al.* 1995). It includes a wide range of projects such as erosion control, reforestation, use of genetically adapted local native species, removal of non-native species and weeds, revegetation of disturbed lands, reintroduction of native species, and habitat improvement for certain species (Jackson *et al.* 1995). Restoration activities are thought to be complementary to conservation efforts because it is assumed that environmental degradation are somewhat reversible processes where human intervention can be used to promote ecosystem recovery of habitat, biodiversity, and services (Choi 2007).

Ecological restoration involves two important ecological concepts: disturbance and succession. Disturbances, which are a change in environmental conditions, tend to alter species composition and functioning of natural ecosystems and often reduce available ecosystem services (human benefits supplied by natural ecosystems) (Jackson *et al.* 1995). Succession is the process by which species of a community changes over time (Jackson *et al.* 1995). Following a disturbance, an ecosystem generally progresses, over few generations, from simple organization of a very few species to a more complex community with many interdependent species (Jackson *et al.* 1995). Late-successional stages tend to generate a wider range of ecosystem services than early-successional stages (Jackson *et al.* 1995). In many ecosystems, communities tend to recover quite easily and quickly following mild to moderate natural and anthropogenic disturbances and restoration efforts are usually not needed in these situations (Jackson *et al.* 1995). However, an ecosystem that experiences more severe disturbances with physical, chemical and biological degradation of the environment often require intensive restoration efforts to recreate environmental conditions that favour ecosystem succession and recovery (Jackson *et al.* 1995). Like Choi (2007) mentioned, ecological restoration is essential for the survival and wellbeing of both humans and nature in the industrialized society we are living.

2.2 Bioremediation

One of the most important concepts in ecological restoration of ecosystems altered by mining is bioremediation. Tailings left after mineral extraction often contain high concentration of inorganic and organic contaminants that are known to cause soil degradation and to negate plant growth and survival. Bioremediation is the use of living organisms to manage and restore contaminated ecosystems (Wenzel 2009). These living organisms can restore contaminated soils by degrading, stabilizing, extracting or transforming polluting substances in soil (Wenzel 2009). According to the U.S. Environmental Protection Agency (EPA), bioremediation is the use of microbes (bacteria, yeast, fungi), algae and plants to break down or degrade toxic chemical compounds that have accumulated in the environment into less toxic or non-toxic substances.

Phytoremediation is one of the bioremediation technique, which can be applied for the treatment of contaminated soils, sludges, sediments and ground waters through contaminant removal, degradation or stabilization and which technically includes different operation procedures, such as phytostabilisation, phytodegradation, phytoextraction, and rhizofiltration (Saier and Trevors 2010). Phytostabilisation aims at establishing a plant cover by the use of tolerant plants that help to reduce the mobility and toxicity of organic or inorganic pollutants through adsorption or precipitation and, at the same time, may increase soil fertility and improve plant establishment (Wenzel 2009). Phytostabilisation reduces contaminant transfer to other ecosystem compartments and in the food chain (Wenzel 2009). Some plant species have the capacity to convert harmful substances into useful sources of nutrients or take up harmful organic contaminants and degrade/transform them into neutral ones (Saier and Trevors 2010). This phenomenon is called phytodegradation and the plants usually chelate the contaminants in the soil in inactive forms using secreted organic compounds (Wenzel 2009; Saier and Trevors 2010). Phytoextraction refers to the removal of contaminants in soil and the storage of these contaminants in plant tissues by species that have the ability to hyper-accumulate metals/metalloids in their shoots and roots (Wenzel 2009). These plants can sequester contaminants in their cell walls or vacuoles away from the sensitive cell cytoplasm where most metabolic processes occur (Saier and Trevors 2010). They can also store the pollutants in their tissues after transporting them into specialized cells and cell compartments (Saier and Trevors 2010). Phytoremediation provides a wide range of advantages over other mechanical and chemical restoration practices: the restoration costs are much cheaper, plants require little care, plants absorb CO_2 instead of producing it, plants do not require the consumption of a huge quantity of fossil fuels, and plants have the potential to yield a wide variety of commercially valuable products such as biomass and timbers (Saier and Trevors 2010; Vodouhe and Khasa 2015).

Only a small portion of plant species possess the tolerance and resistance characteristics needed to be used in phytoremediation (Marmiroli 2007). Genetic variation within a plant species can also play a very important role in the ability of plant to grow in harsh contaminated environments (Marmiroli 2007). Some genes in plants are known to control the resistance and tolerance of plants to contaminated soils (Marmiroli 2007). Therefore, some individuals and varieties within a species have better growth and survival success in polluted soils than others. Plant breeding and selection for heavy metal tolerance may be a promising field for the restoration of mined sites (Khasa et al. 2002). When we look at the restoration of ecosystems located in the boreal forest biome, not too many local native species of trees can be considered to be used in reforestation of mine tailings due to low species richness. The tolerance or resistance of boreal plant species to harsh soil conditions of mine tailings can be improved with the association between valuable tree species and their microsymbionts (mycorrhizal and/or nitrogen-fixing symbionts). Phytobial remediation is based on the use of plants and their associated microorganisms to remove, stabilise, or detoxify pollutants. Mycorrhizoremediation, is a phytobial remediation technology in which plants are associated with their mycorrhizal fungal symbionts to restore and revegetate disturbed lands (Khan 2006). Mycorrhizal fungi can immobilise metals/metalloids in soil by taking up the elements and accumulating them in their biomass via intracellular sequestration, by precipitating them, or by absorbing them onto their chitin cell walls (Wenzel 2009). They can also efficiently explore soil micro-pores that are not accessible to plant roots due to their relatively small size, impede soil contaminant transport through increased soil hydrophobicity, and protect plant roots from direct interaction with the pollutant via the formation of ectomycorrhizal sheath (Wenzel 2009).

3. Terrestrial Land Reclamation: Traditional Practices in Canada

In the past, most mine tailing restoration efforts have focused on agricultural practices such as fertilization, soil mechanical improvement, hydro-seeding of herbaceous species and use of organic amendments. Warman (1988) studied the success of different herbaceous species and fertilizers in the revegetation of lead-zinc mine tailings in Nova Scotia, Canada. In a potting experiment and in the field, the growth of 12 different grass and legume species alfalfa (*Medicago sativa*), buckwheat (*Fagopyrum esculentum*), couchgrass (*Agropyron repens*), ladino clover (*Trifolium repens*), meadow fescue (*Festuca elatior*), orchard grass (*Dactylis glomerata*), red clover (*Trifolium pratense*), reed canary grass (*Phalaris arundinacea*), annual ryegrass (*Lolium multiflorum*), sweet clover (*Melitotus offinalis*), timothy (*Phleum pratense*), and yellow foxtail (*Setaria glauca*)) on mine tailings was evaluated with or without fertilizers (Warman 1988). It was found that the use of fertilizers every year was essential (high level of N-P-K) for plant growth because none of the 12 species was able to mature without fertilization (Warman 1988). Hydro-seeding was very efficient with fertilizers and also a lot cheaper than transplanting herb seedlings (Warman 1988). Alfalfa, couchgrass, and red clover were the three most successfully introduced species capable of vegetating the tailing site (Warman 1988).

Furthermore, Renault *et al.* (2001; 2002) studied seed germination and seedling survival of different herbaceous species planted on gold mine tailings of central Manitoba mined sites. In greenhouse, seeds of Indian mustard (*Brassica juncea*), white mustard (*Sinapis alba*), slender wheatgrass (*Agropyron trachycaulum*), altai wildrye (*Elymus angustus*), reed canary grass, creeping foxtail (*Alopecurus arundinaceus*), streambank wheatgrass (*Agropyron riparium*), and tall fescue (*Festuca alatior*) were seeded on tailings without peat, mixed with peat, and mixed with peat and sand (1:1) (Renault *et al.* 2001; Renault *et al.* 2002). Indian mustard, white mustard, tall fescue seeds had the highest germination rates, but survival after three months was low (Renault *et al.* 2002). Addition of peat to the tailings increased significantly germination rates and greatly improved seedling survival rate (Renault *et al.* 2001). Tall fescue and reed canary grass were the two species that produced the highest biomass when grown on tailings with peat (Renault *et al.* 2001). In the field, a peat layer of 5 cm was added on the tailing surface and red-osier dogwood (*Cornus stononifera*), yellow willow (*Salix lutea*), white spruce (*Picea glauca*), jack pine (*Pinus banksiana*), tamarack (*Larix laricina*), and bog birch (*Betula glandulosa*) were planted on site (Renault *et al.* 2001). Seedling survival rate was relatively low in treatments without peat (Renault *et al.* 2002). Most tree seedlings were able to survive in treatments with peat (Renault *et al.* 2001). Tamarack and white spruce had the highest survival rate followed by bog birch and jack pine (Renault *et al.* 2002). In the field again, seeds of the same herbaceous species used in greenhouse experiment were hydro-planted on tailings with and without peat (Renault *et al.* 2002). Wheatgrass species had the highest survival rate followed by tall fescue and altai wildrye on all treatments, but in general it was quite low (Renault *et al.* 2002). Indian mustard and white mustard were able to survive on tailings with peat (Renault *et al.* 2001).

Giasson *et al.* (2006) evaluated the impact of arbuscular mycorrhizal fungi (AMF) on the extraction of different heavy metals (As, Cd, Zn, Se, Pb) in contaminated soil similar to the ones found on many mining sites. A grass mixture of *Festuca rubra* (35%), *F. eliator* (35%), *Agropyron repens* (25%), and *Trifolium repens* (5%) with five different treatments of arbuscular mycorrhizal fungi (*Rhizophagus irregularis, Funneliformis mosseae, Claroideoglomus etunicatum, Gigaspora gigantea*, and no AMF) was grown on heavy metal contaminated soil in greenhouse in order to identify the best AMF for metal extraction (Giasson *et al.* 2006). This type of grass mixture is currently used for vegetating mine tailings in Eastern Canada (Giasson *et al.* 2006). No fertilizers were added during the experiment. It was found that 30 to 70 % of grass roots in the mycorrhizal treatments were colonized (Giasson *et al.* 2006). The four species of AM fungi significantly increased plant survival, growth, and metal extraction (Giasson *et al.* 2006). There was variation in metal translocation to plants among AM fungi and plants inoculated with *R. irregularis* had the highest extraction (Giasson *et al.* 2006).

Soil organic matter plays a very important role in plant growth and nutrition. Microorganisms use it as a carbon source and food (Bot and Benites 2005). As they break down the organic matter through decomposition, they release excess nutrients (N, P, K, *etc.*) into soil in forms available to plants (Bot and Benites 2005). This process of decomposition is called mineralization and is considered the heart of ecosystem nutrient cycling (Bot and Benites 2005). Soil organic matter is composed of humic and non-humic substances. Five to 25 % of the organic matter is made of non-humic substances. These substances are nutritional carbohydrates that are easily decomposed by microorganisms (Van den Driessche 1991). They tend to promote aggregate formation and better soil structure because of their capacity to bind to inorganic soil particles (Bot and

Benites 2005). Humic substances, on the other hand, are decay resistant by-products of the decomposition of organic matter (Van den Driessche 1991). They have variable electromagnetic charges and their negative charges from the carboxyl groups (-COOH) ease the storage of positively charged nutrient cations, which help to retain nutrients available to plants at all-time improving fertilizer efficiency (Bot and Benites 2005; Van den Driessche 1991). These negative charges can also promote the storage of considerable quantities of positively charged toxic pollutants contributing to the reduction of plant toxicity caused by excess trace metals (Bot and Benites 2005; Van den Driessche 1991). Furthermore, the addition of organic matter is beneficial to plants because it tends to increase soil water retention and infiltration capacity, buffer soil pH, protect soil against erosion, reduce evaporation and negate plant desiccation from the soil surface, supply nutrients for plant uptake and soil biological activity, and ameliorate soil microbial diversity. The application of organic amendments is often utilized to improve soil organic matter content.

Many different organic amendments have been incorporated on mine tailings in order to improve soil conditions and increase success rate of restoration activities. Some organic and inorganic amendments that have been used in the past include sewage and paper sludge, compost (De Coninck and Karam 2008; Warman 1988), peat (Renault et al. 2001), lime (Reid et al. 2009), limestone, and sawdust (Renault et al. 2002). Organic amendments are known to improve plant survival on tailings through the incorporation of a layer of organic matter that can furnish important nutrients to plant, improve rooting medium, facilitate vegetation establishment, and ameliorate fertility, humidity and temperature of soil surface (De Coninck and Karam 2008). De Coninck and Karam (2008) tested the growth and survival of maize (*Zea mays*) on tailings, amended with compost (made of peat moss and shrimp wastes) and a chelating solution (EDTA), in the Gaspé Copper mine in Quebec. The addition of compost increased significantly both shoot and root biomass of maize (De Coninck and Karam 2008). Chelators also had a positive effect on plant biomass but not as much as the compost treatment (De Coninck and Karam 2008). The height of maize plant was lower in tailings treated with chelators alone compared to tailings amended with both compost and chelators (De Coninck and Karam 2008) suggesting that the use of both organic amendments together was beneficial to plant growth.

In the mining industry, it exists many different mine closure methods. Reid et al. (2009) have outlined the tailing site management and closure methods that were developed for a copper zinc underground mine located in Quebec (Canada). Two options are usually considered during mine operation: tailing may be sent to a disposal area or used for backfilling of the mine (Reid et al. 2009). The first option is generally chosen due to its cheaper cost (Reid et al.

2009). Three methods are currently used for the mine closure phase (Reid *et al.* 2009). Firstly, tailings are kept submerged in water to limit oxygen contact with heavy metals and sulphidic minerals. Secondly, the top one metre of tailings is desulphurized in order to decrease environmental contamination and the tailing surface is then stabilized with 30 cm of granular soil. It may be very expensive to add granular soil over the whole tailing area. In the last and more environmentally friendly method, the tailing disposal site is reclaimed using three layers of geological materials: a supporting layer of waste rock from the mine, a low permeability layer of silt, and a protective layer of organic soil and grass seeding in order to limit erosion. This method is the most interesting of the three, but it is very expensive to transport all the required substrates over the tailings. As a result, it is our responsibility to find cheaper phytotechnologies for the restoration of disturbed mining ecosystems.

4. Boreal Forest and Candidate Tree Species for Ecological Restoration of Mined Sites

The boreal forest is Canada's largest vegetation zone, which covers 55 % of the country's terrestrial area. Forest stands are mainly composed of balsam fir (*Abies balsamea*), white spruce (*Picea glauca*), and white birch (*Betula papyrifera*) on mesic sites (MRN 2012). Black spruce (*Picea mariana*), jack pine (*Pinus banksiana*) and tamarack (*Larix laricina*) are often found on less favourable sites with trembling aspen (*Populus tremuloides*) (MRN 2012). Forest dynamics are mainly controlled by forest fires and spruce budworm outbreak (MRN 2012).

White spruce is one of the most commercially valuable and planted tree species in Canada (Sutton 1973). It is used primarily for pulpwood and as lumber for general construction (Nienstaedt and Zasada 1990). White spruce has a broad native distribution across Canada from Newfoundland to Yukon and grows from sea level to about 1520 m of altitude (Sutton 1973; Nienstaedt and Zasada 1990; Farrar 1995). It has been described as a plastic species because of its ability to repopulate rocky areas at the end of glaciation (Nienstaedt and Zasada 1990). It grows under highly variable conditions, including extreme climates and soil conditions (Nienstaedt and Zasada 1990). Therefore, it has huge potential in ecological restoration of mine spoils in the boreal forest. White spruce grows on a wide variety of soils (glacial, lacustrine, marine, and alluvial origin) and rock formations (granites, gneisses, sedimentaries, slates, schists, shales, and conglomerates) (Nienstaedt and Zasada 1990). It can thrive on both acid and alkaline soils but tends to reach its best growth at pH between 4.7 and 7.0 (Nienstaedt and Zasada 1990). This species tolerates relatively well low fertility conditions and shade; thereby well suited to be planted with pionneer shade intolerant tree species such as poplars and alders (Nienstaedt and Zasada 1990). Only a few species of fungi forming ectomycorrhizas have been found on white spruce (Nienstaedt and Zasada 1990).

Jack pine is a medium-sized coniferous tree (Burns 1990). It is a commercially valuable species in Canada and an important source of pulpwood, lumber, and round timber (Farrar 1995). Jack pine is an early-successional species and invades quite easily areas where mineral soil has been exposed by major disturbances such as fires (Burns 1990). For this reason, it has a tremendous potential to be used in the revegetation of rocky mined sites. Jack pine is usually found on sandy soils of the spodosol and entisol soil orders (Burns 1990). It can also grow on loamy soils, on thin soils over granites and metamorphic rocks of the Canadian Shield, on limestones, and on peat moss (Burns 1990). Although it does not naturally grow on alkaline soil surface, it can grow on calcareous soils (pH ≈ 8.2) in association with alkaliphilic ectomycorrhizal fungi (Burns 1990). Optimal conditions for survival and seedling establishment are supplied by mineral soil and burned seedbeds where competition from other vegetation is not severe and water table is high (Burns 1990). Jack pine is one of the most shade-intolerant trees in its native range. Therefore, it may be better to outplant it in pure stand and avoid mixing it with other plant species. Severe drought may kill many seedlings particularly on coarse soils (Burns 1990).

Speckled alder (*Alnus incana* ssp. *rugosa*) is a tall shrub (3-8 m tall) found widely in the Canada's boreal forest (Farrar 1995; Karolewski *et al.* 2008). It usually grows along streams, rivers, and wetland; however, it does not necessarily require moist soil and can grow on dry shallow stony sites. It is shade intolerant and grows very quickly on poor soils. As a result, it produces a lot of biomass and contribute quickly to soil organic pool. On the other hand, green alder (*Alnus viridis* ssp. *crispa*) tends to grow on dry upland sites with jack pine (Farrar 1995; Karolewski *et al.* 2008). It is the most widespread tall shrub (3-12 m tall) found in the Canadian boreal forest (Farrar 1995; Karolewski *et al.* 2008). Green alder tolerates shade and drought better than any of the other alder species and grows well on poor soils (Karolewski *et al.* 2008). Green alder is known for colonizing avalanche chutes in mountains where larger trees that compete with it are killed regularly by avalanche damage. It survives the avalanches through its ability to regenerate from roots and broken stumps (Karolewski *et al.* 2008). Therefore, it may be a very good candidate for the restoration of open mined sites with steep slope. Both pioneer alder species are often used in the reforestation of non-fertile soils, which they enrich by means of nitrogen fixing *Frankia* found in root nodules (Quoreshi *et al.* 2007). The actinomycetes *Frankia* spp. obtain all sources of carbon and energy from the alder, giving fixed nitrogen in exchange (Roy *et al.* 2007). In exchange, alder trees acquire 70 to 100 % of its nitrogen from the *Frankia* symbionts (Roy *et al.* 2007). They have shallow root system and form vigorous stump and root suckers when cut or damaged (Karolewski *et al.* 2008). They can also form a symbiotic relationship with both ectomycorrhizal (ECM) and arbuscular mycorrhizal (AM) fungi leading to the development of a symbiosis between four different organisms: Alder-*Frankia*-ECM-AM (Roy *et al.* 2007). Alder species tend to develop a closer and more frequent association with ECM fungi compared to AM fungi (Roy *et al.* 2007). However, this association is only formed with a small number of ECM fungal species (high specificity, mainly species of the *Alpova* genus) (Quoreshi *et al.* 2007). These mycorrhizal fungi play an important role in the enhancement of alder nutrition (essentially P and N) (Roy *et al.* 2007). The use of alders has been a success story for the rehabilitation of many disturbed sites in the past due to its rapid growth, easy establishment, huge litter production, and excellent nitrogen fixation capacity that favour colonization of other plant species in the ecosystem (Pregent *et al.* 1987; Quoreshi *et al.* 2007; Roy *et al.* 2007). For these reasons, green alder and speckled alder are excellent candidates for the ecological restoration of mine spoils.

5. Ecology of Ectomycorrhizal Fungi and Their Role in Plant Nutrition

Ectomycorrhizal (ECM) fungi evolved 225 million years ago (Dahlberg 2001). It has been estimated that 7750 species of fungi (of the subdivisions *Basidiomycotina*, *Ascomycotina* and *Zygomycotina*) form ECM symbioses. However, based on estimates of knowns and unknowns macromycte diversity, a final estimate of ECM species richness would likely be between 20 000 and 25 000 (Rinaldi *et al.* 2008). ECM fungi have been well adapted to rocky materials for a very long time and have permitted the colonization of trees on harsh environments such as mountain bedrocks (Dahlberg 2001). Associated with bacteria, they have developed the ability to alter minerals in order to acquire essential elements (P, K, Ca, Mg, Fe, S, Mn, Mo, Cu, Zn, B and Cl) for their development (Dahlberg 2001). Elements such as P, Ca, and K tend to be rarely found under ionic form in soil (Dahlberg 2001). ECM fungi associated with bacteria are able to extract these important elements from rock particles and transform them into soluble forms improving the availability of these elements in soil for plant nutrition (Dahlberg 2001). Although ECM plant partners (phytobionts) represent only about 8 000 species or about 5% of vascular plants (mostly in the families *Pinaceae, Betulaceae, Fagaceae, Dipterocarpaceae, Salicaceae* and *Myrtaceae*), these species are of global importance because of their disproportionate occupancy and domination of terrestrial ecosystems in boreal, temperate and subtropical forests (Smith and Read 2008). The symbiosis Tree-ECM fungus is very complex. A single plant may be associated with many different ECM fungal species while a single fungus may connect with many plants at the same time (Allen *et al.* 2003). Most ECM fungal species can form a symbiotic association with a broad range of tree species (Dahlberg 2001). They may even form mycorrhizas with both angiosperm and gymnosperm species (Dahlberg 2001).

These ECM fungi do not colonize cortical root cells but rather form an intercellular interface, consisting of highly branched hyphae forming a latticework between epidermal and cortical root cells, known as the Hartig net (Dahlberg 2001). Furthermore, they form an extracellular layer of mycelium called fungal mantle (sheath), which surrounds the tips of fine roots (Dahlberg 2001). Transfer of materials between fungus and plant take place in the Hartig Net (Landeweert *et al.* 2001). The mantle protects fine roots from pathogens and all materials must go through the fungus first before reaching the root cells (Dahlberg 2001). The association plant-ECM is a mutualistic relationship because ECM fungi obtain most of its carbon from their host plant in exchange to soluble nutrients and minerals (Smith and Read 2008). ECM mycelia growing in soil are smaller than tree fine roots, so they can reach areas and micro-pores where roots cannot (Smith and Read 2008). Therefore, ECM fungi have access to a bigger water and nutrient pool (better absorbing surface) in soil than plants. When low P and N concentrations limit plant photosynthesis, there is an excess of carbon in plants (Smith and Read 2008). At this moment, fungal hyphae explore soil for both P and N and transport them into plant roots in exchange for excess plant C (energy as simple sugars and amino acids) (Smith and Read 2008). Photosynthetic rate depends mainly on N (for RuBP carboxylate), P (for ATP and ADP), Fe and Mg (for chlorophyll), water (to keep stomata open in order to fix CO_2), and internal CO_2 content (Smith and Read 2008). ECM fungi can create a C sink in plants by increasing P and N uptake which enhances photosynthesis (Smith and Read 2008). ECM fungi can also improve plant water uptake opening stomata in the process, which increases again photosynthesis (Smith and Read 2008). Some species form rhizomorphs (hyphal aggregates) that can transport water and nutrients more effectively to roots over long distances in soil (Lamhamedi and Fortin 1991; Allen *et al.* 2003). ECM fungi can reduce the negative effects of drought by providing water sources normally not accessible to plant roots (Allen *et al.* 2003).

Furthermore, ECM fungi can acquire K and P in saline soil and transfer them to plants to balance high levels of Na (Allen *et al.* 2003; Smith and Read 2008). They can also improve the Ca:Mg ratio in bedrock (Allen *et al.* 2003; Smith and Read 2008). This means that ECM fungi can render available to plants the most limited elements in soil, which may improve the overall plant growth and survival. It has been proven that roots colonized by ECM fungi are a lot more efficient in taking up P, N, Zn, Cu, Ni, S, Mn, B, Fe, Ca, and K in soil than roots without their symbionts, especially in low fertility soil (Allen *et al.* 2003; Smith and Read 2008). External hyphae contribute up to 25% of N, 80% of P, and 10% of K absorbed by plants (Allen *et al.* 2003; Smith and Read 2008). ECM fungi influence plant nutrient uptake through enzyme production and mineral alteration in order to mobilize nutrients from organic N and P, amino acids, peptides, proteins, amino sugars, chitin, and nucleic acids that are unavailable to plants (Smith and Read 2008). Most of the P content in forest soils is only found in organic forms such as phytates, nucleic acids, and phospholipids (Allen *et al.* 2003; Smith and Read 2008). ECM fungi produce a wide variety of enzymes called phosphatases that degrade and transform organic P into inorganic forms available to plants (Allen *et al.* 2003; Smith and Read 2008). When N sources are absorbed by the fungus, it is rarely transported into the same form to host plant (Allen *et al.* 2003; Smith and Read 2008). NO^{-3} is transported directly to plants, but NH_4^+ cannot be transported within the fungal tissue because it is toxic. For that reason, NH_4^+ is transformed into glutamine first and then transferred through fungal tissues to plant roots (Allen *et al.* 2003; Smith and Read 2008). Certain ECM species, such as *Thelephora americana* and *Pisolithus tinctorius*, are specialized in acquiring NH_4^+ by scavenging large areas in soil while others, such as *Hebeloma crustuliniforme* are adept at acquiring organic N (Allen *et al.* 2003). Proteins and amino acids containing N can also be taken up by ECM fungi, transformed into non-toxic forms and then finally transferred to the plant for nutrition (Dahlberg 2001; Read *et al.* 2004).

6. Role of Ectomycorrhizal Fungi in Weathering

Fungi have the ability to penetrate solid materials through physical and chemical mechanisms (Hoffland *et al.* 2004). Their hyphal tips tend to produce organic anions and protons when growing which help break down weak spots in solid rock particles (Hoffland *et al.* 2004). The production of organic anions has substantial carbon costs. Because mycorrhizal fungi receive high amount of carbon from the host plant, they are the most important living organisms involved in weathering (Hoffland *et al.* 2004). Fungal hyphae usually grow following scratches, ridges, and grooves, and penetrating cracks, pores, and tunnels that were formed in rocks during previous abiotic weathering (Hoffland *et al.* 2004). The expansion and contraction of fungal hyphae during wetting, drying, freezing, and thawing periods accelerate physical weathering of rocks (Hoffland *et al.* 2004).

Fungi produce two different groups of chemical weathering agents: proton-based and ligand-based agents (Hoffland *et al.* 2004). Proton-based agents (carbonic acids) are produced right alongside the fungal hyphal tips and ECM mantle (Hoffland *et al.* 2004). Ligand-based agents take into account organic anions, siderophores, and other polyphenolic acids (Hoffland *et al.* 2004). Low molecular mass organic acids (LMMOAs) belong to both groups and are the most important agents involved in biological weathering (Landeweert *et al.* 2001; Hoffland *et al.* 2004). They include oxalic, citric, and malic acids (Landeweert *et al.* 2001; Drever and Stillings 1997). These acids tend to bind easily to metal cations (such as Al^{3+}, Fe^{3+}, Ca^{2+}, Mg^{2+}, Mn^{2+}, Zn^{2+}, and Cu^{2+}) reducing free cation activity in soil solution which leads to a decrease in the saturation state and promotes further mineral dissolution and weathering (Landeweert *et al.* 2001; Drever and Stillings 1997). More than 50 % of the respiration in soil is done by mycorrhizal fungi (Landeweert *et al.* 2001). The dissolution of respiratory CO_2 in water generates carbonic acids, which leads to a decrease in pH and thus increases mineral solubilisation and weathering (Landeweert *et al.* 2001; Taylor *et al.* 2009). Uptake of ammonium by ECM fungi also leads to soil acidification due to excess cation uptake that causes an efflux of H^+ (Hoffland *et al.* 2004). Weathering caused by the decrease in pH increases cation exchange capacity and concentrations of soluble and exchangeable K^+, Ca^{2+}, and Mg^{2+} in soil improving plant nutrition (Hoffland *et al.* 2004). Weathering performance varies greatly among ECM fungal species because some species produce more organic acids than others and nutrient release has been positively correlated to oxalic acid concentration in soil (Cairney 1999). Therefore, ECM fungal species that produce a lot of oxalic acids may be more efficient in transferring nutrients to plant roots than others. Recent studies have shown that, by producing organic acids, ECM fungi associated with plants increase weathering and alter a wide range of minerals such as apatite, feldspath, biotite, and hornblende to obtain essential nutritional elements for their development (Wallander and Hagerberg 2004; Balogh-Brunstad *et al.* 2008; Berner *et al.* 2012; Smits *et al.* 2012).

7. Ectomycorrhizal Fungal Community Response to Heavy Metal in Soil

Weathering by ECM fungi often triggers the formation of secondary minerals (Hoffland *et al.* 2004). Secondary minerals include metal oxalates, iron and aluminum oxides, and carbonates (Hoffland *et al.* 2004). These minerals play a very important role in the immobilization of excess calcium and toxic heavy metals (Hoffland *et al.* 2004). As stated earlier, nutrients and water must go through the ECM fungus first before entering root cells (Dahlberg 2001). This phenomenon can be quite useful in soil that contains high toxicity of certain elements, because the ectomycorrhizal fungal mantle can protect plant roots against metal toxicity (Dahlberg 2001; Dixon and Buschena 1988). ECM fungi can absorb toxic elements in soil and sequester them in their cell walls or vacuoles, which reduces considerably the toxicity of heavy metals to plants (Khan 2006). However, heavy metals in high concentration can also decrease ECM fungal development (growth and survival) in soil. *Suillus luteus* growth on inoculated jack pine and white spruce seedlings was significantly reduced by the addition of heavy metals in soil (Khan 2006). ECM fungal formation on white spruce roots was almost completely eliminated with high levels of Pb, Ni, Cd, and Cu in soil (Khan 2006). ECM fungal formation on jack pine was significantly reduced by concentrations of Ni and Cd above 10 ppm (Khan 2006). Yet, root and shoot biomass of ECM-inoculated white spruce and jack pine seedlings was relatively greater than non-inoculated seedlings (Khan 2006). It is also important to not forget that heavy metal tolerance can vary greatly among ECM fungal species and strains (Khan 2006). For that reason, it is very important to identify ECM fungal species and strains that are adapted to soil conditions in which reforestation activities have to be carried out.

8. Ectomycorrhizal Diversity in Boreal Forest

The boreal forest is known to have very high ECM fungal species diversity. Buée *et al.* (2009) have identified between 600 and 1000 different uncultured fungi (between 249 and 408 taxonomic groups) in each of the forest soil samples they studied by pyrosequencing analyses. ECM fungal species from the orders Boletales, Agaricales, Thelephorales, Russulales, Cantharellales, and Sebacinales were the most abundant species in the analyzed boreal forest soil samples (Buée *et al.* 2009). Species from genera such as *Cortinarius, Tomentella, Thelephora, Russula, Lactarius, Clavulina, Descolea*, and *Laccaria* were present in all communities studied and they may be considered as generalists because of their capacity to colonize most boreal ecosystems (Buée *et al.* 2009). Gagné *et al.* (2006) found that ECM fungal species richness in different clearcut plantations in Alberta was considerably similar (11 to 13 species found on tree roots on each site). The community had relatively high diversity, but again the variability among sites was quite low because many species were found in all plantations (Gagné *et al.* 2006). Kernaghan *et al.* (2003) discovered that fertilizers may decrease considerably ECM fungal species diversity on seedling roots in nurseries. It may also decrease root colonization by ECM fungal species such as *Amphinema byssoides* that are known to improve seedling growth and health (Kernaghan *et al.* 2003). Mature forests contain usually both generalist and specialist species, while disturbed ecosystems are mainly dominated by generalists such as *Paxillus involutus* and *Laccaria lacata* (Villeneuve *et al.* 1989). For the ecological restoration of mine tailings, it may be very important to identify generalist ECM fungal species that are adapted to the disturbed site.

ECM fungi not only have high species diversity, but they also show high genetic diversity within species. This high genetic diversity is often displayed by intraspecific physiological variation among ECM fungal strains (Cairney 1999; Lamhamedi and Fortin 1991; Kernaghan *et al.* 2002). Different strains of the same species can have different temperature and pH optimums for growth (Cairney 1999). Furthermore, some strains may produce more mycelia and rhizomorphs than others (Cairney 1999). Lamhamedi and Fortin (1991) found that mycelial development (biomass) varied greatly among *Pisolithus* sp. strains. The authors suggested that mycelial production was under genetic control (Lamhamedi and Fortin 1991). ECM fungal strains of the same species may also exhibit variations in their ability to colonize host plant roots. It was demonstrated that some dikaryons of *Laccaria bicolor* have a better capacity to form Hartig net between root cortex cells and mantle sheaths around fine roots than others (Cairney 1999). It is believed that ECM fungal colonization is controlled by the production of hormones and the activity of enzymes, which varies greatly within ECM fungal species (Cairney 1999). Strains may vary in their ability to stimulate host plant growth (Cairney 1999). Furthermore, they may differ in their ability to improve availability of nutrients in soil and protect the host plant against water stress (Cairney 1999). Mycelial strands are the main channels for the translocation of water and nutrients to host plants (Lamhamedi and Fortin 1991). Nutrient and water uptake tends to be greater in ECM fungal strains that produce mycelial strands in large quantities (Lamhamedi and Fortin 1991). Furthermore, NH_4^+ absorption and organic acid production abilities vary significantly among ECM fungal strains (Cairney 1999). Therefore, some ECM fungal strains may be more efficient in altering minerals in soil than others because they can release higher content of organic acids. In addition, there is intraspecific variation in the sensitivity of ECM fungi to toxic metals such as Al, Zn, and Cd (Cairney 1999; Colpaert *et al.* 2000). Some strains are more adapted and tolerant to high concentration of toxic elements than others (Cairney 1999; Colpaert *et al.* 2000). For all these reasons, it is important to identify and use strains that are adapted to conditions of the restoration site in order to maximize inoculation benefits for plant growth and survival.

9. Role of Free-Living Soil Bacteria in Plant Growth and Nutrient

Soil bacteria have been used in agricultural practices for many decades and they have tremendously helped enhance plant productivity worldwide (Hayat *et al.* 2010). Free-living soil bacteria that are beneficial to plant growth are commonly referred to as plant growth promoting rhizobacteria (PGPRs) (Hayat *et al.* 2010). These PGPRs can promote plant growth by synthesizing particular compounds, such as hormones, beneficial to plants, easing nutrient uptake by plants, and preventing plant diseases (Chanway 1997; Hayat *et al.* 2010). Free-living N-fixing bacteria such as *Azotobacter* and *Azospirillum* species have shown the ability to increase yield of many crops (rice, cotton, wheat, etc.) (Hayat *et al.* 2010). It is thought that this increase in yield is associated with enhanced root development thanks to the increase in biological N_2 fixation and water and mineral uptake (Hayat *et al.* 2010). Other free-living diazotrophic bacteria sometimes capable of N_2 fixation include species of the genus *Acetobacter, Bulkholderia, Enterobacter, Citrobacter*, and *Pseudomonas* (Hayat *et al.* 2010). Phosphorous is one of the three most essential macronutrients for plant growth and development (Hayat *et al.* 2010). Bacteria from the genera *Pseudomonas, Bacillus, Rhizobium, Azotobacter, Burkholderia, Achromobacter, Agrobacterium, Microccocus, Aerobacter, Flavobacterium,* and *Erwinia* are known to have the capacity to solubilize insoluble inorganic phosphate (tricalcium phosphate, dicalcium phosphate, hydroxil apatite, and rock phosphate) and render it available to plants (Hayat *et al.* 2010). PGPR have been used in bioremediation in the past in order to remove complex contaminants and heavy metals from iron, copper, silver, and uranium mines (Hayat *et al.* 2010). Important bacterial genus used in bioremediation includes *Bacillus, Pseudomonas, Methanobacteria,* and *Deinococcus* (Hayat *et al.* 2010). PGPR involved in plant pathogen suppression belongs especially to the genus *Bacillus* and *Pseudomonas* (Hayat *et al.* 2010). These bacterial strains have the ability to produce antibiotic and induce systemic resistance to diseases in plant host (Hayat *et al.* 2010). For example, *Pseudomonas* strains suppress fungal pathogens by producing antifungal metabolites and by sequestering iron in the rhizosphere rendering it unavailable to other organisms (Hayat *et al.* 2010). For all these reasons, PGPR play a very important role in plant health, growth, and nutrition.

Weathering of soil minerals is a very important nutrient source for trees in boreal forest ecosystems (Calvaruso *et al.* 2006). Calvarulo *et al.* (2006) wanted to determine the contribution of tree roots and specific PGPR to mineral weathering. Scots pine seedlings (*Pinus sylvestris*) were grown with or without PGPR in a substrate containing quartz and biotite (Calvaruso *et al.* 2006). The authors found that pine root inoculation with *Bulkholderia glathei* significantly increase biotite weathering by 40% for Mg and 50% for K (Calvaruso *et al.* 2006). Furthermore, two of the three bacterial strains used in the experiment had a positive effect on pine growth and root size and this phenomenon was caused by improved plant nutrition (Calvaruso *et al.* 2006). Weathering rate of biotite was highly variable among *B. glathei* strains (Calvaruso *et al.* 2006). This study suggests that certain PGPR species have developed great abilities to weather soil mineral. Furthermore, some strains within the same species may be more effective in weathering and favouring plant growth and nutrition than others.

The effect of PGPR on coniferous tree growth in temperate forests has been fairly investigated in the last four decades. Many studies have demonstrated that tree seedling height and biomass were increased by the inoculation of *Arthrobacter spp.* and *Agrobacterium spp.* (for *Pinus sylvestris*), *Pseudomonas putida* (for *Pinus banksiana*), *Bacillus polymyxa* (for *Pinus contorta*), *Arthrobacter citreus* and *Pseudomonas fluorescens* (for *Picea mariana* and *Picea glauca*), *Pseudomonas putida*, *Hydrogenophaga pseudoflava*, *Bacillus polymyxa*, and *Staphylococcus hominis* (for *Picea glauca x engelmanii*) in laboratory, greenhouse and/or field trials (Chanway 1997). In these studies, inoculated pine and spruce biomass was increased, on average, from 32 % to 49 % after one growing season on reforested sites (Chanway 1997). Furthermore, certain PGPR strains (*Pseudomonas* and *Bacillus* species) have the ability to enhance considerably conifer root colonization by ECM fungi (Chanway 1997). The inoculation of many bacterial strains from the genus of *Bacillus* and *Paenibacillus* on Loblolly pine (*Pinus taeda*) and Slash pine (*Pinus elliotti*) had mixed effect on seedling growth and biomass under greenhouse conditions (Enebak *et al.* 1997). Some strains improved seedling growth and biomass while others did completely the opposite (Enebak *et al.* 1997). These findings suggest that the effects of PGPR inoculation on tree growth are species-specific. In another study, many *Bacillus* and *Paenibacillus* strains were found thriving in high abundance in root tissues of lodgepole pine (*Pinus contorta* var. *latifolia*) and western red cedar (*Thuja plicata*) (Bal *et al.* 2012). These two tree species are known for their ability to grow in soils severely limited in nitrogen sources (Bal *et al.* 2012). Three *Paenibacillus* strains were showing abilities to fix atmospheric nitrogen (Bal *et al.* 2012). Furthermore, in growth chamber experiments, inoculated pine seedlings were receiving 30 % to 66 % of their foliar N from bacterial N fixation (Bal and Chanway 2012). Similar results were obtained with diazotrophic PGPR thriving inside *Suillus tomentosus/Pinus contorta* tuberculate ectomycorrhizae (Paul *et al.* 2013). These findings suggest that some diazotrophic PGPR may favour coniferous tree growth and nutrition in N-poor soils through atmospheric N fixation.

10. Phytobial Restoration and Direct Reforestation of Mined Sites in the Boreal Forest

Phytobial restoration refers to the use of plants and their symbionts for the restoration, revegetation, and direct reforestation of degraded lands and ecosystems (Nadeau and Khasa 2015). Many studies have been conducted in Alberta, Canada, on the potential use of trees and their symbionts for the restoration of disturbed oil sand tailings. These oil sand tailings are formed after the extraction of the bitumen from the tar sand and are saline-alkaline, with high content of sodium, sulfate, and calcium (Kernaghan *et al.* 2002). Kernaghan *et al.* (2002) studied the growth of many different ECM fungal species native to Canadian boreal forest in water released by the tailings and in medium with different levels of alkalinity, under aseptic laboratory conditions, to identify promising fungi for the reclamation of saline-alkaline habitats. It was found that the ECM fungal species *Suillus brevipes, Rhizopogon rubescens, Paxillus involutus*, and *Amphinema byssoides* were very sensitive to alkaline soils and were not able to growth with the presence of water released by oil sand tailings (Kernaghan *et al.* 2002). On the other hand, different strains and species of *Laccaria* and *Hebeloma* in addition to *Wilcoxina mikolae* were more tolerant to alkaline treatments and had better overall growth in the presence of tailing-released water than other species (Kernaghan *et al.* 2002). *Laccaria* strains displayed the highest mycelial growth and the authors suggested that these species may be excellent candidates for use in rehabilitation of degraded saline-alkaline sites (Kernaghan *et al.* 2002). These *Laccaria* strains were collected from roots of white spruce, tamarack, Douglas-fir, jack pine, lodgepole pine, and balsam fir trees (Kernaghan *et al.* 2002). *Hebeloma* strains came from roots of white spruce, lodgepole pine, and Norway spruce trees (Kernaghan *et al.* 2002).

Bois *et al.* (2005) investigated the mycorrhizal status of pure reclamation materials and revegetated tailing sands from the Canadian oil sand industry. This study is the first crucial step in the use of microbial inoculants in the real world (Quoreshi 2008). The authors found that the composite tailing sands material was deprived of active mycorrhizal propagules while all other materials showed some level of inoculum potential (tailing sands, deep overburden, muskeg peat, three reclaimed sites) (Bois *et al.* 2005). Arbuscular mycorrhizal fungi were observed on roots of clover and poplar. Pine roots were also colonized by vesicle forming hyphae of an unidentified fine endophyte and by dark septate fungi (Bois *et al.* 2005). Ectomycorrhizas were observed on both pine and poplar (Bois *et al.* 2005). Using morphological and molecular analyses, six ectomycorrhizal (ECM) fungi were identified to the genus or species level: *Laccaria* sp., *Thelephora americana* Lloyd, *Wilcoxina* sp. (E-strain), *Tuber* sp. (I-type), a Sebacinoid, and a Pezizales species. Fungi of the genera *Laccaria* and *Wilcoxina* were the most frequently observed ECM species on the roots of jack pine and hybrid poplar (Bois *et al.* 2005). The researchers indicated that planting grass species such as barley many years before tree planting may favour the development of AM fungi over ECM fungi and explain the low ECM status on conifer species of the reclaimed sites (Bois *et al.* 2005). Because of poor status of ECM on disturbed sites, the study suggested the need to inoculate tree seedlings with suitable mycorrhizal fungi before outplanting in order to improve their survival and growth.

Furthermore, Bois *et al.* (2006a) compared *in vitro* growth development of two salt-resistant ECM fungal strains (*Laccaria bicolor* and *Hebeloma crustuliniforme*) identified by Kernaghan *et al.* (2003) with three ECM fungal species (*Suillus tomentosus, Hymenoscyphus* sp., and *Phialocephala* sp.) that were isolated from Syncrude's oil sand tailings in Alberta, Canada. This experiment was done on modified Melin-Nokrans medium (MMN) containing different concentrations of NaCl (Bois *et al.* 2006a). The authors found that the two Ascomycota species (*Hymenoscyphus* sp. and *Phialocephala* sp.) were more resistant to NaCl treatments than the other three basidiomycota species (Bois *et al.* 2006a). *L. bicolor* had the highest decrease in growth and biomass yield out of the three basidiomycota species under increasing NaCl concentrations (Bois *et al.* 2006a). *H. crustuliniforme* was the most resistant of the three basidiomycota species to water stress while *S. tomentosus* showed the best biomass yield under all NaCl treatments (Bois *et al.* 2006a). This study suggests that native strains of ECM fungi growing naturally on oil sand tailings may be better adapted and resistant to the extreme conditions and yield better growth on disturbed oil sand tailings than allochthonous species from natural forests.

In another research, Bois *et al.* (2006b) looked at the effects of ECM fungal inoculation on white spruce and jack pine seedlings grown in salt-affected oil sand tailings in greenhouse. The authors inoculated the seedlings with three different ECM fungal strains (*Hebeloma crustuliniforme, Laccaria bicolor*, and *Suillus tomentosus*) that were selected for their tolerance to high salinity in soil (Bois *et al.* 2006b). It was found that inoculation improved considerably the growth and decreased stress caused by salinity in both white spruce and jack pine seedlings (Bois *et al.* 2006b). The highest biomass production was encountered on white spruce seedlings inoculated by *S. tomentosus* (Bois *et al.* 2006b). These seedlings were also the least affected by saline treatments (Bois *et al.* 2006b). Thus, it was suggested that white spruce seedlings inoculated by *S. tomentosus* may be the most suited treatment for the reforestation of saline oil sand tailings in Alberta (Bois *et al.* 2006b). Once again, better growth results were obtained from ECM fungi that came directly from the disturbed sites. In treatments where salinity was not too high (< 200mM NaCl), high biomass production was also obtained with trees inoculated by *L. bicolor* (Bois *et al.* 2006b). Other results showed that white spruce seedlings inoculated with *H. crustuliniforme* were not affected negatively and biomass yield was not reduced significantly by salinity treatments compared to controls (Bois *et al.* 2006b). Sodium accumulation was higher in jack pine seedlings compared to white spruce indicating better salinity tolerance in white spruce (Bois *et al.* 2006b).

Onwuchekwa *et al.* (2014) examined the effect of white spruce and jack pine seedling inoculation with ECM fungi on their growth and survival on oil sand tailings in a field trial. Seedlings were inoculated with *Hebeloma crustuliniforme*, *Suillus tomentosus*, and/or *Laccaria bicolor* (Onwuchekwa *et al.* 2014). The inoculation increased the growth rate of both tree species after two growing seasons, but this increase was greater on jack pine (Onwuchekwa *et al.* 2014). White spruce and jack pine seedlings inoculated with *H. crustuliniforme* showed the greatest increase in height growth rate (60 % and 100 %, respectively) compared to controls (Onwuchekwa *et al.* 2014). Inoculation did not affect the survival of jack pine seedlings (Onwuchekwa *et al.* 2014). On the other hand, white spruce seedlings inoculated with *S. tomentosus* only and with treatments of mixed ECM fungal species (HC+ST, HC+LB, ST+LB, and HC+ST+LB) had a greater survival (up to 70 % increase) compared to controls (Onwuchekwa *et al.* 2014). These results demonstrated very well the importance of ectomycorrhizal association on white spruce and jack pine for seedling survival and growth on oil sand tailings. Further research was conducted on oil sand spoils of three other sites in Alberta (Nadeau and Khasa 2015). After four growing seasons, white spruce seedling productivity (plot volume index) was enhanced between 100 % and 115 % while jack pine productivity was improved between 45 % and 60 % on the three sites when inoculated with *H. crustuliniforme*, *L. bicolor*, *S. tomentosus*, or *Amphinema byssoides* (Nadeau and Khasa 2015). In addition, alder seedling volume growth was increased by 65 % and 76 % on oil sand spoils of two different sites in Alberta after two growing seasons when inoculated with N-fixing *Frankia* (Nadeau and Khasa 2015).

Lefrançois *et al.* (2010) evaluated the effect of growing *Frankia*-inoculated alders on soil quality of oil sand tailings in Alberta. Biomass and nitrogen content of alder seedlings were measured to assess the impact of *Frankia* inoculum on alder growth and health (Lefrançois *et al.* 2010). The plantation of *Frankia*-inoculated alders on the oil sand tailings improved considerably soil quality after two years through the increase of both organic matter content (leaf litter, six times greater) and cation exchange capacity in soil (Lefrançois *et al.* 2010). Furthermore, growth of alders with its *Frankia* symbiont led to a decrease in soil pH (Lefrançois *et al.* 2010); thereby, reducing by 70 % the negative effect of tailing salinity (Na^+) on plants. The process of decreasing pH is thought to be linked to the nitrification activity associated with nitrogen fixation by the actinomycetes *Frankia* spp. (Lefrançois *et al.* 2010). Improvement of soil conditions by alders and its *Frankia* symbiont demonstrated that this symbiotic interaction can create a better environment for plant growth in extreme site conditions of degraded ecosystems through nitrogen fixation and fast accumulation of organic matter (leaf litter) on soil surface. *Frankia*-inoculated alders also had a positive impact on native microbial community activity (Lefrançois *et al.* 2010). This study suggests that alders associated with its *Frankia* symbiont are able to grow and perform well without the addition of fertilizers in harsh and nutrient-limiting environment. Furthermore, Nadeau *et al.* (2016) showed that inoculating alders with specific ECM fungal strains can improve seedling health by up to 17.5 %, height by up to 100 %, and total dry biomass by up to 90 % when exposed to oil sand-like stressors (NaCl, Na_2SO_4, Naphthenic acids) under *in vivo* bipartite symbiotic conditions. Alders have a tremendous potential to be used for the restoration and rehabilitation of boreal degraded ecosystems.

In another project, Nadeau *et al.* (2014) studied the role and importance of soil microorganisms – PGPR and ECM fungi – in promoting the health, growth, and nutrition of white spruce on waste rocks and fine tailings of Sigma-Lamaque gold mine in Quebec, Canada. The project was divided into three studies. First, the community structure of ECM fungi associated with white spruce roots was analyzed on four different locations near the mining site (Nadeau *et al.* 2014). Results demonstrated that the mining site had a significantly different ECM fungal community compared to the surrounding forest edge, natural forest, and nursery ecosystems (Nadeau *et al.* 2014). Second, a laboratory experiment was performed for *in vitro* selection of promising ECM fungi that were ecologically adapted to ground waste rocks (Nadeau *et al.* 2014). Results demonstrated that the two ECM fungi isolated from roots of healthy white spruce on the mining site (*Cadophora finlandia* and *Tricholoma scalpturatum*) yielded 13-fold and five-fold greater growth (ergosterol content), respectively, on waste rocks than ECM fungus from the natural forest (*H. crustuliniforme*) (Nadeau *et al.* 2014). Third, a glasshouse experiment to grow white spruce seedlings on waste rocks and fine tailings was conducted and the performance of different treatments of ECM fungi and PGPR was evaluated (Nadeau *et al.* 2014). After 32 weeks of growth, inoculation with *C. finlandia* and *T. scalpturatum* significantly improved the health of seedlings by 130 % and 60 %, respectively, whereas *H. crustuliniforme* did not (Nadeau *et al.* 2014). *C. finlandia* enhanced seedling health by increasing foliar nitrogen (25 %) and root water (4 %) uptake and reducing foliar iron toxicity (50 %) (Nadeau *et al.* 2014). *Pseudomonas putida*, a PGPR isolated from the mining site, considerably enhanced by 25 % seedling aerial growth, whereas *Azotobacter chroococcum* and *Rhizobium radiobacter* did not (Nadeau *et al.* 2014). *P. putida* improved seedling growth by reducing foliar Ca toxicity (20 %) (Nadeau *et al.* 2014). These results suggest that site-adapted ECM fungi and PGPR play a very important role in the health, growth, and nutrition of white spruce on mine spoils (Nadeau *et al.* 2014). However, *in vitro* and *in vivo* selection of microorganisms is fundamental for determining the most promising and efficient strains. Quoreshi

(2008) has proposed a valuable flow chart for the use of microbial (fungal and bacterial) inoculants in the real world (Fig. 1).

Application of Microbial Inoculants to Enhance Reclamation Success

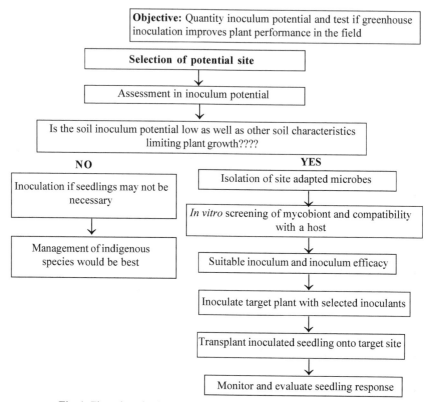

Fig. 1: Flow chart for the use of microbial inoculants in the real world.

Conclusion

New green technologies using symbiotic microorganisms have a huge potential in enhancing the practices of ecological restoration, surface bioremediation, direct reforestation, and rapid reconstruction of healthy ecosystems on mined sites in the Canadian boreal forest. Tree and microbe genetics may play a very important role in plant ability to colonize and establish on newly-formed mineral ecosystems following mining activities. Future studies should focus on (1) the identification of genes more expressed in tree seedlings and their microsymbionts highly efficient in phytoremediation and (2) the development of genomic markers for assisted selection of promising tree and symbiotic microorganisms to be used in phytobial restoration and direct reforestation of mined sites.

References

Allen MF, Swenson W, Querejeta JI, Egerton-Warburton LM, Treseder KK (2003) Ecology of mycorrhizae: A conceptual framework for complex interactions among plants and fungi. Ann. Rev. Phytopath. 41, 271-303.

Bal A, Anand R, Berge O, Chanway CP (2012) Isolation and identification of diazotrophic bacteria from internal tissues of *Pinus contorta* and *Thuja plicata*. Can. J. For. Res. 42, 807-813.

Bal A, Chanway CP (2012) Evidence of nitrogen fixation in lodgepole pine inoculated with diazotrophic *Paenibacillus polymyxa*. Bot. 90, 891-896.

Balogh-Brunstad Z, Keller CK, Dickinson JT, Stevens F, Li CY, Bormann BT (2008) Biotite weathering and nutrient uptake by ectomycorrhizal fungus, *Suillus tomentosus*, in liquid-culture experiments. Geo. Cos. Acta 72, 2601-2618.

Berner C, Johansson T, Wallander H (2012) Long-term effect of apatite on ectomycorrhizal growth and community structure. Mycorrhiza 22, 615-621.

Bois G, Piché Y, Fung MYP, Khasa DP (2005) Mycorrhizal inoculum potentials of pure reclamation materials and revegetated tailing sands from the Canadian oil sand industry. Mycorrhiza. 15, 149-158.

Bois G, Bertrand A, Piché Y, Fung M, Khasa DP (2006a) Growth, compatible solute and salt accumulation of five mycorrhizal fungal species grown over a range of NaCl concentrations. Mycorrhiza 16, 99-109.

Bois G, Bigras FJ, Bertrand A., Piché Y., Fung MYP, Khasa D (2006b) Ectomycorrhizal fungi affect the physiological responses of *Picea glauca* and *Pinus banksiana* seedlings exposed to an NaCl gradient. Tree Physiol., 26, 1185-1196.

Bot A, Benites J (2005) The importance of soil organic matter: key to drought-resistant soil and sustained food production. Food and Agriculture Organization of the United Nations, Rome, Italy.

Buée M, Reich M, Murat C, Morin E, Nilsson RH, Uroz S, Martin F (2009) 454 pyrosequencing analyses of forest soils reveal an unexpectedly high fungal diversity. New Phytol. 184, 449-456.

Burns RM (1990) *Pinus banksiana* Lamb. Silvics of North America. Volume 1 Conifers. USDS, USA. Retrieved October 27th 2012.

Cairney JWG (1999) Intraspecific physiological variation: implications for understanding functional diversity in ectomycorrhizal fungi. Mycorrhiza 9. 125-135.

Calvaruso C, Turpault M, Frey-Klett P (2006) Root-associated bacteria contribute to mineral weathering and to mineral nutrition in trees: a budgeting analysis. Appl. Environ. Microbiol. 72(2), 1258-1266.

Chanway CP (1997) Inoculation of tree roots with plant growth promoting soil bacteria: an emerging technology for reforestation. For. Sci. 43(1), 99-112.

Choi YD (2007) Restoration ecology to the future: a call for new paradigm. Restor. Ecol. 15(2), 351-353.

Colpaert JV, Vandenkoornhuyse P, Adriaensen K, Vangronsveld J (2000) Genetic variation and heavy metal tolerance in the ectomycorrhizal basidiomycete *Suillus luteus*. New Phytol. 147, 367-379.

Dahlberg A (2001) Community ecology of ectomycorhizal fungi: an advancing interdisciplinary field. New Phytol. 150, 555-562.

De Coninck AS, Karam A (2008) Impact of organic amendments on aerial biomass production, and phyto availability and fractionation of copper in a slightly alkaline copper mine tailing. Int. J. Min. Rec. Env. 22(4), 247-264.

Dixon RK, Buschena CA (1988) Response of ectomycorrhizal *Pinus banksiana* and *Picea glauca* to heavy metals in soil. Plant Soil 105, 265-271.

Drever JI, Stillings LL (1997) The role of organic acids in mineral weathering. Colloids Surfaces A: Physicochem. Eng. Aspects 120, 161-181.

Enebak SA, Wei G, Kloepper JW (1997) Effects of plant growth-promoting rhizobacteria on loblolly and slash pine seedlings. For. Sci. 44, 139-144.

Farrar JL (1995) Trees in Canada. Natural Resources Canada, Canadian Forest Service, Headquarters, Ottawa, Copublished by Fitzhenry and Whiteside Limited, Markham, Ontario.

Gagné A, Jany J, Bousquet J, Khasa DP (2006) Ectomycorrhizal fungal communities of nursery-inoculated seedlings outplanted on clear-cut sites in northern Alberta. Can. J. Bot. 36, 1684-1694.

Giasson P, Jaouich A, Cayer P, Gagné S, Moutoglis P, Massicotte L (2006) Enhanced phytoremediation: a study of mycorrhizoremediation of heavy metal-contaminated soil. Remediation 17, 97-110.

Hayat R, Ali S, Amara U, Khalid R, Ahmed I (2010) Soil beneficial bacteria and their role in plant growth promotion: a review. Ann. Microbiol. 60, 579-598.

Hoffland E, Kuyper TW, Wallander H, Plassard C, Gorbushina AA, Haselwandter K, Holmstrom S, Landeweert R, Lundstrom US, Rosling A, Sen R, Smits MM, Van Hees PAW, Van Breemen N (2004) The role of fungi in weathering. Front. Ecol. Environ. 2, 258-264.

Jackson LL, Lopoukhine N, Hillyard D (1995) Ecological restoration: a definition and comments. Restor. Ecol. 3, 71-75.

Karolewski P, Oleksyn J, Giertych MJ, Zytkowiak R, Reich PB, Tjoelker MG (2008) Primary and secondary host plant differ in leaf-level photosynthetic response to herbivory: evidence from *Alnus* and *Betula* grazed by the alder beetle, *Agelastica alni.* New Phytol. 140, 239-249.

Kernaghan G, Hambling B, Fung M, Khasa D (2002) *In vitro* selection of boreal ectomycorrhizal fungi for use in reclamation of saline-alkaline habitats. Restor. Ecol. 10, 43-51.

Kernaghan G, Sigler L, Khasa D (2003) Mycorrhizal and root endophytic fungi of containerized *Picea glauca* seedling assessed by rDNA sequence analysis. Microbial Ecol. 45, 128-136.

Khan AG (2006) Mycorhizoremediation – an enhanced form of phytoremediation. J. Zhejiang Uni. Sci. B. 7(7), 503-514.

Khasa DP, Hambling B, Kernaghan G, Fung M, Ngimbi E (2002) Genetic variability in salt tolerance of selected boreal woody seedlings. For. Ecol. Manage. 165, 257-269.

Lamhamedi MS, Fortin JA (1991) Genetic variations of ectomycorrhizal fungi: extrametrical phase of *Pisolithus sp.* Can. J. Bot. 69, 1927-1934.

Landeweert R, Hoffland E, Finlay RD, Kuyper TW, Van Bremen N (2001) Linking plants to rocks: ectomycorrhizal fungi mobilize nutrients from minerals. Trends Ecol. Evol. 16(5), 248-254.

Lefrançois E, Quoreshi A, Khasa D, Fung M, Whyte LG, Roy S, Greer SW (2010) Field performance of alder-*Frankia* symbionts for the reclamation of oil sands sites. Appl. Soil Ecol. 46, 183-191.

Marmiroli N (2007) Genetic variability and genetic engineering in phytoremediation. In: Advanced Science and Technology for Biological Decontamination of Sites Affected by Chemical and Radiological Nuclear Agents (Ed) Marmiroli N, Samotokin B and Marmiroli M, NATO Science Series: IV: Earth and Environmental Sciences, Springer, Netherlands, Vol. 75, pp 89-108.

Marshall, B. (2014) Faits et chiffres de l'industrie minière canadienne 2014. L'Association minière du Canada (AMC). http://mining.ca/fr/documents/faits-et-chiffres-2014

Ministère des Ressources Naturelles (MRN) du Québec (2012) Vegetation zones and bioclimatic domains in Québec. Retrieved October 30th 2012 from www.mrn.gouv.qc.ca/english/ publications/forests/publications/zone-a.pdf

Nadeau MB, Azaiez A, Roy S, Greer C, Khasa D (2014) Development of a new green technology for the revegetation of abandoned gold mine tailings using specific symbiotic microorganisms associated with *Picea glauca*. In: Mine Closure 2014 (Ed) Weiersbye IM, Fourie AB, Tibbett M and Mercer K, University of the Witwatersrand, Johannesburg, ISBN 978-0-620-62875-4.

Nadeau MB, Khasa DP (2015) The plant-soil-microbes, an ecofriendly way to reconstruct healthy ecosystems on mining sites: a global perspective. In: Mine Closure 2015 (Ed) Fourie AB, Tibbett M, Sawatsky L and van Zyl D, InfoMine Inc., Vancouver, Canada, pp 1061-1071.

Nadeau MB, Gagné A, Bissonnette C, Bélanger PA, Fortin AJ, Roy S, Greer C, Khasa D (2016) Performance of ectomycorrhizal alders exposed to Canadian oil sands tailing stressors under in vivo bipartite symbiotic conditions (in press).

Nienstaedt H, Zasada JC (1990) *Picea glauca* (Moench) Voss. Silvics of North America, Volume 1: Conifers. United States Forest Service, USA. Retrieved October 27th 2012. http://www.na.fs.fed.us/spfo/pubs/silvics_manual/table_of_contents.htm

Onwuchekwa NE, Zwiazek JJ, Quoreshi A, Khasa DP (2014) Growth of mycorrhizal jack pine (*Pinus banksiana*) and white spruce (*Picea glauca*) seedlings planted in oil sands reclaimed areas. Mycorrhiza 24, 431-441.

Quoreshi AM, Roy S, Greer CW, Beaudin J (2007) Inoculation of green alder (*Alnus crispa*) with *Frankia*-ectomycorrhizal fungal inoculant under commercial nursery production conditions. Nat. Plant J. 8(3), 271-281.

Quoreshi AM (2008) The Use of Mycorrhizal Biotechnology in Restoration of Disturbed Ecosystem In: Mycorrhizae : Sustainable Agriculture and Forestry (Ed) Siddiqui ZA, Akhtar MS and Futai K, Springer, Ottawa, Canada, pp 303-320

Paul LP, Chapman WK, Chanway CP (2013) Diazotrophic bacteria reside inside *Suillus tomentosus/ Pinus contorta* tuberculate ectomycorrhizae. Bot. 91, 48-52.

Pregent G, Camiré C, Fortin JA, Arsenault P, Brouillette JG (1987) Growth and nutritional status of green alder, jack pine, and willow in relation to site parameters of borrow pits in James Bay territory, Quebec. Rec. Rev. Res. 6, 33-48.

Read DJ, Leake JR, Perez-Moreno J (2004) Mycorrhizal fungi as drivers of ecosystem processes in heathland and boreal forest biomes. Can. J. Bot. 82: 1243-1263,

Reid C, Bécaert V, Aubertin M, Rosenbaum RK, Deschênes L (2009) Life cycle assessment of mine tailings management in Canada. J. Clean. Prod. 17, 471-479.

Renault S, Sailerova E, Fedikow MAF (2001) Phytoremediation of mine tailings and bio-ore production: results from a study on plant survival at the central Manitoba (AU) minesite (NTS 52L/13); In: Report of Activities 2001, Manitoba Industry, Trade and Mines, Manitoba Geological Survey, pp 138-149.

Renault S., Sailerova E, Fedikow MAF (2002) Phytoremediation of mine tailings and bio-ore production: progress report on seed germination, survival and metal uptake of seedlings planted at Central Manitoba (Au) minesite (NTS 52L13); In: Report of Activities 2002, Manitoba Industry, Trade and Mines, Manitoba Geological Survey, pp 255–265.

Roy S, Khasa DP, Greer CW (2007) Combining alders, frankiae, and mycorrhizae for the revegetation and remediation of contaminated ecosystems. Can. J. Bot. 85, 237-251.

Rinaldi AC, Comandini O, Kuyper TW (2008) Ectomycorrhizal fungal diversity: separating the wheat from the chaff. Fungal Divers. 33, 1-45.

Saier MH, Trevors JT (2010) Phytoremediation. Water Air Soil Pollut. 205 S61-S63.

Smith SE, Read DJ (2008) Myc. sym., 3rd ed. Academic Press, London, UK.

Smits MM, Bonneville S, Benning LG, Banwart SA, Leake JR (2012) Plant-driven weathering of apatite – the role of an ectomycorrhizal fungus. Geobiology 10, 445-456.

Sutton RF (1973) Histoire naturelle de l'Épinette blanche (*Picea glauca* (Moench) Voss). Ministère de l'Environnement, Service Canadien des Forêts, Publication # 1250f, Ottawa, Canada.

Taylor LL, Leake JR, Quirk J, Hardy K, Banwart SA, Beerling DJ (2009) Biological weathering and the long-term carbon cycle: integrating mycorrhizal evolution and function into the current paradigm. Geobiology 7, 171-191.

Van den Driessche R (1991) Mineral nutrition of conifer seedlings. CRC Press, Boca Raton, Florida, USA.

Villeneuve N, Grandtner MM, Fortin JA (1989) Frequency and diversity of ectomycorrhizal and saprophytic macrofungi in the Laurentide mountains of Quebec. Can. J. Bot. 67, 2616-2629.

Vodouhe GF, Khasa, DP (2015) Local communities' perception of mine sites restoration in Abitibi (Quebec). Int. J. Phytoremediat. 17, 962–972.

Wallander H, Hagerberg D (2004) Do ectomycorrhizal fungi have significant role in weathering of minerals in forest soils? Symbiosis 37, 249-252.

Warman PR (1988) The Gays river mine tailing revegetation study. Land. Urban Plan. 16, 283-288.

Wenzel WW (2009) Rhizosphere processes and management in plant-assisted bioremediation (phytoremediation) of soils. Plant Soil 321, 385-408.

11

Microbial Restoration in Mined-out Areas Under Reduced Ecosystem - An Overview

Chaubey, O.P., Bohre Priyanka and Jamaluddin

Abstract

Opencast mining destroys the original land ecosystem and microbial community due to mining and stockpiling of mine rejects on adjoining land, which is a matter of great environmental concern. Mining creates formation of wastelands, formation of gullies, ravines, overburdens, reduced water infiltration and increased runoff. Reclamation of mined out areas is often difficult due to its chemical, physical and biological traits. Absence of topsoil is the most common feature of the mine spoils dumps. Arbuscular mycorrhizal fungi (AMF), the synonym of VAM fungi, are important microbes of soil that form symbiotic association with most of the terrestrial plants on the earth. Observations led to use of legume microsymbionts as biological substitutes for fertilizers. AMF diversity increases with species diversity of plants as the potential number of association increases. Dominant arbuscular mycorrhizal fungi can prevent the invasion of non-mycorrhizal plants on land where they have established symbiosis and promote their mycorrhizal host. Fungal and bacterial colonies were very less while actinomycetes were negligible. The four genera of bacteria were recorded in different unexcavated soil from different species. They were *Rhizobium, Azotobacter, Azospirillum and Pseudomonas fluorescens.* The potential of soil microorganisms has been recognized widely in improvement of soil quality, soil formation, aggregation and re-vegetation through their activities in litter decomposition and nutrient cycling. Their activities such as phosphate solubilization, nitrogen fixation, oxidation of various inorganic components of soil or mineralization of inorganic components and mycorrhizal symbiosis are major beneficial activities that play very important role in soil system functioning. An active soil microbial biomass is an essential factor in the long term fertility of soils. The nutritional characteristics like organic carbon, available nitrogen and available phosphorus maintained significant positive correlation with the density of AMF and microbial biomass in different dominant tree species. There is an urgent need to develop and extend technological measures to rehabilitate mined out lands.

Keywords: Arbuscular mycorrhizal fungi, Microbial restoration, Mineralization Reduced Ecosystem

1. Introduction

Mining is regarded as an industrial operation of withdrawing resources from the earth. Opencast mining destroys the original land ecosystem and microbial community due to mining and stockpiling of mine rejects on adjoining land, which is a matter of great environmental concern. Due to stockpiling of mine rejects, the landscapes that emerge are devoid of supportive and nutritive capacity for biomass development. After mining huge overburden dumps are left, the soil is disturbed; the humus rich soil gets buried deep down leaving the sterile sandy and boulder subsoil on to the top. The site, thus, made available for reclamation and re-vegetation poses inhospitable conditions. The restoration to its original state is a very daunting task. Mining creates formation of wastelands, formation of gullies, ravines, overburdens, reduced water infiltration and increased runoff. The life support ability of the habitat is reduced. Pits are formed which during rains get filled up. There is a general pollution of soil, water and the atmosphere. Natural plant communities get disturbed and the habitats become impoverished due to mining presenting a very rigorous condition for plant growth. To combat the consequences of land degradation due to mining some remedial measures are required for the reclamation of degraded land with a view to restore its productivity and fertility. Reclamation of mined out areas is often difficult due to its chemical, physical and biological traits. Absence of topsoil is the most common feature of the mine spoils dumps. If present, it is very poor in nitrogen, which is essential for plant growth. This is due to the absence of soil organic matter provided by decay of dead plant material. In addition, to this the stony nature of mine spoil wastes aggravates the situation further for vegetation establishment by developing low infiltration rates, unfavorable porosity and high bulk density due to massive compaction.

Since the progress of natural vegetation process is very slow on mine spoils, selective plantation of suitable native species is desired in most cases. Restoration of these recalcitrant mined out sites with tree plantation along with suitable organic amendment is important for developing the vegetation canopy and to improve the properties of mine spoil such as its structural aggregation, nutrient cycling and availability, thereby improving its productivity and fertility. Application of ecological principles and concepts is the foundation of sustainable development.

2. Arbuscular Mycorrhizal Fungi

Arbuscular mycorrhizal fungi (AMF), the synonym of VAM fungi, are important microbes of soil that form symbiotic association with most of the terrestrial plants on the earth. These mycorrhizae can grow inside the roots of plants and serve them in a number of ways. These fungi are able to thrive well on poor physical substrates by extending their hyphal network to the establishing plants. These fungi improve availability of nutrients in utilizable forms. Their role in removing heavy metals from overburden soils, retention in their cell wall, and ability to bind soil particles proved to be excellent in reclamation efforts. The arbuscular mycorrhizal fungi (AMF) have been reported to moderate the adverse effects of higher temperature, improve plant–water relations and increase stomatal conductance, besides preventing various soil-borne diseases. Restoration is defined as a tactic employed to return degraded lands to its original condition. Of the total earth's surface, 78% of the land area is unsuitable for agriculture. Out of the remaining land, about 9% suffers from physical, chemical and biological constraints requiring special management practices. Since, land degradation processes involve loss of vegetation and accelerated run off and soil loss; they require assistance for their restoration and regeneration. Mycorrhizal fungi and nitrogen-fixing bacteria are among the major beneficial components of soil microbial community, which contribute to plant growth and survival by reducing stresses through symbiosis (Timmer and Leyden 1980; Sylvia and Williams 1992; Simon et al. 1993; Remy et al. 1994, Douds and Nagahashi 2000; Eriksson 2001; Mozafar et al. 2000, Rillig et al. 2003, Smith and Read 2002; Smith et al. 2003; Rillig, 2004; Wright 2005; Yamato 2005). These plants have special nutritional relationships between the two symbionts: the high phosphorus requirement of the nitrogen fixing root nodule and the high nitrogen requirement of the chitin walled AMF and the high carbon requirement of both. Since each symbiont can supply the other's need in excess, the endophytes can bring about the requirements of the plant when the association is grown in nutrient-deficient soil (Norris et al. 1994). In addition, AMF mycelia can extend to a long distance and link the rhizosphere and mycorrhizosphere of different plant species. It can make common pool of the available nutrients of the other plant species (Norris et al. 1994). This way, the nutrients released into the overlapping mycorrhizosphere by plant root exudation or by root and nodule decay become available for non-nitrogen fixing plants (Simard et al. 1997), which can play important role and facilitates the survival and growth of other plant species. Such observations led to use of legume microsymbionts as biological substitutes for fertilizers. Suggested roles of mycorrhizal fungi in natural and managed ecosystems are well established (Brundrett 2002; Gianinazzi 1996; Matekwor *et al.* 2005).

The demand for a particular mineral nutrient depends on plant internal requirements, while the supply of that nutrient primarily depends on its availability and mobility in soils (Allen *et al.* 1979; Marschner 1995, Marschner and Timonen 2004). Mineral nutrients such as phosphorus have very limited mobility in soils (Parfitt and Russell 1977; Insam and Domsch 1988; Marschner 1995; Booze *et al.* 2000; Boswell *et al.* 1998, Bucking and Shachar 2005). Thus, to obtain more phosphorus, plants must bypass the depletion zones by further root activity elsewhere in the soil. The outcome of this quest for phosphorus (and other relatively immobile soil resources) should largely be determined by the surface area of a plant's root system. The most important role of mycorrhizal fungal hyphae is to extend the surface area of roots. The capacity of plants to influence nutrient availability in soils will also depend on the extensiveness and activity of their root system, since young roots are the primary source of exudates (Harley and Smith 1983; Curl and Truelove 1986; Uren and Reisenaur 1988; Warcup 1990; Jeffries *et al.* 2003).

AMF diversity increases with species diversity of plants as the potential number of association increases. Dominant AMF can prevent the invasion of non-mycorrhizal plants on land where they have established symbiosis and promote their mycorrhizal host (Pham 2007). Research has shown that AMF release an unidentified diffusional factor, known as the mycofactor, which activates the nodulation factor's inducible gene MtEnod11. This is the same gene involved in establishing symbiosis with the nitrogen fixing, rhizobial bacteria (Bolan 1991; Kosuta *et al.* 2003; Xie *et al.* 1995; Sara 2003). There is also evidence to suggest that AMF may play an important role in mediating the plant species' specific effects on the bacterial composition of the rhizosphere. The use of AMF in ecological restoration has been shown to enable host plant establishment on degraded soils by improving its quality and health (phytoremediation), such as physical and biological properties, nitrogen, organic matter, and restoration of native AMF (Sandra Brawn 2002), and production of soil protein known as glomalin in incubated soils from forested, afforested, agricultural, and grasslands (Kabir and Koide 2000; Tran 2001; Hai 2009). Chaubey *et al.* (2012) reported that population of AMF spores, microbial biomass and infection of roots with AMF spores increased with the age of plantations of dominant tree species raised on the overburden soils due to rehabilitation activities at Jhingurda, open cast coal mine of Singrauli, Madhya Pradesh, India.

In a study of flag stone mines at Shivpuri district, microbial status in rhizosphere of important tree species occurring in flagstone mines such as *Diospyros melanoxylon, Holarrhina antidysenterica, Butea monosperma, Solanum nigrum, Lagerstroemia parviflora, Madhuca latifolia, Ziziphus xylopyrus, Carica opaca, Cassia tora, Aristida adscensionis, Acacia catechu, Gymnosporia montana, Flacourtia indica, Alysicarpus indica, Hyptis suaveolens, Anogeissus pendula, Boswellia sevrala, Achyranthes aspera, Cenchrus cilaris, Acacia leucophloea* and *Prosopis juliflora* were assessed. The colonies of fungi in the most of samples were followed by bacteria. However, the colonies of actinomycetes were less and recorded in few samples. The representative samples of mine soil from flagstone area showed fewer colonies of all microbes. Fungal and bacterial colonies were very less while actinomycetes were negligible. The four genera of bacteria were recorded in different unexcavated soil from different species. They were *Rhizobium, Azotobacter, Azospirullum and Pseudomonas fluorescens.* There were considerable variations in the bacterial species. *Rhizobium* was in more number in rhizosphere of leguminous plants. However, the bacterial species were represented by *Azotobacter* present in all samples of unexcavated soil. But *Pseudomonas* was present in few samples while *Azosporillum* species could be recorded rarely. The AMF population varieds to a considerable extent in all plant species. The unexcavated forest soil of different species showed species of *Glomus* which is represented by *Glomus mosseae, Glomus intraradices* and *Glomus deserticola* while one species of *Acaulospora* i.e., *Acaulospora scrobiculata* could be noticed. All AMF species varied in frequency in different species. Grasses harbored more AMF. The spore number in most of the sample varied from 86 to 145 per100g soil. The mined soil hardored only *Glomus* and *Acaulospora* species. The spore number was less in mined soil i.e. from 78 to 83 per 100g soil. The diversity of AMF species was very less in mined soil. Comparatively the unexcavated forest soils had more bacterial, fungal, AMF and actinomycetes population compared to mined soil. On the basis of species examination and spore population AMF were less in mined soil. Afforestation of the mined area needs seedlings boosted up with the representative microbial species of the area. The seedlings raised in the nursery can be boosted up by such microbial consortium for undertaking the plantation (Chaubey 2011).

3. Potential of Soil Microorganisms

The potential of soil microorganisms has been recognized widely in improvement of soil quality, soil formation, aggregation and re-vegetation through their activities in litter decomposition and nutrient cycling. Their activities such as phosphate solubilization, nitrogen fixation, oxidation of various inorganic components of soil or mineralization of inorganic components and mycorrhizal symbiosis are major beneficial activities that play very important role in soil system functioning. Soil microorganisms inhabitating the rhizosphere environment interact with plant roots and mediate nutrient availability to the plants. Implications of plants and their symbionts like mycorrhizal fungi, N-fixing bacteria and free living rhizosphere population of bacteria promote plant establishment and growth. Thus role of soil microorganisms in nutrient cycling is unique. In addition to their effects on soil fertility, they also enhance soil structure by binding together soil particles. An active soil microbial biomass is an essential factor in the long term fertility of soils. Soil microorganisms play significant role in soil fertility and ecosystem functioning. Without microbes and their functions, plant species cannot be supported by the soil alone (Kennedy and Smith 1995). It controls the major processes involved in nutrient transformation, cycling, organic matter, management and macro-aggregation for favorable water and aeration characteristics (Singh and Goel 2009; Hargreaves *et al.* 2003). In a biomass and carbon sequestration study at northern coal field (NCL) Singrauli, there was gradual increase in microbial biomass from younger to older plantations in different species. It ranged from 40.2 (2 years old plantation) to 51.5 mg kg^{-1} (18 years old plantation) in *Tectona grandis* Linn. f., from 32.5 (2 years old plantation) to 66.6 mg kg^{-1} (18 years old plantation) in *Dalbergia sissoo* Roxb., from 21.2 (2 years old plantation) to 52.7 mg kg^{-1} (18 years old plantation) in *Azadirachta indica* A. Juss., from 35.5 (2 years old plantation) to 50.3 mg kg^{-1} (19 years old plantation) in *Cassia siamea* Lamk, from 24.2 (2 years old plantation) to 54.7 mg kg^{-1} (18 years old plantation) in *Pongamia pinnata* (Linn) Pierre and from 31.7 (6 years old plantation) to 42.6 mg kg^{-1} (10 years old plantation) in *Gmelina arborea* Roxb. It was primarily due to increase in microbial population in older plantations as the availability of organic matter and humus had increased over a period of time (Bohre *et al.* 2013a,b). It was primarily due to the fact that older plantations had huge population of AMF spores in different plant species. In younger plantations, the colonization by AMF gradually decreased due to less available carbon and other nutrients. Similarly, colonization of AMF in the roots was noticed. In older plantations, all five genera of AMF including *Acaulospora* sp., *Glomus* sp., *Gigaspora* sp., *Scutellospora* sp. and *Sclerocystis* sp. were recorded. Out of these, *Acaulospora* and *Glomus* were in high frequency. The number of spore population was more in *Acaulospora* and *Glomus*, while *Gigaspora* and *Scutellospora* species were less in number. *Sclerocystis* was recorded in few samples especially in older plantations. In terms of AMF and root colonization with AMF, the *Dalbergia sissoo* proved most promising species, followed by *Pongamia pinnata* and other species. These species were good for nitrogen fixation. With the increasing age of species, the microbial biomass also increased in different species. The findings were comparable with the observations recorded by Daft and Nicholson (1974), Gupta and Shukla (1991), Jamaluddin and Chandra (2009), Chaubey *et al.* (2012) in different studies regarding AMF status and plant succession. The role of AMF was found to recover the nutrient status of coalmine spoil soil.

The nutritional characteristics like organic carbon, available nitrogen and available phosphorus maintained significant positive correlation with the density of AMF and microbial biomass in different dominant tree species. Their quantity may, therefore, be the good indices of microbial biomass and density of AMF. However, the best positive correlation between microbial biomass and nutritional characteristics was found with available phosphorus followed by available nitrogen and carbon in different dominant species. This can be attributed to the fact that to obtain more phosphorus plants establish symbiotic relationship with the AMF and increased the surface area of root (Curl and Truelove, 1986; Uren and Reisenaur, 1988; Chaubey *et. al.* 2013). Banerjee *et al.* (2000) reported a significant positive correlation between the number of organisms and organic carbon in coal mine spoil of Gevra colliery, Chattisgarh, India where number of organisms varied from species to species with significant differences. Soil microbial biomass and AMF were recognized as the driving force behind nutrient transformation in soil and thus, have a major role in soil fertility and ecosystem functioning. Soil microbial biomass was useful in determining the degree of recovery of degraded ecosystem and nutritional budget. Microbial activity improved gradually during restoration of mine spoils (Stroo and Jeneks 1982). Moreover, these tree species proved to be promising to improve the fertility status of degraded soil through development of soil microflora, microbial biomass and nutritional status of mine spoils.In a study on reclamation of mine spoil dump using integrated biotechnological approach at Sasti Coal Mine, Maharashtra, and amendment of FYM application @50T/ha along with biofertilizer application to mine spoil helped to improve the physico-chemical and microbiological characteristics of the coal mine spoil dump. The organic matter level and nutrient status with respect to N, P, and K also increased, thereby improving the fertility and productivity of mine spoil dump. Application of organic amendment and biofertilizers helped in reducing the toxicity of heavy metals due to the increased organic matter content of FYM which contains high proportion of humid organic matter. Due to incorporation of organic matter the rhizospheric productivity was increased. This ultimately resulted in higher below ground biomass production as compared to control. The above ground biodiversity (AGB) also improved thereby showing more plant girth and biomass in *Cassia siemea* and *Tectona grandis.* The results of the study indicated that integrated biotechnological approach restored the productivity and fertility of mine spoil dump within three years leading to the development of self sustainable ecosystem and turned the barren mine spoil dump into lush green mat of vegetation cover, thereby mitigating the environmental hazards and conserving the biodiversity of the area. Thus, using Integrated Biotechnological Approach (IBA) technology the degraded ecosystem was converted into a source of revenue in a short span. The studies conducted open a new avenue to implement the technology in different types of wastelands (Juwarkar and Jambhulkar 2009).

In a study on ecological approach in rehabilitation of lignite mined wastelands in the Indian Arid Zone, by extending ecological principles, success in rehabilitation could be achieved in the desired time frame. Survival of species planted at lignite mine spoil was 82.82%. Maximum growth was observed in *Parkinsonia aculeata* followed by other species. Mid February to March has proved to be better plantation season in lignite backfill. Direct seeding of Senna (*Cassia angustifolia*) and intercropping perennial grasses (*Cenchrus ciliaris* and *Lasiurus sindicus*), cereal crop (*Pennisetum glaucum*) as well as vegetables has proved successful. Role of local people and NGO's for participatory mine rehabilitation has been highlighted as a major future requirement of these programmes (Kumar and Kumar 2009).

In a study on bio-restoration of coal mine spoil with fly ash and biological amendments in Eastern Jharia of Bharat Coking Coal Limited (BCCL), out of 17 species tried, *Acacia auriculiformis* gave better performance in terms of photosynthesis and soil conservation efficiencies on coal mine spoil dumps. The article incorporated the attempts made to investigate the utilization of fly ash in reclamation of coal mine spoil for agriculture and forestry and recommended that standard leaching and sub scale tests on release of metals, and efficiency of suitable amendments in immobilizing the metals, specific to the mine spoil and fly ash in question, should be carried out before implementing at large scale field observation. Soil amendment can be established after repeated field experiments. It is desirable to (i) have an understanding of the response of crops and plant species to ash amendments of various types of mine spoil, (ii) determine the changes that take place in the physicochemical and biological characteristic of the spoils as a result of fly ash application, (iii) evaluate potential of site end results of sustained used of fly ash and (iv) develop an appropriate policy/ recommendation for the sustainable use of fly ash in mine spoil reclamation or degraded forest lands. Fly ash may also contain naturally occurring radio-nuclides from the U and Th series, as well as K. Trace metals and radio-nuclides may have an impact on the environment surroundings and may affect the decision to dispose of fly ash in ash ponds or to apply fly ash in agriculture and forestry (Srivastava and Ram 2009).

Of the 4 bio-productive systems, *viz.* forests, croplands, fisheries, and grasslands, on which the humanity depends for all the matter and energy it needs, grasslands are the most neglected in India. Natural grasslands do not occur in India except in the Alpine zone. And though millions of hectares are recorded as pasture lands in the revenue records, they are not productive, and do not qualify to be called grasslands. They have neither been maintained, nor artificially generated. Therefore, planting of grasses on coal - mined surfaces would serve not only ecological and environmental purposes, but also socio – economic purposes. Vetiver grass (*Vetiveria zizaniodes*) which is known for its effectiveness in erosion and sediment control, and highly tolerant to adverse soil conditions is one of the most suited species for vegetating coal - mined lands. Experimental studies show that when adequately supplied with nitrogen and phosphorus for fertilizing, *Vetiveria* could grow in highly adverse conditions (Lal 2009).

4. Rehabilitation of Mined Out Lands

Increased environmental awareness and improvement in the knowledge compels environmentalists and technocrats to accept the modern approach of mined out land reclamation on the basis of the interpretation of the whole ecology of a site and its intended after-use. Preparation of an eco-restoration plan prior to the commencement of mining is essential. Once the mining is over, the land should be rehabilitated immediately. The physical and chemical properties of the spoils should be thoroughly investigated prior to rehabilitation (Nath 2009).

There is an urgent need to develop and extend technological measures to rehabilitate mined out lands for potential use since mining gives rise to sharp changes in the landscape and causes adverse ecological impact through depletion of flora and fauna. In this context, an ecosystem approach for rehabilitation of mine spoils and overburden areas would be adopted. On the basis of above review, following recommendations/suggestions are emerged out:

1. Coal mined out areas should be tackled on an ecosystem approach basis which incorporates land preparation, soil and water conservation on watershed basis and including different elements of biodiversity, preferably the grasses and other herbaceous flora in the restoration programme.

2. There is a need to utilize the available land resources for maximum production of economic biomass for mitigating the effect of Green House Gases (GHG), and to encourage the plantation of socio-economic importance, it is essential to have plant species suitable for soil conservation (soil binding), yielding Non-timber forest products (NTFP) including wood fuel, fodder and bio-energy, oil bearing species such as *Jatropha curcas*, *Pongamia pinnata*, *Schleichera oleosa*, etc.

3. Soil and mine spoil should be studied specifically with a view to introducing suitable culture of AMF and other micro fauna and micro flora to enrich the quality of that soil to yield better and accelerated results of reclamation measures.

4. Leguminous tree and other plants inhabiting nitrogen fixing bacteria in their root system should be given priority in afforesting mined overburdens.

5. Instead of using chemical weedicides which may have adverse effects on the environment, biological control of weeds should be adopted. There is a vast potential of research in this area.

6. Fly ash with biological amendments may be gainfully used for enhanced rehabilitation of coalmine spoils.

7. Reclamation improves ecosystem attributes of coal mine wastes with passage of time and hence they need to be better looked after.

8. Mining operation deteriorates the water qualities by various ways but mine drainage water should be utilized after amendment as a resource instead of throwing out.

9. Already, a lot of researches have been carried out on plant species selection and their characteristics but still there is a lot of scope to manage post plantations techniques and management. The reclamation and revegetation process can be enhanced many folds, if post mining operators and plantations managers take proper management practices into considerations.

10. Plantation of native species of high productivity and soil conserving species based on high photosynthetic ability and soil conserving efficiency need to be preferred and encouraged.

11. Surface soil should be utilized in agriculture and can be used in reclamation of the over burden.

12. Follow up legal action should be ensured.

References

Allen MF, Moore TS, Christensen M and Stanton N (1979) Growth of vesicular-arbuscular mycorrhizal and non- mycorrhizal *Bautelona gracilus* in a defined medium. Mycologia 71, 666-669.

Banerjee SK, Das PK and Mishra TK (2000) Microbial and nutritional characteristics of coal mine overburden spoils in relation to vegetation development. J. Ind. Soc. Soil Sci. 48, 63-66.

Bohre P, Chaubey OP and Singhal P K (2012) Biomass accumulation and carbon sequestration in *Dalbergia sissoo* Roxb. Int. J. Bio-Sci. Bio-Technol. 4 (3), 29-44.

Bohre P, Chaubey OP and Singhal PK (2013a) Biomass production and carbon sequestration by *Cassia siamea* Lamk in degraded ecosystem. In: Sustainable Bio-diversity Conservation in the Landscape (Ed) Chaubey OP, and Ram Prakash, Aavishkar publishers, Jaipur, pp 15-34.

Bohre P, Chaubey OP and Singhal PK (2013). Biomass accumulation and carbon sequestration in *Tectona grandis* Linn. f. and *Gmelina arborea* Roxb. Int. J. Bio-Sci. Bio-Technol. 5 (3), 153-173.

Bolan NS (1991) A critical review of the role of mycorrhizal fungi in the uptake of phosphorus by plants. Plant Soil 134, 189–207.

Booze-Daniels JN, Daniels WL, Schmidt RE, Krouse JM and Wright DL (2000) Establishment of low maintenance vegetation in highway corridors In: Reclamation of Drastically Disturbed Lands (Ed) Barnhisel RI, Agronomy Series No. 41, Am. Soc. Agron., Madison, WI, pp 887-920

Boswell EP, Koide RT, Shumway D L and Addy HD (1998) Winter wheat cover cropping, VA mycorrhizal fungi and maize growth and yield. Agric. Ecosyst. Environ. 67, 55–65.

Brundrett, MC (2002) Coevolution of roots and mycorrhizas of land plants. New Phytol. 154, 275–304.

Bucking H and Shachar-Hill Y (2005) Phosphate uptake, transport and transfer by arbuscular mycorrhizal fungus is increased by carbohydrate availability. New Phytol. 165(3), 889–912.

Chaubey, O. P., Priyanka Bohre and P. K. Singhal (2012) Impact of Bio-reclamation of Coal Mine Spoil on Nutritional and Microbial Characteristics - A Case Study. Int. J. Bio-Sci. Bio-Technol. 4 (3): 69-79.

Chaubey, OP, Bohre P, Jamaluddin and Singhal PK (2013) Microbial succession and restoration of degraded ecosystem under different tree cover. Mycorrhiza News, 25(2), 2-13.

Chaubey OP (2011) Preparation of reclamation plan for flagstone mines of shivpuri district Microbial succession and restoration of degraded ecosystem under different tree cover. Final report submitted to M.P. State Mining Corporation, Bhopal (M.P.).

Curl, E A and Truelove B (1986) The Rhizosphere. Springer-Verlag, Berlin. Daft, MJ and Nicholson TH (1974) AM in plants colonizing coal waste in Scotland. New Phytol. 73, 1129-1138.

Douds DD and Nagahashi G (2000) Signalling and recognition events prior to colonisation of roots by arbuscular mycorrhizal fungi. In: Current Advances in Mycorrhizae Research. (Ed) Podila GK and Douds DD, APS Press. Minnesota, pp 11-18.

Eriksson, A (2001) Arbuscular mycorrhizae in relation to management history, soil nutrients and plant diversity. Plant Ecol. 155, 129–137.

Gianinazzi-Pearson V (1996) Plant cell responses to arbuscular mycorrhizae fungi: getting to the roots of symbiosis. The Plant Cell, 8(10), 1871–1883.

Gupta OP and Shukla RP (1991) The composition and dynamics of associated plant communities of sal plantations. Trop. Ecol. 82, 296-309.

Hai Vo Dai (2009) Research on capacity of carbon sequestration in Urophylla plantation in Vietnam. Magazine of Agriculture and Rural Development, No 1/2009, Ha Noi, pp 102 – 106.

Hargreaves RP, Brookes PC Ross GJS and Poulton PR (2003) Evaluating soil microbial biomass carbon as an indicator of long term environmental change. Soil Boil. Biochem. 35, 401-407.

Harley JL and SE Smith (1983) Mycorrhizal Symbiosis. Academic Press: London.

Insam H and Domsch KH (1988) Relationship between Soil Organic Carbon and Microbial Biomass on Chronosequences of Reclamation Sites. Microbial Ecol. 15, 177-188.

Jamaluddin and Chandra KK (2009) Mycorrhizal establishment and plant succession in coal mine overburden. In: Sustainable Rehabilitation of Degraded Ecosystems (Ed) Chaubey OP, Vijay Bahadur and Shukla PK, Aavishkar Publishers and Distributors Jaipur, India, pp 157-163.

Jeffries P, Gianinazzi S, Perotto S, Turnau K, Barea J (2003) The Contribution of arbuscular mycorrhizal fungi in sustainable maintenance of plant health and soil fertility. Biol. Fertil. Soils 37: 1–16.

Juwarkar A and Jambhulkar HP (2009) Reclamation of mine spoil dump using integrated biotechnological approach at Sasti coal mine, Maharashtra. In: Sustainable Rehabilitation of Degraded Ecosystems (Ed) Chaubey OP, Vijay Bahadur and Shukla PK), Aavishkar publishers, distributors Jaipur, India, pp 92-108.

Kabir Z and Koide RT (2000) The effect of dandelion or a cover crop on mycorrhiza inoculum potential, soil aggregation and yield of maize. Agri. Ecosyst. Environ. 78, 167–174.

Kennedy AC and Smith KL (1995) Soil microbial diversity and sustainability of agriculture soils. Plant Soil 170: 75-86.

Kosuta S, Chabaud M , Lougnon G, Gough C, Denarie J, Barker D and Bacard G (2003) A diffusible factor from arbuscular mycorrhizal fungi induces symbiosis-specific method expression in roots of Medicago truncatula. Plant Physiol. 131(3), 952–962.

Kumar S and Kumar P (2009) Ecological Approach in Rehabilitation of Lignite Mined Wastelands in the Indian Arid Zone. In: Sustainable Rehabilitation of Degraded Ecosystems (Ed) Chaubey OP, Vijay Bahadur and Shukla PK, Aavishkar Publishers and Distributors, Jaipur, India.

Lal JB (2009) Vetiveria the Miracle Grass for Post – Mining Rehabilitation. In: Sustainable Rehabilitation of Degraded Ecosystems (Ed) Chaubey OP, Vijay Bahadur and Shukla PK, Aavishkar Publishers and Distributors, Jaipur, India.

Marschner P and Timonen S (2004) Interactions between plant species and mycorrhizal colonization on the bacterial community composition in the rhizosphere. Appl. Soil Ecol. 28, 23–36.

Marschner H (1995) Mineral nutrition of higher plants, 2nd edn. Academic Press, London.

Matekwor A, Nakata EM and Nonaka M (2005) Arum - and Paris-type arbuscular mycorrhizas in a mixed pine forest on sand dune soil in Niigata Prefecture, central Honshu, Japan. Mycorrhiza 15(2), 129–36.

Mozafar A, Anken T, Ruh R and Frossard E (2000) Tillage intensity, Mycorrhizal and non mycorrhizal fungi and nutrient concentrations in maize, wheat and canola. Agron. J. 92, 1117–1124.

Norris JR, Read D and Verma AK (1994) Techniques for mycorrhizal research: methods in microboilogy. Academic Press, Harcourt Brace and Company, Publishers, London.

Pham, Tuan Anh (2007) Forecasting CO_2 sequestration in natural forests in Tuy Duc District, Dak Nong Province, Vietnam. Master Thesis, Forestry University Vietnam 25.

Romain Pirard (2005) Pulpwood plantations as carbon sinks in Indonesia: Methodological challenge and impact on livelihoods. Carbon Forestry, Center for International Forestry Research, CIFOR.

Remy W, Taylor T, Hass H and Kerp H (1994) Four hundred-million-year-old vesicular arbuscular mycorrhizae (abstract). PNAS, 91(25), 11841-11843.

Rillig M, Ramsey P, Morris S and Paul E (2003) Glomalin, an arbuscular-mycorrhizal fungal soil protein, responds to land-use change. Plant Soil 253, 293–299.

Rillig M (2004) Arbuscular mycorrhizae, glomalin and soil aggregation. Can. J. Soil Sci. 84, 355–363.

Sandra Brown (2002) Measuring carbon in forests: current status and future challenges. Environ. Pollut. 116, 363–372.

Sara, Beth Gann (2003) A Methodology for Inventorying Stored Carbon in An Urban Forest, Falls Church, Virginia 27.

Schimel DS (1995) Terrestrial ecosystems and the carbon cycle. Global Change Biol. 1, 77–91.

Simard SW, Perry DA, Jones MD, Myrolds DD, Durall DM and Molina R (1997) Net transfer of carbon between ectomycorrhizal tree species in the field. Nature 388, 579-582.

Simon L, Bousquet J, Levesque C, Lalonde M(1993) Origin and diversification of endomycorrhizal fungi and coincidence with vascular land plants. Nature 363, 67–69.

Singh B and Goel VL (2009) Rehabilitation of degraded soil sites through afforestration programmes: A case study. In: Sustainable Rehabilitation of Degraded Ecosystems (Ed) Chaubey OP, Vijay Bahadur and Shukla, Aavishkar Publishers and Distributors, Jaipur, India, pp 67-76.

Smith SE and Read DJ (2002) Mycorrhizal Symbiosis. Academic Press: London.

Smith S, Smith A, Jakobsen I (2003) Mycorrhizal fungi can dominate phosphorus supply to plant irrespective of growth response. Plant Physiol. 133 (1), 16–20.

Srivastava NK and Ram LC (2009) Bio-restoration of coal mine spoil with fly ash and biological amendments. In: Sustainable Rehabilitation of Degraded Ecosystems (Ed) Chaubey OP, Vijay Bahadur and Shukla, Aavishkar Publishers and Distributors, Jaipur, India, pp 77-99.

Stroo HF and Jenks EM (1982) Enzymatic activities and respiration in mine spoils. Soil Sci. Soc. Am. J. 46, 548-553.

Sylvia DM and Williams SE (1992) Vesicular-arbuscular mycorrhizae and environmental stress. In: Mycorrhizae and sustainable Agriculture (Ed) Bethlenfalvay GJ, Linderman RG, ASA Special Publication 54, Madison, pp 101-124.

Timmer L and Leyden R (1980) The relationship of mycorrhizal infection to phosphorus-induced copper deficiency in sour orange seedlings. New Phytol. 85, 15–23.

Tran, Van Con (2001) Define some species for production plantation in the north central highlands. Scientific Report. FSIV.

Uren NC and Reisenaur HM (1988) The role of root exudates in nutrient acquisition. In: Advances in Plant Nutrition Vol. 3 (Ed) Tinker B and Lauchli A, Praeger, New York, pp 79-114.

Warcup JH (1990) The mycorrhizal associations of Australian Inuleae (Asteraceae). Muelleria, 7, 179-187.

Wright (2005) Roots and Soil Management: Interactions between roots and the soil. In: Management of Arbuscular Mycorrhizal Fungi (Ed) Zobel RW and Wright SF, Am. Soc. Agron. pp183-197.

Xie Z, Staehelin C, Vierheilig H, Weimken A, Jabbouri S, Broughton W, Vogeli-Lange R and Thomas B (1995) Rhizobial nodulation factors stimulate mycorrhizal colonization of ndulating and nonnodulating soybeans. Plant Physiol. 108 (4), 1519-1525.

Yamato, Masahide (2005) Morphological types of arbuscular mycorrhizas in pioneer woody plants growing in an oil palm farm in Sumatra, Indonesia. Mycoscience 46, 66.

12

Biological Synergism for Reclamation of Mined Lands

Anuj Kumar Singh and Jamaluddin

Abstract

Mine spoils pose adverse conditions for soil microbes and plant growth, due to its low organic content and other essential nutrients, unfavourable soil chemistry and poor soil texture. Poor microbial population inhibits nutrient transformation, consequently the establishment of the plants and ultimately affects the process of ecological succession. A synergy among different groups of beneficial microbes accompanied by plantation of suitable tree species on mined over area has proved to be effective for speedy and efficient revitalization of the mined affected lands. Inoculation of plantation with efficient strains of native microbes accelerates the soil redevelopment and helps to establish nutrients cycle readily which attracts native flora thereby the pace of natural regeneration is increased. It is therefore imperative to consider the application of effective microflora in restoration of mined lands along with the plantation.

Keywords: Biological synergism, Microbes, Mines restoration.

1. Introduction

Mining causes massive damage to landscapes and biological communities of the earth. Natural plant communities get disturbed and the habitats become impoverished due to mining, presenting a very rigorous condition for plant's growth, possesses a very serious threat to the environment, resulting in the reduction of forest cover, erosion of land in a greater scale, pollution of air, water and land and reduction in biological diversity. The problem of waste rock dumps has become devastating to the landscape around mining areas.

Mining and forestry comes together with rehabilitation. Kopp (1978) viewed rehabilitation as an early dominant and desired activity in the development of forestry. Forest researchers and foresters have been involved in rehabilitation of mine spoils on a massive scale from the turn of this century (Sheail,1974) and still the technologies and approaches are being developed and updated from time to time to accelerate the development process in different mine environments. The large-scale land disturbances associated with mining activities and related concerns about the environmental deterioration have triggered the increasing number of rehabilitation strategies, which aim for the restoration of the disturbed land and ecosystem.

It has long been recognized that soil microorganisms are the major driving force behind nutrient transformation in soil, thus they have a major role in soil fertility and ecosystem functioning. Microbial biomass is both the agent of biochemical changes in soil and a repository of plant nutrients that are more labile than the bulk of the soil organic matter. Nitrogen fixing microbes have a pivotal role in soil fertility and plants growth on disturbed lands. Nitrogen fixing microbes can exist as free-living organisms in associations of different degrees of complexity with other microbes and plants. Positive effects of *Rhizobium* sp. inoculation in combination with *Azotobacter* sp. or *Azospirillum* sp. inoculants have been reported for a variety of legumes and forage species. Other significant groups including phosphate solubilising microorganisms (PSM) and mycorrhizae have been applied in restoration of disturbed areas like mined lands. Development of effective microbial strains of beneficial microbes have also been another significant aspect in the process of restoration of disturbed soil ecosystems (Singh and Jamaluddin 2011b). Microbial or biological approach including the application of microbial biomass along with the plantation of suitable tree species primarily leguminous one have long been practised with necessary modifications based on prevailing site conditions. This technique is highly recommended due to its ecosystem friendly nature and comparatively low cost of operation. The present article endeavors to review some of the significant microbial aspects, synergy among biological components and processes involved in rehabilitation of different mined lands.

2. Role of Microorganisms in Soil System

It is universally accepted that the microbes are beneficial in the activity of important functions performed in the ecosystem. Soil microbial biomass, living part of the soil organic matter, is an agent of transformation of organic matter and source of available nutrients. The effective groups of microbes belong to bacteria, fungi, and actinomycetes. Activities of soil microbes can change the soil environment much rapidly and provide key controlling influence on the rate at which nutrient cycling processes take place and play a very important role in the formation of macroaggregates.

The potential of soil microorganisms has been recognized widely in improvement of soil quality, soil formation, aggregation and revegetation through their role in litter decomposition and nutrient cycling. Their activities such as phosphate solubilization, nitrogen fixation, oxidation of various inorganic components of soil or mineralization of inorganic components and mycorrhizal symbiosis are major beneficial activities that play a very important role in soil system functioning. Soil microorganisms inhabiting the rhizosphere environment interact with plant roots and mediate nutrient availability to the plants. Implications of plants and their symbionts like mycorrhizal fungi, N-fixing bacteria and free-living rhizosphere population of bacteria promote plants' establishment and growth. In addition to their effects on soil fertility, they also enhance soil structure by binding together soil particles. An active soil microbial biomass is an essential factor in the long-term fertility of soils.

A number of studies have been previously conducted in order to establish the role of microbes in plants' growth in varying soil and climatic conditions. Microorganisms improve the nutrient status and texture by addition of organic matter (Palaniappan and Natarajan 1993). Free living or symbiotic nitrogen fixers improve the nitrogen status with micronutrients and growth promoters. The microorganisms alter the pH of the habitat making it suitable for the establishment of higher plants. The humic and fulvic acid fractions of humus are known to chelate micronutrients like Cu, Fe, Zn and Mn and also exert buffering action (Relan et al. 1986).

3. Microbes in Different Mine Environment

Overburdens when dumped in un-mined areas in the vicinity of the mines create mine spoils. Nutrient deficient spoils are generally hostile to plant growth and reclamation strategies other than natural colonization of mine spoils are very painstaking process. Some important researches on the study of the impact of mining on the soil, vegetation and associated microbial population are reviewed hereby. These mine spoils represent extremely rigid substrata for plant growth and development. Colonization, establishment and maintenance of vegetation on these spoils are enormously difficult. The physical factors, which limit plant establishment and survival include high temperature, moisture stress (Richardson 1975), soil particle size (Down 1974) and compaction (Hall 1957). Soil fertility is also a major factor regulating the plant growth. The shortage of organic matter is attributed to the absence of litter (Schafer et al. 1980). Power (1978) considers soil physico-chemical characteristics like texture, pH, electrical conductivity, soluble Ca, Mg, Na, B, cation exchange capacity, exchangeable cations, gypsum and calcium carbonate equivalents as being crucial to the prediction of plant growth potential of mine overburdens with water holding capacity and infiltration rates as the other important variables. Bradshaw *et al.* (1975) and Bell and Ungar (1981) found high temperature and low moisture of surface coalmine spoils to be important factors limiting plant growth.

The factors contributing to the early colonization of mine dumps have given considerable attention by various workers. Bradshaw (1983), Chadwick (1973), Byrens and Miller (1973) found natural succession on coal mine spoils a slow process due to surface mining altering physico-chemical properties. These spoils present a special habitat where conditions are extremely unfavourable for plant growth and establishment. Marrs and Bradshaw (1980) and Marrs *et al.* (1981) studied the development of ecosystem of China clay waste. Iron mine tailings were studied by Leisman (1957); and Shetron & Duffek (1970). Floristic diversity of lead mining wastes was studied by Clarke and Clarke (1981), lead and zinc by Kimmerer (1984) and copper mining wastes by Goodman and Gemmel (1978); Veeranjaneyulu and Dhanaraju (1990). Doerr and Guernsely (1956) dealt with the environmental effects of strip mining and underground mining, which create conspicuous landscape features and associated phenomena. Mukherjee (1987,) described about the land degradation associated with surface and sub-surface mining. Chadwick *et al.* (1987) outlined the environmental implications of increased coal mining and utilization. Chaudhury (1992) dealt with the impact on mining activities on environment and also the management and protection of the mined areas.

To the extent possible, mine spoils need to be levelled or terraced in order to provide suitable substratum. Levelling will vary according to the type of mine, methods of mining and the way in which a particular area has been worked. For instance, in case of surface and opencast mines the procedures will be levelling and fencing of the area. In case of shaft and underground mines although overburden can be treated in similar fashion but mined out areas and abandoned mines will have to have different strategy depending upon the context. Mined out areas in hillside slopes may require contour dikes. Levelling will provide a base of coarse material over which to spread sediment. Some of the large mining pits that have developed into reservoirs can be developed as water-bodies aesthetically appealing for ecotourism and simultaneous fish culture to support local livelihoods. The main physical problems with mine spoils are shallow substrate of soil (or often lack of it), large cavities in the very coarse-grained substrate, very high stone content, extremely coarse texture, compaction, and the limited availability of moisture.

Mining activities have a deleterious effect on plant and soil microbial community health that can be ameliorated with the addition of amendments in order to achieve successful biological rejuvenation. Amendments provide a nutrient source and an environment that enhances the development of a fully structured and functional soil microbial community (Tate *et al.* 1985). Functional microbial communities are composed of a complex species that differ in their environmental tolerances, physical requirements and habitat adaptations (Norland *et al.* 1991). The soil microbial community functions in a number of biogeochemical processes that are fundamental to soil development and plant growth.

For reclamation of such problematic mine spoils, microbial inoculants like PSM, *Azospirillum, Rhizobium* and arbuscular mycorrhizal fungi (AMF) can be used because when added to the spoil they provide special advantages. Fixation of atmospheric nitrogen and mobilization of essential micronutrients make them easily accessible to plants. This approach leads to achieve the fertility, thus improving the water holding capacity of soil and creating topsoil to sustain high quality vegetation cover. The rejuvenation of mine spoil dump and mined land productivity and fertility through amendment of these microbial inoculants may enable restoration of the degraded land ecosystem. Thus, the improvement in the physico-chemical and microbial status of soil through organic blending, inoculation with biofertilizers, screening of suitable plant species and establishment of bio-geochemical cycle in the mine spoils are, therefore, essential to achieve the objectives of restoration of land fertility, productivity and biological rejuvenation of limestone mined spoils.

4. Plant Growth Promoting Rhizobacteria (PGPR) and Soil Fertility

The rhizobacteria that exert beneficial effects on plant development are termed as plant growth promoting rhizobacteria (PGPR). Some PGPR can improve nodulation and N_2 - fixation in legume plants (Zhang *et al.* 2004; Lucas-Garica *et al.* 2004). Inoculation of phosphate solubilizing bacteria (PSB) enhanced nodulation and N_2-fixation in Alfalfa plant, in parallel with an increase in the phosphorus content of plant tissues (Toro *et al.* 1998). It is therefore obvious that an improvement in phosphorus nutrition of the plant resulting from the presence of PSB responsible for increased nodulation and N_2 - fixation, as it is well known that these processes are phosphorus dependent (Barea *et al.* 2005).The PGPR are known to participate in many important ecosystem processes, such as the biological control of plant pathogens, nutrient cycling and seedling growth. Selected strains of PGPR have been used as seed inoculants. The PGPR mediated processes involved in nutrient cycling include those related to non-symbiotic nitrogen fixation and those responsible for increasing the availability of phosphate and other nutrients in the soil. Many asymbiotic diazotrophic bacteria have been described and tested as biofertilizers (Barea *et al.* 2004).

Azospirillum sp. are considered as one of the most important PGPR. A significant activity of these bacteria is the production of auxin type hormones that affect root morphology and thereby improve nutrient uptake from the soil. This may be more important than their N-fixing activity. *Azospirillum* are being used as seed inoculants under field conditions (Zahir *et al.* 2004). In disturbed soil conditions where nutrients particularly nitrogen is a constraint, *Azospirillum* can immensely help in nitrogen nutrition of the plants through atmospheric nitrogen fixation.

Azotobacter sp. are known to influence plant growth through their ability to fix nitrogen, excretion of ammonia in the rhizosphere along with root exudates, production of growth promoting substances and phosphate solubilization. The multiple action of *Azotobacter* contributes to better germination percentage of seeds, increase root and shoot length and improve nitrogen nutrition to plants (Shende *et al.* 1977).

Pseudomonas is typical PGPR and their interactions with AM fungi mutually enhanced each other's colonization and achieved additive plant growth enhancement. Another mechanism of action of PGPR on plant growth is the production of siderophores. The siderophores are produced by most fungi and bacteria including *Pseudomonas, Rhizobium* and *Azotobacter* (Meyer and Linderman 1986). The species of *Pseudomonas, Bacillus, Aspergillus, Penicillium* etc. have been reported to be active in the bioconversion of insoluble phosphorus. These organisms produce organic acids like citric, glutamic, succinic, lactic and tartaric which are responsible for solubilization of insoluble forms of phosphorus. Phosphorus solubilizing microorganisms synergistically interact with N-fixing microorganisms. Taking into cognizance, the phosphorus solubilizing microorganisms are being exploited as biofertilizers in agriculture, horticulture, forestry, agroforestry (Gaur 1990) and the same needs to be extended for mined spoils restoration.

5. Interaction Between PGPR and Arbuscular Mycorrhizae

The role of microorganisms in nutrient cycling is unique. An active biomass is an essential factor in the long-term fertility of restored soil. It is therefore essential that microbes beneficial for plant growth have to be introduced to the spoils. Among the different microbes AMF, nitrogen fixers and phosphate solubilizers are very important for any plant. Microbial activity in the rhizosphere affects rooting patterns and the supply of available nutrients to plants thereby modifying the quality and quantity of root exudates (Gryndler 2000; Barea, 2000). Carbon fluxes are crucial determinants of rhizosphere function. The release of root exudates and decaying plant materials provide sources of carbon compound for the heterotrophic soil biota as either growth substrates or structural material for root associated microbiota (Werner 1998). Nitrogen-fixing microbes can exist as free-living organisms in associations of different degrees of complexity with other microbes and plants. The most abundant elements in the atmosphere (N_2) are very often the limiting element for the growth of most organisms. Many soil organisms interact with each other to overcome of the limitation. Positive effects of *Rhizobium* sp. inoculation in combination with *Azotobacter* sp. or *Azospirillum* sp. inoculants have been reported for different forage and grain legumes.

Phosphorus is an important plant nutrient next to nitrogen for plants growth. The most important aspect of phosphorus cycle is microbial mineralization, solubilization and immobilization besides chemical fixation of phosphorus in the soil. Phosphorus solubilising microorganisms convert insoluble inorganic phosphate compounds into soluble form. A considerable higher concentration of phosphate solubilising bacteria is commonly found in the rhizosphere soil. Also the fungal genera with this capacity are *Penicillium* and *Aspergillus* (Suh *et al.* 1995; Whitelaw *et al.* 1999). Singh and Jamaluddin (2010) have reported the beneficial effects of *Pseudomonas fluorescens* in the growth of *P. pinnata and J. curcas* in limestone mined spoils. AMF association enhances nitrogen gain in ecosystems by increasing the N-fixation rates of plant and N_2-fixing bacterial associations. In many ecosystems a large fraction of the available nitrogen in the soil is ammonium and mycorrhizal fungi readily transport ammonium from soil to plant (Ames *et al.* 1983). Nitrogen fixed in association with one plant is transported to an adjacent plant *via* hyphal connections. N fixed by soybeans was transported to maize *via* the mycorrhizal hyphae which significantly increased the growth and nitrogen status of maize plants (Van Kessel *et al.* 1985).

As mycorrhizae may enhance the ability of the plant to cope with water stress conditions associated to nutrient deficiency and drought, mycorrhizal fungi has been proposed as a promising tool for improving restoration success in semi-arid degraded areas. By stimulating the development of beneficial microorganisms in the rhizosphere, the use of AMF infected plants could reduce the amount of fertilizer needed for the establishment of vegetation and could also increase the rate at which the desired vegetation becomes established by stimulating the development of beneficial microorganisms in the rhizosphere. Degraded soils are common targets of revegetation efforts in the tropics, but they often exhibit low densities of AMF fungi. This may limit the degree of mycorrhizal colonization in transplanted seedlings and consequently hamper their seedling establishment and growth in those areas. Inoculation of native and well adapted microbial flora may prove a proficient tool for restoration of heavily degraded mine spoils. Native beneficial microbial flora like AMF along with PSB and N fixing bacteria were isolated, multiplied and re-inoculated in different important plant species viz. *Pongamia pinnata, Jatropha curcas, Withania somnifera* and *Ailanthus excelsa*. All the inoculated plants exhibited enhanced growth and development as compared to uninoculated one. Moreover, inoculation with beneficial PGPR changed the soil characteristics and also allowed increased invasion and natural succession on planted spoil as compared to unplanted sites. Similar results on enhanced growth of planted species with inoculation of *G. mosseae* in limestone mine spoils were also reported by Rao and Tak (2002). It indicated soil rebuilding and alteration of soil conditions from harsh to conducive for plants growth.

6. Phosphorus Nutrition in Disturbed Lands

Phosphorus is an important plant nutrient next only to nitrogen for plant growth. The most important aspect of phosphorus cycle is microbial mineralization, solubilization and immobilization besides chemical fixation of phosphorus in the soil. Phosphorus solubilizing microorganisms convert insoluble inorganic phosphate compounds into soluble form. A considerable higher concentration of phosphate solubilizing bacteria is commonly found in the rhizosphere soil. Also the fungal genera with this capacity are *Penicillium* and *Aspergillus* (Suh *et al.* 1995; Whitelaw *et al.* 1999).

A significant amount of phosphorus is bound in organic forms in the rhizosphere. Phosphatase enzymes produced by plants and microbes are presumed to convert organic phosphorus into available phosphate, which is absorbed by plants. Phosphate availability is one of the major growth limiting factors for plants in many natural ecosystems. Plants absorb phosphorus from the soil as inorganic phosphate ions, but their availability is severely restricted by reactions of inorganic and organic phosphates with soil constituents. These enzymes can significantly import the availability and recycling of phosphorus in and around the rhizosphere. Phosphatases constitute an enzyme group which is presumed to catalyze the hydrolysis of several organic phosphate monoesters and liberating available phosphorus and occurring scattered in all tissue cells of plant organs. Extracellular soil acid phosphatases catalyze the hydrolysis of organic phosphate and thus constitute an important link between biologically unavailable and bio-available phosphorus pools in the soil (Juma and Tabatabai 1988, Krishna and Bagyaraj 1985).

Phosphatase is related to plant's ability to make soil phosphorus available for absorption. The intracellular acid phosphatase is responsible for the P-hydrolysis from organic compounds favouring phosphorus mobilization and translocation from senescent tissues (Lefebvere *et al.* 1990; Duff *et al.* 1994). Therefore, this enzyme activity is a physiological characteristics related to plant efficiency in relation to phosphorus acquisition and utilization (Tadano *et al.* 1993). Plants usually secrete root acid phosphatase when phosphorus availability is low. However, plant species differ in secretion ability and enzyme activity (Yan *et al.* 2001). Studies conducted at the Indian Institute of Horticulture Research, Bangalore, India on phosphatase activity in *Carica papaya* roots showed that the inoculation with arbuscular mycorrhizal fungi enhanced the activity of both acid and alkaline phosphatases on the root surface and soil surrounding the root region. They also observed acid phosphatase activity was much greater than alkaline phosphatase activity. Comparatively acid phosphatase activity has been reported to be more promising in phosphorus solubilization. Acid phosphatases contribute to the mineralization of organic phosphorus compounds in the soil, thus enhancing the biological availability of released inorganic phosphorus. Both the roots and mycorrhizal symbionts produce acid phosphatases. Studies conducted at the International Crops Research Institute for the Semi - arid Tropics (ICRISAT), Telangana, India showed that in sterilized soil, acid and alkaline phosphatase activities in the rhizospheres of mycorrhizal and non-mycorrhizal plants were higher than in non- rhizosphere soil. With acid phosphatase, the activity was highest in the rhizosphere of mycorrhizal plants followed by rhizosphere of non-mycorrhizal plants and non-mycorrhizal soil (Krishna and Bagyaraj 1985). Root and rhizosphere levels of acid phosphatase activity were found to be higher in plants inoculated with AMF (Dodd *et al.* 1987). The studies conducted by Antibus and Dighton (1993) also concluded that the phosphatase activity was linked with phosphorus uptake after carrying out experiment on Red maple and Grey birch. The above reports are sufficient to establish the role of phosphatases in phosphorus solubilization and availability to the plants.

7. Development of Effective Strains of Native Microbes

As it is well known that mine overburden dumps are quite deficient in plant nutrients and due to inhospitable soil conditions, the growth and establishment of plants or trees is nearly impossible. Such soil needs reestablishment of disturbed nutrients cycle and supply of beneficial nutrients. However, in disturbed lands, the microbial community which is largely responsible for nutrients decomposition and mineralization is also disturbed or very scanty in population. Thus the primary objective of land rehabilitation is the restoration of microbial population. The well adapted native microbial flora could be of immense importance to accelerate the restoration process.

The beneficial PGPR isolates from the disturbed soils must undergo laboratory assay for the development of more resistant strains which would be more tolerant towards the harsh soil conditions of the mining area. Microbial isolates, particularly bacterial isolates native to the acidic or alkaline mine overburden dumps inhabiting the soil of unfavourable pH are well adapted to that medium and has enhanced tolerance towards such soil conditions. Such efficient strains may be developed and multiplied in bulk for use in the fields which are deficient in nutrients, phosphorus in particular. Leyval *et al.* (1990) also reported bacterial solubilization of mica mineral which led to the release of unavailable nutrients to Pine and Beech. Watteau and Berthelin (1994) emphasized that the organic acids produced by rock dissolving bacteria enhanced the weathering of primary rock minerals by chelating cat ions from minerals. The rhizosphere of the plant which is rich in microbial diversity especially of bacteria, contribute significantly to the supply of nutrients to plants (Hinsinger and Gilkes 1993).These bacterial population can be developed in more capable strains and can be multiplied for application in the field. Chang and Li (1998) demonstrated induced weathering processes by combined effect of plant root and soil microbes which may be responsible for transformation of mineral structure, removing cations from the minerals for rapid uptake by plants. Seshadri *et al.* (2000) tested three strains of *Azospirillum* sp. to study their capability to solubilize inorganic phosphate *in vitro*. Gaur (1990) reported solubilization of tricalcium phosphate in liquid medium by using bacterial inoculants and demonstrated phosphorus solubilizing bacteria as a significant bacterial biofertilizers. Agnihotri (1970) tested some fungal isolates for efficiency of solubilization of insoluble phosphate and observed the solubilization process to a maximum extent. Similar observations in respect to the activity of *Streptomyces sp.* against rock phosphate solubilization were made by Bardiya and Gaur (1974). The periodic solubilization of ferric and aluminium phosphate was investigated in liquid medium by Gaur and Gaind (1983). Ahmed and Jha (1968) reported the solubilization of rock phosphate and hydroxyapatite by some bacteria and fungi. Wani *et al.* (1979) also confirmed the rock solubilization ability of selected strains of *Pseudomonas striata* and *Bacillus polymyxa* on both solid and liquid media. The investigations carried out in the present study are an indication of the development of new strains of the test organisms. These improved strains can be further multiplied in bulk and can be used for biological solubilization of rock bound minerals in nutrient deficient soils like mine overburden dumps.

For instance, in reclamation of limestone mine overburden dumps, the phosphate solubilizing bacteria (PSB) should be screened out natively from the mined dumps and more calcium tolerant strains should be developed through rigorous laboratory assays and multiplied in bulk and applied at the restoration sites. Such tolerant strains might more effectively break down or solubilize calcite rocks or rock bound minerals than the organisms inhabiting in normal soil. It has been previously reported that the bacterial isolates which were native to calcareous soil were capable to solubilize calcite rock more effectively than the introduced one (Singh and Jamaluddin 2011a). It is recommended that the bacterial strains, which are native to such calcareous soil system should be screened out and multiplied in bulk so that they can be applied for the purpose of calcite rock solubilization and release of rock bound minerals in the nutrient deficient soils like limestone mined over burden dumps and other similar disturbed sites. Such efficient strains of PGPR are earlier reported to be capable enough to solubilize rock bound phosphates and play a very important role in nutrients supply particularly of phosphorus in nutrients deficit soils. Such efficient strains may be of vital importance for vegetation of mined out areas.

8. Plantation Accelerates Ecosystem Recovery

Plantation is the oldest technology for the restoration of lands damaged by human activity. A primary objective for achieving satisfactory rehabilitation of a mined landscape is to establish a permanent vegetation cover. There is increasing evidence that forest plantations can play a key role in harmonizing long-term forest ecosystem restoration goals with near-term socioeconomic development objectives. Plantations can play a critical role in restoring productivity, ecosystem stability and biological diversity to degraded area. Relative to unplanted sites, plantations have a marked catalytic effect on native forest development (succession) on severely degraded sites. Numerous studies have demonstrated that land rehabilitation benefits from plantations because it allows jump-starting succession. The catalytic effects of plantations are due to changes in under storey microclimatic conditions (increased soil moisture, reduced temperature, *etc.*), increased vegetational- structural complexity, and development of litter and humus layers that occur during the early years of plantation growth. The development of a plantation canopy can alter the understorey microclimate and the soil physical and chemical environment to facilitate recruitment, survival and growth of native forest species. Singh and Jamaluddin (2011b) reported 14 species including trees, herbs and shrubs growing naturally in limestone mined spoils in Central India. Those included *Lantana camara, Calotropis procera, Acacia nilotica, Acacia catechu, Butea monosperma, Ocimum gratissimum, Tridax procumbens, Ailanthus excelsa, Ipomoea* sp.*, Zizyphus mauritiana, Phyllanthus niruri, Parthenium* *hysterophorus, Azadirachta indica* and *Argemone mexicana* among the prominent species growing naturally on calcite soils.

The colonization of plant species on coal mine spoils is influenced by the particle size of the soil derived from the overburden and coal mine wastes. With high clay content, the soil becomes water logged, whereas with silt content, the soils become compact forming crust which often restrict seedling growth and entry of water and air into the soil system (Richardson *et al.* 1975). Soil pH has been a major determinant in controlling plants' growth on impoverished lands such as mine spoils. Some of the previous studies have suggested different tree species suitable to the land intended for restoration. However, native and nitrogen fixing tree species have been a constant recommendation for most of the degraded lands. Revegetation of iron-ore mine areas of Madhya Pradesh was studied by Prasad (1989) who observed better growth performance of *Dalbergia sissoo, Albizia procera, Pongamia pinnata*, etc. in the manured pits.

Revegetation of mine spoils with perennial species is important for accelerating the natural succession. Plant species with higher tolerating capacity often produces extensive ground cover, plant biomass and productivity that prevent soil from erosion and increases the infiltration rate of the water in the soil. Increase of plant biomass and productivity leads to pleasing appearance of overburden dump. Thus, plantations may act as 'foster ecosystems', accelerating development of genetic and biochemical diversity on degraded sites. Plantations have an important role in protecting the soil surface from erosion and allowing the accumulation of fine particles. They can reverse degradation process by stabilizing soils through development of extensive root systems. Once they are established, plants increase soil organic matter, lower soil bulk density, and moderate soil pH and bring mineral nutrients to the surface and accumulate them in available form. The plants accumulate these nutrients and re-deposit them on the soil surface in organic matter, from which nutrients are much more readily available by microbial breakdown. Most importantly, some species can fix and accumulate nitrogen rapidly in sufficient quantities to provide a nitrogen capital, where none previously existed, more than adequate for normal ecosystem functioning. Once the soil characteristics have been restored, it becomes easier to restore a full-fledged vegetation.

In the dolomite mine spoils, Prasad (1989) reported the successful plantations of *Gmelina arborea, A. auriculiformis, E. tereticornis* and *P. pinnata* and it also helped in acceleration of natural successional process in mined spoils. Earlier, Prasad and Pandey (1985) recommended *P. pinnata, D. strictus, A. lebbek, A. procera* and *E. officinalis* for the restoration of limestone mined out areas. Goswami and Jamaluddin (2006) have described successful establishment of *Eucalyptus hybrid, Casia siamea, Acacia auriculiformis,*

Dalbergia sissoo and *Dendrocalamus strictus* in iron ore mine spoils in central India. Much similarly, Mukhopadhyaya and Maiti (2010) have reported that tree species like *A.auriculiformis, Casia siamea, Dalbergia sissoo, Gmelina arborea, Leucaena leucocephala, Dendrocalamus strictus and Terminalia arjuna* have been performing well in coal over burden dumps in eastern India. The successful establishment and growth of *P.pinnata, Withania somnifera, Jatropha curcas, Acacia nilotica and Ailanthus excelsa* in limestone mine overburden dumps have been reported by Singh and Jamaluddin (2011b). They have also studied the catalytic effects of plantation supplemented by microbial inoculation in natural regeneration of native species. The frequency and density of regenerated species have been recorded in greater numbers in inoculated area than uninoculated one. This suggested that the inoculation helped plantation to attract surrounding flora on mine spoils.

Fig. 1: Limestone mined spoils heaped around mines

Fig. 2: Different strains of *Pseudomonas fluorescens* isolated from calcite mine spoils

Fig. 3: Growth of *P. pinnata* in calcite mine spoil

Fig.4: Growth of *J. curcas* in calcite mine spoil

Conclusion

The application of native microbial population which is well adapted and stress tolerant and plantation of fast growing tree species may ensure primary goal of re-establishment of the soil's natural biogeochemical cycles. Such progress, in turn, would allow the natural invasion of multiple herbs, shrubs and tree species that would not only help in soil stabilization but would enhance the soil's physico-chemical and nutritive properties. In addition, integrating such microbial biofertilizers as *Rhizobium, Azospirillum* and AMF along with phosphate solubilizing microbes at the early stage of the plants' growth would enable the plant to become more tolerant to stress by ensuring continuous supplies of nutrients during their early stages of growth. The use of microbial biofertilizers would reverse the damage and will subsequently enhance the nutritive capacity of soils and the plantation will catalyze the restoration process. A synergy between the above ground and below ground flora would thus lead to accelerated recovery of the ecological balance of the mine degraded land.

References

Agnihotri UP (1970) Solubilization of insoluble phosphates by some soil fungi isolated from nursery seedbeds. Can. J. Microbiol. 6, 877-880.

Ahmed N, Jha KK (1968) Solubilization of rock phosphate by microorganisms isolated from Bihar soils. J. Gen. Appl. Microbiol. 14, 89-95.

Ames RN, Reic CPP, Proter LK, Cambardella C (1983) Hyphal uptake and transport of nitrogen from two N-labeled sources by *Glomus mosseae* a vesicular arbuscular mycorrhizal fungus. New Phytol. 95, 381-396.

Antibus RK, Dighton J (1993) Acid phosphatase activities and oxygen uptake in mycorrhizas of selected deciduous forest trees. In: Proceedings of the 9th North American Conference on Mycorrhizae, 8-12 August, 1993,University of Guelph, Guelph, Ontario, Canada.

Barea JM (2000) Rhizosphere and mycorrhizae of field crops. In: Biological Resource Management, Connecting Science and Policy (OECD) (Ed) Toutant P, Balazs E, Galante E, INRA and Springer, Berlin, Heidelberg, New York, pp110-125.

Bradshaw AD, Dancer WS, Handley JF, Sheldon JC (1975) Biology of land revegetation and reclamation of china-clay wastes, In: The Ecology of Resource Degradation and Renewal. The 15th Symposium of the British Ecological Society, 10-12 July 1973 (Ed) Chadwick, MJ and Goodman, GT, Blackwell Scientific Publication, Oxford, pp 363-384.

Byrens WR, Miller JH (1973) Natural revegetation and cast over burden properties of surface-mined coal lands in southern Indian, In: Ecology and Reclamation of Devastated Land (Ed), Hutnik RJ and Davis G, Gordon and Breach, New York, pp 285-306.

Chadwick KJ (1973) Methods of assessment of acid colliery spoils as a medium for plant growth, 81-91. In: Ecology and Reclamation of Devastated Land. Vol. I (Ed) Hutnik RJ and Davis G, Gordon and Breach, New York.

Chang TT, Li CY (1998) Weathering of limestone, marble and calcium phosphate by ectomycorrhizal fungi and associated microorganisms. Taiwan J. Forestry Sci. 13(2), 85-90.

Chaudhury AB (1992) Mine Environment and Management (An Indian Scenerio). Ashish Publishing House. New Delhi.

Doerr A, Guernsely L (1956) Man as a geomorphological agent: an example of coal mining. Ann. Assoc. Am. Geog. 46, 197-210.

Down CG (1974) The relationship between colliery waste particle sizes and plant growth. Environ. Conserv.1, 29-40.

Duff SMG, Sarath G, Plaxton WC (1994) The role of acid phosphatase in plant phosphorus metabolism. Physiol. Planatarum, 90, 791-800.

Gaur AC, Gaind S (1983) Microbial solubilization of phosphates with particular reference to iron and aluminium phosphate. Sci. Cult. 49, 110-112.

Goodman GT, Gemmel RP (1978) The maintenance of grassland on smelter wastes in the lower Swansea Valley. II. Copper smelter waste. J. Appl. Ecol. 15, 875-883

Goswami MG, Jamaluddin (2006) Distribution of mycorrhizal fungi in iron mine areas of Chhattisgarh. Indian Phytopath. 59(3), 395.

Gryndler M (2000) Interactions of arbuscular mycorrhizal fungi with other soil organisms, In : Arbuscular Mycorrhizas: Physiology and Function (Ed) Kapulnik Y and Douds Jr DD, Dordrecht, Kluwer academic publishers, The Netherlands, pp 239-262.

Hall IG (1957) The ecology of disused pit heaps in England. J. Ecol.45, 689-720.

Hinsinger P, Gilkes RJ (1993) Root-induced irreversible transformation of a trictahedral mica in the rhizosphere of rape. J. Soil Sci. 44, 535-545.

Hinsinger P, Gilkes RJ (1995) Root-induced dissolution of phosphate rock in the rhizosphere of lupins grown in alkaline soil. Aust. J. Soil Res. 33, 477-489.

Juma NG, Tabatabai MA (1988) Phosphatase activity in corn and soybean roots: conditions for assay and effects of metals. Plant Soil 107, 39-47.

Kimmerer RW (1984) Vegetation development on a dated series of abandoned lead and zinc mines in southwestern Wisconsin. Am. Midl. Natl. 111, 332-341.

Kopp H (1978) Forestry in the federal republic of Germany, with particular reference to the rehabilitation of forest ecosystems In: The Breakdown and Restoration of Ecosystems (Ed) Holdgate MW and Woodman MJ, Plenum press, New York, pp 279-292.

Krishna KR, Bagyaraj DJ (1985) Phosphatases in the rhizospheres of mycorrhizal and non-mycorrhizal groundnut. J. Soil Biol. Ecol. 5(2), 81-85.

Lefebvre DD, Duff SMG, Fife Julien-Inal, Singh C, Plaxton WC (1990) Response to phosphate deprivation in Brassica nigra suspension cells. Enhancement of intracellular cell surface and secreted phosphatase activities compared to increases in Pi-absorption rate. Plant Physiol. 93, 504 -511.

Leisman GA (1957) A vegetation and soil chronosequence on the Mesabi iron range spoil banks, Minnesota. Ecol. Monogr. 27, 221-245.

Leyval C, Turnau K, Haselwandter K (1997) Effect of heavy metal pollution on mycorrhizal colonization and function: physiological, ecological and applied aspects. Mycorrhiza 7, 139-153.

Lucas-Garica JA,Probanza A,Ramos B,Colon-Flores JJ,Guitierrez-Manero FJ (2004) Effects of plant growth promoting rhizobacteria (PGPRs) on the biological nitrogen fixation, nodulation and growth of Lupinus albus I.cv.Multolupa. Eng. Life Sci. 4, 71-77.

Marrs RH, Roberts RD, Skeffington RA, Bradshaw, AD (1981) Ecosystem development on naturally-colonized china clay wastes. II. Nutrient compartmentation. J. Ecol.69, 163-169.

Marrs RH, Bradshaw AD (1980) Ecosystem development on reclaimed china-clay wastes. III. Leaching of nutrients. J. Appl. Ecol. 17, 727-736.

Meyer RJ, Linderman RG (1986) Response of subterranean clover to dual inoculation with vesicular arbuscular mycorrhizal fungi and plant growth promoting rhizobacterium Pseudomonas putida. Soil Biol. Biochem.18(2), 185-190.

Mukherjee R (1987) Surface mining and land degradation in Raniganj coal field, Barddman, Geog. Rev. of India, Vol 49, 1-69.

Mukhopadhyay S, Maiti SK (2010) Natural mycorrhizal colonization in tree species growing on the reclaimed coalmine overburden dumps: Case study from Jharia Coalfields, India. The Bioscan 3, 761-770.

Norland MR, Veith DL and Dewar SW (1991) Initial vegetative cover on coarse taconite tailing using organic amendments on Minnesota's Iron Range. In Proc. 8th Meet. Amer. Soc. Surf. Min. Reclam. Durango, CO., pp 263-277.

Palaniappan SP, Natarajan (1993) Practical aspects of organic matter maintenance in soils. In: Organics in Soil Health and Crop Production (Ed) Thompson PK, Tree crops development foundation, India, pp 23-41.

Prasad R (1989) Removal of forest cover through mining and technology for retrieval. Proceedings of National Seminar on depletion of soil and forest cover. J. Trop. Forest. 5, 109-116.

Prasad R, Pandey RK (1985) Natural plant succession in the rehabilitation of bauxite and coal mine overburden in Madhya Pradesh. J. Trop. Forest. 1, 79-84.

Rao AV, Tak R (2002) Growth of different tree species and their nutrient uptake in limestone minespoil as influenced by arbuscular mycorrhizal (AM) fungi in Indian arid zone. J. Arid Environ. 51, 113-119.

Relan PS, Tekchand SS, Kumari R (1986) Stability constraints of Cu, Pb, Zn, Fe and cadmium complexes with humic acids from manure. J. Indian Soc. Soil. Sci. 34, 250.

Richardson JA (1975) Physical problems of growing plants on colliery wastes. In: Ecology of Resource Degradation and Renewal (Ed) Chadwick MJ and Goodman GT, Blackwell Scientific Publication, Oxford, England, pp 275- 285.

Sheail J (1974) Early experience in wasteland planting. Town Country Plan. 42, 224-226.

Sheshadri S, Muthukumarasamy, Lakshminarsimhan C, Ignacimuthu S (2000) Solubilization of inorganic phosphates by *Azospirillum halopraeferans*. Curr. Sci. 79(5), 565-567.

Shetron SG, Duffek R (1970) Establishing vegetation on iron mine tailings. J. Soil Water Conserv. 25, 227-230.

Singh AK, Jamaluddin (2010) Role of Microbial inoculants on growth and establishment of plantation and natural regeneration in limestone mined spoils. World J. Agr. Sci. 6(6), 707-712.

Singh AK, Jamaluddin (2011a) Development of resistance and enhanced mineral solubilization capacity of plant growth promoting rhizobacteria native to calcareous soils of limestone mined spoils. Indian Forester 37(12), 1366-1370.

Singh AK, Jamaluddin (2011b) Status and diversity of arbuscular mycorrhizal fungi and its role in natural regeneration on limestone mined spoils. Biodiversitas 12(2), 107-111.

Suh JS, Lee SK, Kim KS, Seong KY (1995) Solubilization of Insoluble Phosphates by *Pseudomonas putida, Penicillium sp. and Aspergillus niger* isolated from Korean Soils. J Korean Soc. Soil Sci. Fertilizer 28(3), 278-286.

Tate RL III ,Klein DA (1985) Soil Reclamation Processes. In: Microbiological Analysis and Applications (Ed) Tate RL, Klein DA and Dekkar M, New York and Basel, pp 1-349.

Toro M, Azcon R, Barea JM (1997) Improvement of arbuscular mycorrhizal development by inoculation with phosphate solubilizing rhizobacteria to improve rock phosphate bioavailability (^{32}P) and nutrient cycling. Appl. Environ. Microbiol. 63, 4408-4412.

Vankessel C, Singleton PW, Hoben HJ (1985) Enhanced nitrogen transfer from a soybean to maize by vesicular-arbuscular mycorrhizal fungi. Plant Physiol. 79, 562-563.

Veeranjaneyulu K, Dhanaraju RM (1990) Geobotanical studies on Nallkonda complex mine. Trop. Ecol. 33, 59-65.

Wani PV, More BB, Patil PL (1979) Physiological studies on the activity of phosphorus solubilizing microorganisms. Indian J. Microbiol. 19, 23-25.

Watteau F, Berthelin J (1994) Microbial dissolution of iron and aluminium from soil minerals: efficiency and specificity of hydroxamate siderophores compared to aliphatic acids. Eur. J. Soil Biol. 30, 1-9.

Werner D (1998) Organic signals between plants and micro-organisms, In: The Rhizosphere: Biochemistry and Organic Substances at the Soil-Plant Interfaces (Ed) Pinton R, Varanini Z and Nannipieri P, New York, Marcel Dekker, pp 197-222.

Whitelaw MA (1999) Growth promotion of plants inoculated with phosphate solubilizing fungi. Adv. Agron.69, 99-151.

Yan X, Liao H, Trull MC, Beebe SE, Lynch, JP (2001) Induction of a major leaf acid phosphatase does not confer adaptation to low phosphorus availability in common bean. Plant Physiol. 125, 1901-1911.

Zahir ZA, Arshad M, Frankenberger WT (2004) Plant growth promoting rhizobacteria: Applications and perspectives in agriculture. Adv. Agron. 81, 97-168.

Zhang S, Reddy MS, Kloepper JW (2004) Tobacco growth enhancement and blue mold disease protection by rhizobacteria: relationship between plant growth promotion and systemic disease protection by PGPR strain 90-166. Plant Soil. 262, 277-288.

13

Status of Beneficial Microbes and their Role as Bio-inoculants for Reclamation of Saline Areas in Tamil Nadu

Mohan, V., Saranya Devi, K., Srinivasan, R. and Sangeeth Menon

Abstract

Alleviating plant salt stress and remediating saline soils are of great economic interest. Beneficial microbes such as arbuscular mycorrhizal (AM) fungi, plant growth promoting rhizobacteria (PGPR), etc. are associated with many plant species including trees. These beneficial microbes can cope up with salinity and help the plants to survive in such soils. Diversity status of beneficial microbes from the samples collected from different salt affected areas was investigated by many researchers. They screened and identified the efficient PGPR isolates isolated from saline areas for plant growth hormone (IAA) production and phosphate solubilization. The significance and highlights of the recent research as well as the past understanding of beneficial microbes and their role in reclamation of saline areas is discussed in this paper.

Key words: Arbuscular mycorrhizal fungi, *Azotobacter, Azospirillum, Bacillus Frankia,* Plant growth promoting rhizobacteria, Saline soil

1. Introduction

Salinity affects more than 7% of the Earth's land area (Parida and Das, 2005). Most of the salinity is natural but the extent of saline soils is increasing in a significant proportion of cultivated agriculture lands because of land clearing or irrigation (Munns 2005). Salinity is considered one of the most significant environmental factors limiting plant growth, productivity (Tian *et al.*, 2004) and survival of glycophytes (Munns 2005). Since salinity problems are commonly associated with agricultural areas, salt resistance of tree species has received relatively little attention. However, salinity can also be of major concern in natural boreal ecosystems (Purdy *et al.* 2005) as well as in urban areas (Lait *et al.* 2001) and industrial reclamation sites (Renault *et al.* 2001). Salt adversely affects plants by inducing osmotic imbalance, ionic toxicity, nutrient deficiencies, or a combination of the above factors (Shannon 1977). It also upsets water balance by interfering with the activity of water channel proteins (aquaporins) which results in an increase in water flow resistance of the root system (Carvajal *et al.* 2000; Martinez-Ballesta *et al.* 2000; Apostol *et al.* 2002 and Lopez-Berenguer *et al.* 2006).

Salinity is a major environmental stress which has been studied extensively and has its impact on agriculture in the past, present and future. The global importance of salt affected soils can be explained by their wide distribution on all continents covering about 10% of the total land surface. At present, there are nearly 954 million hectares of saline soils on earth's surface. It is extremely difficult to quantify the social and economic costs of salt prone land and water resources. High salt concentrations in the root zone soil limit the productivity of nearly 8.11 million hectares otherwise productive lands in India. Plant species vary in how well they tolerate salt-affected soils. Some plants will tolerate high levels of salinity while others can tolerate little or no salinity. The relative growth of plants in the presence of salinity is termed their salt tolerance. Since soil salinity is a major abiotic stress in plant agriculture worldwide. It has led to research into salt tolerance with the aim of improving crop plants. However, salt tolerance might have much wider implications because transgenic salt-tolerant plants often also tolerate other stresses including chilling, freezing, heat and drought (Naseri and Reycroft 2002).

Salinity refers to the dissolved concentration of major inorganic ions (Na, Ca, Mg, K, HCO_3, SO_4, Cl). Total salt concentration is expressed either in terms of the sum of either the cat ions or an ions in mM/l or sum of cat ions plus an ions in mg/l. For analytical reason practical index of salinity is calculated using electrical conductivity (EC). It is expressed in units of dS/m. An approximate relation between EC and salinity concentration is 1dS/m=10mmol/l=700mg/l. All soil contains some water soluble salts. Plants absorb essential plant nutrients in the form of soluble salts, but the excessive accumulation of soluble salts called soil salinity, suppresses plant growth. Saline or salt affected soils are common in arid and semi-arid regions. Salts in the soil occur as ions (electrically charged from of atoms or compounds). Ions contribute to the soil salinity include Cl^-, SO_4^{-2}, HCO_3^-, Na^+ Ca^{2+}, Mg^{2+} and rarely NO_3^- or K^+. The salts of these ions occur in highly variable concentration and proportions. They may be indigenous, but more commonly they are brought in to an area in the irrigation water or in waters draining from adjacent areas. Natural drainage is often so poorly developed in arid regions that salts collect inland basins rather than being discharged in to sea. Salinity in urban areas often results from the combination of irrigation and groundwater processes. Irrigation is also now common in cities (Rhoades *et al.* 1994).

Salt affected soils are caused by excess accumulation of salts, typically most pronounced at the soil surface. Salts can be transported to the soil surface by capillary transport from a salt laden water table and then accumulate due to evaporation. They can also be concentrated in soils due to human activity, for example the use of potassium as fertilizer, which can form sylvite, a naturally occurring salt. As soil salinity increases, salt effects can result in degradation of soils and vegetation (Blaylock 1994). Nutrient deficiencies and excess salinity can inhibit plant growth. Plants are stressed in three ways by salinity: (1) low water potential in the root medium leads to water deficits in plants, (2) the toxic effects of ions, mainly Na and Cl and (3) nutrient imbalance caused by depression in uptake or transport (Leon 1975). There are three important methods to reclaim salt affected areas which are as under:

(i)   Leaching: Salts must be leached out of the plant root zone by applying additional water. The water in excess of plant needs is called the leaching fraction.

(ii) Phytoremediation: Phytoremediation (from the Ancient Greek φυτο (phyto, plant), and Latin remedium (restoring balance or remediation) describes the treatment of environmental problems. Phytoremediation consists in mitigating pollutant concentrations in contaminated soils, water, or air, with plants able to contain, degrade, or eliminate metals, pesticides, solvents, explosives, crude oil and its derivatives, and various other contaminants from the media that contain them (Meagher 2000).

(iii) Microbes: Microbes play an important role in reclaiming process. The important microorganisms employed are: (a) PGPR, (b) mycorrhizal fungi (ecto- and endomycorrizal fungi) and (c) actinomycetes (*Frankia*). To overcome salt-stress problems, it is possible to select salt-tolerant plants, use biological processes such as mycorrhizal interactions or beneficial microbial interactions or desalinate soil by leaching excessive salts (Munns 2005). The desalination of soils is not economically viable for sustainable agriculture. This paper highlights of the recent research as well as the past understanding of different beneficial microbes such as mycorrhizal fungi and PGPR and their efficacy in phytohormone production, salt tolerance and growth enhancement of plants and their establishment in saline areas is discussed in this paper.

2. Status of PGPR in Saline Areas

PGPR is a group of bacteria that actively colonize plant roots and increase plant growth and yield. The mechanisms by which PGPR promote plant growth, include the ability to produce phytohormones, asymbiotic N_2 fixation, antagonistic ability against phytopathogenic microorganisms by production of siderophores, the synthesis of antibiotics, enzymes and/or fungicidal compounds and also solubilization of mineral phosphates and other nutrients (Gholami *et al.* 2009). PGPR are associated with plant roots and augment plant productivity and immunity; however, recent work by several groups shows that PGPR also elicit so-called 'induced systemic tolerance' to salt and drought (Yang *et al.* 2009). Enormous numbers of microbial populations and species in the soil, especially in the rhizosphere, intensive and extensive interactions have been established between soil microorganisms and various other soil organisms, including plant roots, and plant growth promotion by rhizosphere microorganisms is well established (Bashan 1998). Rhizobacteria play an important role within the interaction between soil and plant. As plants grow on marginal soils such as saline soils, the importance of the rhizobacteria increases as they mobilize nutrients and provide tolerance ability to the plants. Earlier reports indicated that isolation of PGPR organism, *Azospirillum* from the rhizosphere soils by Caballero-Mallado and Valdes (1983) and also from many tropical trees by Subba Rao (1984). *Pseudomonas pulifori* isolates were isolated from root free soil, rhizosphere and rhizoplane of wheat growing in alkaline soil and *Bacillus subtilis* isolates from the saline tract by Gaur (2004). Tilak *et al.* (2005) have reported that the species of *Pseudomonas* such as *P. striata, P. cissicola, P. fluorescens, P. pinophillum, P. putida, P. aeruginosa, P. stutzeri* have been isolated from rhizosphere of brassica, chickpea, maize, soya bean and other crops, desert soils and Antartica lake. The reported *Bacillus* species include B. *brevis, B. cereus, B. circulans, B. flimus, B. megaterium, B. polymyxa* and B. *subtilis* from the rhizosphere of legumes, cereals, jute, chilli and oat. Trembekar *et al.* (2009) have reported fifty strains of PSB *viz., Bacillus subtilis* from the saline tract of Vidharba which are currently used as bio-fertilizers by local farmers. Kloepper *et al.* (2007) reported IST (induced systemic resistance) to salt tolerance by *Arabidopsis* using *Bacillus subtilis* GBO3, a species that has previously been used as a commercial biological control agent. Recently, Mohan and Sangeetha Menon (2015) have reported the diversity status of beneficial microflora in saline soils of Tamil Nadu and Pudhucherry in Southern India and in this study, total of 51 PGPR isolates (phosphate solubilizing bacteria – 18; *Azotobacter* spp. - 16 and *Azospirillum* spp.-17) and 25 different AM fungi were reported by them.

2.1 Indole Acetic Acid Production and Phosphate Solubilization by PGPR Isolates

There are considerable populations of phosphate-solubilizing bacteria in soil and in plant rhizosphere. Earlier researchers have examined the ability of different bacterial species to solubilize insoluble inorganic phosphate compounds, such as tricalcium phosphate, dicalcium phosphate, hydroxyapatite and rock phosphate. The bacterial genera having P solubilizing capacity *viz.*, *Pseudomonas, Bacillus, Rhizobium, Burkholderia, Achromobacter, Agrobacterium, Microccocus, Aerobacter, Flavobacterium* and *Erwinia* were also reported (Sperberg 1958; Goldstein 1986). A considerably higher concentration of phosphate-solubilizing bacteria is commonly found in the rhizosphere in comparison with non-rhizosphere soil. Contrary to the earlier reports, Reinhold *et al.* (1986) reported salt sensitivity of *A. lipoferum* strains from the roots of kallar grass and suggested that osmotolerance in *A. lipoferum* may also depend on the source of soil or plant.

The role of *Azotobacter chroococcum* in producing plant growth promoting substance and nitrogen fixation might be due to its role in (1) decreasing the absorption of sodium and (2) increasing both nitrogen and magnesium concentration. Tripathi *et al.* (1998) stated that production of the phytohormone indole acetic acid was not affected by salinity as strongly as nitrogenase activity. Lower levels of salinity, in fact stimulated indole acetic acid production, but at above 100 mM NaCl, IAA production was inhibited. At low levels of salinity, increase in IAA production may cause proliferation of roots and possibly aid in the uptake of proline or betaines released by the plant roots. This might be one of the strategies for the better survival of both plants and the associated *Azospirillum*. Nautiyal (2000) reported that the strain NBR12601 isolated from the rhizosphere of chickpea and alkaline soils could solubilize phosphorus in presence of 10% salt, pH 12 at 45°C suggesting that extensive diversity in appropriate habitats may lead to recovery of effective bacteria.

The IAA producing PGPR have been isolated from Kallar grass [*Leptochloa fusca* (L.) Kunth] grown in salt affected soil of Pakistan by Perrig *et al.* (2007) and their growth promoting effects have been documented on rice by Mirza *et al.* (2006). Total of 150 bacterial isolates belonging to *Bacillus, Pseudomonas, Azotobacter* and *Rhizobium* from different rhizosphere soils of chick pea in the vicinity of Allahabad was recorded by Joseph *et al.* (2007). These test isolates were biochemically characterized and screened *in vitro* for their plant growth promoting traits like production of Indole Acetic Acid (IAA), ammonia (NH$_3$), hydrogen cyanide (HCN), siderophore and catalase and Guang-Can *et al.* (2008) have screened and isolated inorganic P-solubilizing bacteria (IPSB) and organic P-mineralizing bacteria (OPMB) in soils taken from subtropical flooded and temperate non-flooded soils and compared inorganic P-solubilizing and organic P-solubilizing abilities between IPSB and OPMB. Ten OPMB strains were isolated and identified as *Bacillus cereus* and *B. megaterium*, and five IPSB strains as *B. megaterium, Burkholderia caryophylli, Pseudomonas cichorii* and *P. syringae*. A total of 72 bacterial isolates belonging to *Azotobacter*, fluorescent *Pseudomonas, Mesorhizobium* and *Bacillus* were isolated from different rhizospheric soil and plant root nodules in the vicinity of Aligarh (Farah *et al.* 2008). These test isolates were biochemically characterized. These isolates were screened *in vitro* for their plant growth promoting traits like production of Indole Acetic Acid (IAA), ammonia (NH$_3$), hydrogen cyanide (HCN), siderophore, phosphate solubilization and antifungal activities.

A total of 20 isolates of bacteria (*Azotobacter* sp., *Acinetobacter* sp., *Bacillus* sp., *Citrobacter* sp., *Flavobacterium* sp., *Klebsiella* sp., *Nitrosomonas* sp., *Pseudomonas* sp., *Rhizobium* sp., *Thiobacillus* sp., *Azospirillum* sp., *Azotobacter chroococcum, Bacillus panthothenticus, Chromobacterium violaceum, C. lividum, Escherichia coli, Flavobacterium breve, Klebsiella aerogenes, Spaerotillus natans* and *Staphylococcus epidermidis*); 9 isolates of fungi (*Aspergillus niger, Bisporomyces* sp., *Monilia* sp., *Cephalosporium* sp., *Verticillum* sp., *Gliocladium* sp., *Penicillium* sp., *Nelicocephalum* sp. and *Cunninghamella* sp.) and 7 isolates of actinomycetes (*Streptomyces, Streptosporangium, Nocardia, Thermomonospora, Thermoactinomyces, Micromonospora, Mycobacterium*) from rhizosphere at Wamena Biological Garden, Jayawijaya, Papua (Widawati *et al.* 2005). Recently, Mohan (2013) tested all the selected PGPR isolates isolated from salt affected areas in Tamil Nadu and Puducherry for IAA production and phosphate solubilisation ability under *in vitro* conditions and got positive results.

2.3. Salt Tolerant Ability of PGPR Isolates

Rhizobacteria play an important role within the interaction between soil and plant. As plants grow on marginal soils such as saline soils, the importance of the rhizobacteria increases as they mobilize nutrients and provide tolerance ability to the plants. In one of the studies, Reinhold *et al.* (1986) and Bilal *et al.* (1990) have pointed out that occurrence of three species of *Azospirillum* viz., *A. brasilense, A. lipoferum* and *A. halopraferens* in the rhizosphere and rhizoplane on Kallar grass, grown as pioneer in high salinity soils indicate that these isolates should be salt tolerant. Glassker (1996) reported that the inorganic and organic compounds play an important role to maintain internal osmolarity of bacteria. The bacteria employ the adaptive strategy of accumulating a broad spectrum of osmotically active solutes by modulating their biosynthesis, catabolism, uptake of efflux. The ability of cells to grow under conditions of elevated osmolarity is determined both by the salt tolerance of their enzymes and by their capacity to accumulate compatible solutes in place of salts. Hartmann and Zimmer (1994) observed that in *A. amazonense* DSM2787 acetylene reduction was very sensitive to the addition of salt showing 90% inhibition with only 40 mM NaCl but in *A. halopraferens* 24% nitrogenese activity of unstressed cells could still be at 250 mM NaCl. Several reports suggest that osmo-tolerance in *Azospirillum* is strain specific and depends on the extent of salinity of the habitat from where the strain is isolated. Tripathi *et al.* (1998) studied that species of *Azospirillum* are widely distributed in salt affected soils in Pakistan, India, and Brazil. Occurrence of these species of *Azospirillum* i.e., *A. brasilense, A. lipoferum, A. halopreferens* in the rhizosphere and rhizoplane of Kallar grass grown on pioneer in high salinity soils indicated that these isolates should be salt tolerant. A variety of *Azospirillum* strains isolated from other habitats showed adaptation to their saline environments. Nautiyal (2000) has reported that the strain NBR12601 isolated from the rhizosphere of chickpea and alkaline soils could solubilize phosphorus in presence of 10% salt, pH 12 at 45^0 C suggesting that extensive diversity in appropriate habitats may lead to recovery of effective bacteria.

PGPR found in association with plants grown under chronically stressful conditions, including high salinity, may have adapted to the stress conditions, and could provide a significant benefit to the plants (Mayak *et al.* 2004). Tilak *et al.* (2005), studied phosphate solubilizing microorganisms which can grow in media containing tricalcium ion and aluminium phosphate hydroxyapatate, bone meal, rock phosphate and similarly in soluble phosphate compounds as a sole phosphate source. The most efficient phosphate solubilizing microorganisms belongs to genera *Bacillus* and *Pseudomonas* amongst bacteria and *Aspergillus* and *Penicillium* in fungi. The salt tolerance of PGPR might be due to osmotolerance mechanism where de-novo synthesis of osmolites and over production of salt stress proteins effectively nullifies the detrimental effects of high osomolarity (Paul and Sudha 2008). Upadhay *et al.* (2009) reported that rhizobacterial isolates from wheat rhizosphere were tolerant to 8% NaCl. Recently, Mohan and Saranya devi (2015) screened potential isolates of PGPR in conferring salt tolerance under *in vitro* condition and they observed that the absorbance of the PGPR culture (*Azotobacter, Azospirillum,* and PSB) broth grown in salt stress to sodium chloride decreased with increasing concentration of the salt (Figs.1-3). Similarly, decrease in growth pattern of PGPR in terms of the absorbance was also observed in salt stress to sodium citrate and sodium sulphate.

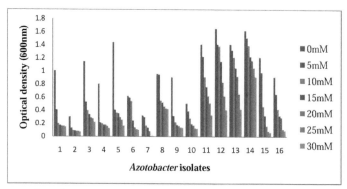

Fig. 1: Effect of various concentrations of sodium chloride (NaCl) on the growth of *Azotobacter* isolates (*Source:* Mohan and Saranya devi, 2015)

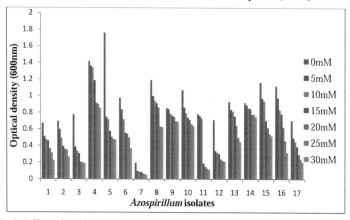

Fig. 2: Effect of various concentrations of sodium chloride (NaCl) on the growth of *Azospirillum* isolates (*Source:* Mohan and Saranya devi, 2015)

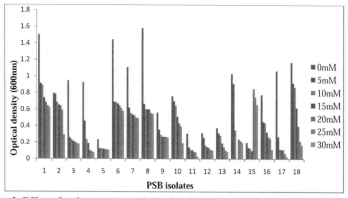

Fig. 3: Effect of various concentrations of sodium chloride (NaCl) on the growth of PSB isolates (*Source:* Mohan and Saranya Devi, 2015)

3. Status of Mycorrhizal Fungi from Different Saline Soil Samples

The term "mycorrhiza" is literally meaning "Fungus Root" to denote the beneficial association between certain soil fungi and plant roots. The mycorrhizal colonization is known to be significantly beneficial for the nutritional uptake of host plants (especially phosphorus, zinc and other essential elements) from deficient soils. There are different kinds of mycorrhizal fungi. Among them, the most widespread and relevant to agriculture, horticulture and forest trees are endomycorrhizal fungi and ectomycorrhizal (ECM) fungi which are very much used in forestry crops.

3.1 Ectomycorrhizal (ECM) Fungi in Saline Soils

This type of symbiotic association occurs mostly on Gymnosperms and few Angiosperm families. The fungi forming ECM fungi mostly belong to the higher fungi (Basidiomycetes and Ascomycetes). It protects the host plant from soil and root-borne pathogens. Few examples of these fungi are *Amanita muscaria, Laccaria laccata, L. fraterna, Pisolithus albus, Rhizopogon luteolus, Scleroderma citrinum, Suillus brevipes, S. subluteus* etc and they are reported in India (Vijayakumar *et al.* 1999; Mohan, 2002, 2005, 2008; Mohan *et al.* 2011). Among these, the most widespread one is the *Pisolithus albus.* The ECM fungus, *Pisolithus* was found very effective in reclamation of many problem areas like mined overburden, saline and other drought prone areas in different parts of the world. Benefits of ECM association to plants used in reforestation and revegetation may include improved survival in many problem areas like mined overburden, saline and other drought prone areas in different parts of the world (Marx 1991; Marx *et al.* 1977) and enhanced growth (Browing and Whitney 1992).

Natarajan *et al.* (1988) reported the ECM fungus *Thelephora remarioides* in association with *Casuarina equisetifolia* in Madras Coast. The fungi involved in the ectomycorrhizas of the Casuarinaceae are, however, not well documented. The only experimental ectomycorrhizas obtained in the Casuarinaceae with an identified fungus are those of *Allocasuarina distyla* with *Pisolithus tinctorius* (Bamber *et al.* 1980). Dixon *et al.* (1993) reported that *in vitro* growth and *in situ* symbiosis of ECM fungi generally declined with increasing substrate salinity. However, salt tolerance of the tested fungi varied significantly between species and between isolates within a species. The ECM fungal genera *Laccaria, Pisolithus* and *Suillus* appeared more tolerant of sodium salts than *Thelephora* and *Cenococcum.* Reddell (1986) and Dixon (1988) observed that *Frankia* and *Suillus* species compartmentalized salt and toxic metals in vacuoles and cell walls, thus partially excluding these agents from metabolic pathways.

Chen and Ellul (2001) determined the efficacy of eighteen isolates of Australian *Pisolithus* species to be resistant to NaCl concentrations of high saline soils. Bandou *et al.* (2006) showed that ECM results in the decrease in Na^+ and Cl^- uptake together with a concomitant increment in P and Kalium (K) absorption and a higher water status in ECM plants may be important salt-alleviating mechanism for the growth of sea grape seedlings in saline soils. The use of controlled mycorrhization may increase the survival of coniferous seedlings used for revegetation of salt-affected sites (Bois *et al.* 2006a,b).

Broadly, salt stress-tolerant ECM fungi having a symbiosis with plants enhanced plant growth and alleviated the negative effects of excess NaCl, however, different fungi use different ways to mitigate the toxic influence of salt stress such as improving plant nutrient uptake and ion balance, protecting enzyme activity, and facilitating water uptake. Ishida (2009) studied ECM fungal community in alkaline-saline soil in northeastern China. They found 11 ECM fungi, including species of *Inocybe*, *Hebeloma*, and *Tomentella* of the Basidiomycota and three Ascomycota species. They also observed that the phylogenetic analysis showed that the phylogroup including *Geopora* sp. is found mostly in alkaline soil habitats, indicating its adaptation to high soil pH and these ECM fungal species detected in this study may be useful for restoration of alkaline-saline areas.

Few studies have been made for the introduction of mycorrhizal fungi for their role in improvement of tree growth in the unfavorable saline conditions (Reddell, 1986; Dixon *et al.* 1993). Extreme soil salinity may adversely affect mycorrhizal fungi in the rhizosphere soil (Reddell 1986). Preliminary studies suggested that rhizosphere microorganisms may be influenced by salt accumulation in soils. Salinity tolerance is an important character to long term survival, reproduction and spread of mycorrhizal fungi in salt affected soils (Perry *et al.* 1987; Jain *et al.* 1989).

Plants were grown over a range (0mM, 200mM, 350mM and 500mM) of NaCl levels for 12 weeks, after 4 weeks of non-saline pre-treatment under greenhouse conditions. Growth and mineral nutrition of the seagrape seedlings were stimulated by *S. bermudense* regardless of salt stress. Although ECM colonization was reduced with increasing NaCl levels, ECM dependency of seagrape seedlings increased. Tissues of ECM plants had significantly increased concentrations of P and K but lower Na and Cl concentrations than those of non-ECM plants. Higher K concentrations in the leaves of ECM plants suggested a higher osmoregulating capacity of these plants. Moreover, the water status of ECM plants was improved despite their higher evaporative leaf surface. The results suggest that the reduction in Na and Cl uptake together with a concomitant increase in P and K absorption and a higher water status in ECM plants may be important salt-alleviating mechanisms for seagrape seedlings growing in saline soils (Bandou *et al.* 2006). The growth models, diameter growth rates, biomass yield and Na^+ contents of three ectomycorrihzal (ECM) fungi, *Suillus bovinus* and *Boletus luridus* were investigated at nine NaCl levels (0, 0.1, 0.2, 0.3, 0.4, 0.5, 0.6, 0.7, 0.8 mol/L). The results showed that the growth models of the three ECM fungal species were not affected by the NaCl concentration, but the growth rates reduced with the increasing NaCl concentration. The growth rates of *B. luridus* and *S. bovinus* were significantly higher than that of *S. luteus* at the same NaCl level; the biomass yields of three ECM fungal species were different, *S. bovinus* < *S. luteus* < *B. luridus*. Of the three species, *B. luridus* exhibited the highest growth rates, best biomass yield, and greatest Na^+ concentration in the mycelia over the NaCl gradient tested, indicating *B. luridus* has the most tolerance to NaCl stress and assimilation to Na^+ under salt stress. The growth rate of *S. luteus* was the lowest, but the biomass yield and Na^+ concentration in the mycelia were only lower than those of *S. luridus*. *S. bovinus* was the most sensitive to NaCl stress and its growth rate was faster than that of *S. luteus*, but the biomass yield and Na^+ concentration in the mycelia were the lowest (Tang *et al.* 2009).

Eighteen isolates, comprising of three putative *Pisolithus* species from eastern and central Australia, were screened for resistance to NaCl during growth in axenic culture. Biomass yield for most isolates showed a decline with increasing NaCl up to 200 mM; however, only five isolates reached an EC_{50} point (effective concentration inhibiting growth by 50%) below 200 mM NaCl. Most *Pisolithus* isolates were thus found to be resistant to NaCl at concentrations found in very saline soils. Of five isolates screened for Na_2SO_4 resistance, none reached an EC_{50} point below 100 mM. While intraspecific variation was evident in the resistance of isolates to both salts, no obvious patterns of interspecific variation were observed. The data thus indicate that isolates of the three Australian *Pisolithus* species are broadly resistant to salinity and may represent useful ectomycorrhizal inoculants for out planting of compatible, salt-resistant host trees at saline sites (Chen and Ellul 2001).

Four ECM fungi, *Cenococcum geophilum*, *Pisolithus tinctorius*, *Rhizopogon rubescens*, and *Suillus luteus*, were grown in liquid MMN media with five different concentrations of NaCl for 30 days, and their mycelial weights were determined. Mycelial weights of *P. tinctorius* and *R. rubescens* were not significantly different between 0 mM and 200 mM, whereas those of *C. geophilum* and *S. luteus* decreased with increasing NaCl concentration, indicating that the former two species were more tolerant to higher NaCl concentrations than the latter species. They further studied the intraspecific differences in NaCl tolerance of nine *P. tinctorius* isolates. They were grown on MMN agar media with six different concentrations of NaCl for 21 days, and their radial growth was measured. In total, the hyphal growth at 25 mM NaCl was significantly higher than those at the other NaCl concentrations, and EC_{50} values were confirmed at between 50 mM and 200 mM. Among the isolates, Pt03 and Pt21 showed measurable growth at 200 mM; the growth of Pt03 was not significantly different between 0 mM and 200 mM. The results indicate that there are intraspecific variations in NaCl tolerance of *Pisolithus* species (Yosuke *et al.* 2006).

3.2 Endomycorrhizal Fungi

This group is one of the major types of mycorrhizal fungi which differ from ECM fungi, in structure. Unlike ECM fungi, which form a system of hyphae that grow around the cells of the root, the hyphae of the endomycorrhizal fungi not only grow inside the root of the plant but penetrate the root cell walls and become enclosed in the cell membrane as well and it makes for a more invasive symbiotic relationship between the fungi and the plant. The penetrating hyphae create a greater contact surface area between the hyphae of the fungi and the host plants. Endomycorrhizal fungi have further been classified into five major groups: Vesicular Arbuscular (presently it has been referred as Arbuscular), Ericoid, Arbutoid, Monotropoid and Orchid mycorrhizae.

3.2.1 Status of AM Fungi in Salt Affected Areas

AM fungi belong to endomycorrhizal group. They are most prevalent and distributed in varied ecosystems. On a global scale, AM fungi are virtually ubiquitous being present in tropical, temperate and arctic regions. Within the different global regions AM fungi have a broad ecological range (Harley and Smith 1983; Gianinazzi and Schuepp 1994). They are found in most ecosystems including dense rain forests, open woodlands, scrub, savanna, grasslands, heaths, sand-dunes and arid and semiarid deserts. Their occurrence within these systems varies according to localized environmental conditions and plant cover. Jeffries and Barea (1994) reported that AM fungi play a key role in ecosystems. AM fungi interact with a wide range of other soil microorganisms, both synergistically and antagonistically, at all stages of their life cycle and some of the interactions may have large effect in agricultural and natural ecosystems. The environmental factors play a crucial role in root microorganisms and microbe-microbe interactions. AM colonization units develop within the root cortex, with hyphae growing longitudinally between the cells and intracellular development of arbuscules. Vesicles, containing large amounts of lipid, are formed later in the maturation of an infection unit. At the same time as colonization spreads within root, extramatrical hypha grows out into the soil (Smith 1988) and absorbs the nutrients from the soil. Therefore, in flow (rate of uptake per unit length of root) of P from soil into the roots of mycorrhizal plants is faster, than into non-mycorrhizal plants (Hale and Sanders 1982; Smith *et al*. 1986). The mechanism underlying the increased rate of P uptake is the efficiency with which mycorrhizal roots exploits the soil profile, with hyphae extending beyond the depletion zone surrounding the absorbing root and its root hairs (Owusu-Bennoah and Wild 1979; Clarkson 1985). The distribution and occurrence of AM fungi differ both qualitatively as well as quantitatively with the change in edaphic factors and the type of vegetation. AM colonization is most common in soils with moderate or low fertility. Changes in soil fertility due to amendments with mineral fertilizers or organic matter markedly affect the activity and survival of AM fungi. AM fungi are able to grow saprophytically around decaying root fragments and other organic matter (Warner and Mosse 1980; Hepper and Warner 1983). Nearly all soils contain indigenous AM fungi (Abbott and Robson 1982) but all of them cannot be considered efficient. Though efficient strains are found infrequently (Powell 1982), their influence on plant growth and yield, resistant to drought and salinity, tolerance to pathogens have been studied in the last few decades (Smith and Read 2000).

Many earlier studies revealed the association of AM fungi associated with plants in saline areas. The mycorrhizae of some inland halophytes were reported by Ballard (1989) and the results of his study on 3 saline sites indicated that moderate to high levels of AM fungi present in several perennial species. He also found that colonization was more sensitive to salinity and temperature than to moisture and nutrient levels. Also, high inoculum potential was found and inoculation increased growth and survival of *Hordeum jubatum*, *G. fasciculatum* spores germinated at a wide range of salinity and temperature. Gupta *et al*. (2002) reported the occurrence of AM fungi in three tree mangrove species namely *Agalia cuculata*, *Heritiera fomes* and *Sonneratia caseolaris* whereas no such association was found in four herbaceous mangroves namely *Acanthus ilicifolius*, *Acrosticum aurenum*, *Derris heterophylla* and *Myristachya wighitiana*. Kannan and Lakshminarasimhan (1988) screened forty eight plant species from 32 families for mycorrhizal association in samples collected from 4 different sites in 1985 and 1986, in Point Calimere, a village on the east coast of India bordered by Bay of Bengal on the east and Palk straight on the south. Spore types of *Glomus fasciculatum*, *G. macrocarpum*, *Sclerocystis rubiformis*, *Sclerocystis sinuosa*, *Gigaspora* sp. and *Acaulospora birecticulata* were recovered from the rhizosphere of all the samples screened. Among them, spores of *Glomus* species were most common and *G. fasciculatum* was most abundant in samples of different sites screened.

Mohan and Natarajan (1988) studied the occurrence of AM fungi in the roots of 26 plants belonging to 16 families of Angiosperms growing in the beach between Ennore and Mahabalipuram in the Coromandel Coast. It was found that 20 plant species belonging to 13 families had AM fungal colonization. AM fungal spores were observed in the sand samples of all the 26 plant species. They reported 8 species of the genus *Glomus viz., G. clarus, G. claraoideum, G. fulvus, G. intraradices, G. microcarpus, G. monospora, G. occultum* and *G. pubescens* for the first time from India. Mohan (1995) also studied the occurrence and importance of AM association in sand dune and desert ecosystems in Rajasthan. The study has shown that the genus *Glomus* was predominantly found in the rhizosphere of both sand dune and desert colonizing plant species including certain tree species such as *Azadirachta indica, Acacia tortilis, Prosopsis cineraria* and *Tecomella undulata.* In another study, Mohankumar *et al.* (1988) studied the distribution of AM fungi in the sandy beach soils of Madras Coast. The study has shown that most plants harboured AM fungi and the AM species reported by them were *Entrophosphora shenckii, Glomus claroideum, G. clarum, G. intraradices, G. microcarpus, G. monospora, G. occultum, G. pubescens* and *G. pustulatum.* They suggested that soil temperature and moisture status of the soil influenced the colonization of AM fungi in the coastal soils. Selvaraj and Subramanian (1991) have found that the number of spores was higher during rainy season than in summer in mangrove ecosystem of Pichavaram and Muthupet estuary. Recently, Mohan (2013) reported the diversity status of AM fungi in saline samples collected from different study locations in Tamil Nadu and Puducherry and a total of twenty five different AM fungi belonging to 3 genera *viz., Acaulospora, Gigaspora* and *Glomus* were recorded. Among them, the genus *Glomus* was found predominant with 17 species, whereas, the genera *Acaulospora* and *Gigapsora* had four species each during the period under investigation. It was also observed that all the samples had percent root colonization and soil spore population but varied with the location and period of collection.

4. Effect of Beneficial Microbes for Enhancing Plant Growth and Reclamation of Saline Areas

Plant – microbe interaction is gaining more attention of agronomists and microbiologists in an area of applied research and halo tolerant or halophilic bacteria are being utilized in the field and they can flourish in hypersaline conditions and interacting with plants may restore these lands to productivity. Salt tolerant beneficial microbes from different sources *i.e.,* saline soils, rhizosphere, rhizoplane and phylloplane of different plants growing in saline areas have been isolated (Yasmin and Hasnian 1997). These microbes were able to resist high levels of sodium chloride stress. Bacteria from different sources could promote seedling growth under saline conditions (Siddique, 1997). Stimulated seedling growth was ascribed to the decreased uptake of Na^+ or increased auxin level in the cells accompanied by decreased Na^+ uptake. This paper highlights the status of different beneficial microbes and their role as bio-inoculants for reclamation of salt affected areas. Various crops in India have been inoculated with diazotrophs particularly *Azotobacter* and *Azospirillum* (Tilak and Saxena 2005). Application of *Azotobacter* and *Azospirillum* has been reported to improve yields of both annual and perennial grasses. Saikia and Bezbaruah (1995) reported increased seed germination of *Cicer arietinum, Phaseolus mungo, Vigna catjung* and *Zea mays*. However, yield improvement is attributed more to the ability of *Azotobacter* to produce plant growth promoting substances such as phytohormone IAA and siderophore azotobactin, rather than to diazotrophic activity.

The PGPR strain induced a higher increase in these antioxidant enzymes in response to severe salinity. Inoculation with selected PGPR could serve as a useful tool for alleviating salinity stress in salt- sensitive plants (Kohler *et al.* 2009). They observed the stress adaptations in PGPR with increasing salinity in the coastal agricultural soils. Their studies revealed that the root colonization potential of the strain was not hampered with higher salinity in soil. Recently, Mohan (2013) studied the effect of different beneficial microbes on growth and establishment of 16 different clones of *Casuarina equisetifolia* in saline areas in Tamil Nadu in nursery and field. The overall outcome of the nursery and filed trial was a positive response of 16 different clones of *C. equisetifolia* plants to inoculation with saline tolerant beneficial microbes as bio-inoculants as compared to uninoculated control. Statistically significant increase in plant growth parameters for most of the clones under nursery and field conditions, confirmed the positive effect of these beneficial microbes as bio-inoculants. This study revealed that the selected salt tolerant N fixing organism (*Frankia + Azotobacter chroococcum*) and P solubilizing/mobilizing organisms (AM fungi) improved the plant growth (primarily plant height, collar diameter, and biomass) and also improved the soil nutrient parameters. The application of PGPRs and mycorrhizal fungi in agriculture is well established, however, a considerable scope exists to carry out extensive research in forestry particularly in saline and salt affected areas.

Conclusion

The health and growth of seedlings in nursery are very critical for their out planting performance in the field especially problem areas like saline and salt affected lands. Seedlings usually undergo transplantation shock after their planting in the field and may suffer low survival rate especially in poor and adverse soils. In the adoption process of the seedlings to new and adverse environments in a complex and natural situation, tetra-partite interaction of AM fungi–PGPR–*Frankia*–Plant plays very important role. From a practical viewpoint, raising successful plantation in saline or salt affected land, apart from using superior plant genotype, is dependent upon soil managements by providing sufficient nutrient for plants to grow well and by giving protection from the soil-borne diseases. It is then necessary to focus on rhizosphere soil because AM fungi and PGPR have abilities to alter and regulate such micro-scale of soil environments. Preparation of seedlings inoculated with superior bio-inoculants before their out planting is a environmentally benign, less expensive and relatively easy nursery technology that can be followed in agriculture and forestry.

References

Abbott LK, Robson AD (1982) Infectivity of Vesicular Arbuscular Mycorrhizal fungi in agricultural soil. Aust. J. Agric. Res. 33, 1049-1059.

Apostol KG, Zwiaze JJ, Mackinnon MD (2002) NaCl and Na_2SO_4 alter responses of jack pine (*Pinus banksiana*) seedlings to boron. Plant Soil.240, 321-329.

Ballard BJ (1989) The mycorrhizae of some inland halophytes. Dissertation Abstracts International B (Science and engineering). 49, 2470.

Bamber RK, Mullette K, Mackowski C (1980) Mycorrhizal studies. In: Research Report, Forest Communications, Sydney, Australia, pp 70-74.

Bandou E, Lebailly F, Muller F, Dulormne M, Toribio A, Chabrol J, Courtecuisse R, Plenchette C, Prin Y, Duponnois R, Thiao M, Sylla S, Dreyfus B, Ba AM (2006) The ectomycorrhizal fungus *Scleroderma bermudense* alleviates salt stress in seagrape (*Coccoloba uvifera* L.) seedlings. Mycorrhiza 16, 559–565.

Bashan Y (1998) Inoculants of plant growth promoting bacteria for use in agriculture. Biotechnol. Adv. 16, 729-770.

Bilal R, Ghulam R, Querishi JA, Malik KA (1990) Characterization of *Azospirillum* and related diazotrophs associated with roots of plants growing in saline soils. World J. Microbiol. Biotech. 6, 46-52.

Blaylock AD (1994) Soil salinity, Salt tolerance and Growth potential of horticultural and Landscape plants. University of Wyoming, Cooperative extension service, Dept. of Plant, Soil and Insect Sciences, College of Agriculture.

Bois G, Bertrand J, Piché Y, Fung MYP, Khasa DP (2006a) Growth, compatible solute and salt accumulation of five mycorrhizal fungal species grown over a range of NaCl concentrations. Mycorrhiza 16, 99-109.

Bois G, Bigras FJ, Bertrand A, Piche Y, Fung MYP, Khasa DP (2006b) Ectomycorrhizal fungi affect the physiological responses of *Picea glauca* and *Pinus banksiana* seedlings exposed to an NaCl gradient. Tree Physiol. 26, 1185–1196.

Browing MHR, Whitney RD (1991) Responses of Jack pine and Black spruce seedlings to inoculation with selected species of mycorrhizal fungi. Can. J. For. Res. 21, 701-706.

Carvajal M, Cerda A, Martinez V (2000) Does calcium ameliorate the negative effect of NaCl on melon root water transport by regulating aquaporin activity. New Phytol. 145, 439–448.

Chen DM, Ellul S (2001) Influence of salinity on biomass production by Australian *Pisolithus* spp. isolates. Mycorrhiza. 11, 231-236.

Clarkson DT (1985) Factors affecting mineral nutrient acquisition by plants. Ann. Rev. Plant Physiol. 36, 111-115.

Dixon RK (1988). Response of Ectomycorrhizal *Quercus rubra* to soil cadmium, nickel and lead. Soil Biol Biochem., 20, 555- 559.

Dixon RK, Rao MV, Garg VK (1993) Salt stress affects *in vitro* growth and *in situ* symbioses of ectomycorrhizal fungi. Mycorrhiza. 3, 63-68.

Farah A, Iqbal A, Khan MS (2008) Screening of free-living rhizospheric bacteria for their multiple plant growth promoting activities. Microbiol. Res.. 163, 173-181.

Gaur S (2004) Diacteylphoroglucinol-producing *Pseudomonas* does not influence AM fungi in wheat rhizosphere. Curr. Sci. 86, 453-457.

Gholami A, Shahsavani S, Nezarat S (2009) The Effect of Plant Growth Promoting Rhizobacteria (PGPR) on germination, seedling growth and yield of maize. World Acad. Sci. Eng. Technol. 49, 19-24.

Gianinazzi S, Schuepp H (1994) Impact of arbuscular mycorrhizas on sustainable agriculture and natural ecosystems, Springer, Birkäuser, Basel.

Glassker E (1996) Mechanism of osmotic activation of the quaternary compound transporter (QacT) of *Lactobacillus plantarum*. J. Bacteriol. 180, 5540-5546.

Goldstein AH (1986) Bacterial solubilization of mineral phosphates: historical perspective and future prospects. Am. J. Altern. Agri. 1, 51–57.

Guang-Can TAO, Shu-Jun TIAN, Miao-Ying CAI, Guang-Hui XIE (2008) Phosphate solubilizing and mineralizing abilities of bacteria isolated from soils. Pedosphere 18(4), 515–523.

Gupta N, Routaray S, Basak UC, Das P (2002) Occurrence of arbuscular mycorrhizal association in Mangrove forest of Bhitar Kanika, Orissa. Indian J. Microbiol. 42, 247-248.

Harley JL, Smith SE (1983) Mycorrhizal symbiosis. Academic Press. Inc., London. 483p.

Hale KA, Sanders FE (1982) Effects of benomyl on vesicular-arbuscular mycorrhizal infection of red clover (*Trifolium pratense* L.) and consequences for phosphorus inflow. J. Plant Nutr. 5, 1355-1367.

Hartmann A, Zimmer W (1994) Physiology of *Azospirillum*. In: *Azospirillum* – Plant Association (Ed) Okon Y, Bocon Raton, CRC Press, pp15-39.

Hepper CM, Warner A (1983) Role of organic matter in the growth of a vesicular arbuscular mycorrhizal fungus in soil. Trans. Br. Mycol. Soc. 81, 155-156.

Ishida TA, Nara K, Ma S, Takano T and Liu S (2009) Ectomycorrhizal fungal community in alkaline-saline soil in northeastern China. Mycorrhiza 19, 32-335.

Jain RK, Paliwal K, Dixon RK, Gjerstad DH (1989) Improving productivity of multipurpose trees in substandard soils in India. Ind. J. For. 87, 38-42.

Jeffries P, Barea JM (1994) Biogeochemical cycling and arbuscular mycorrhizas in the sustainability of plant–soil systems. In: Impact of Arbuscular Mycorrhizas on Sustainable Agriculture and Natural Ecosystems (Ed) Gianinazzi S and Schüepp H, Birkhäuser Verlag, Basel, pp 101-115.

Joseph B, Ranjan RR, Lawrence R (2007) Characterization of plant growth promoting rhizobacteria associated with chickpea (*Cicer arietinum* L.). Inter. J. Pl. Produc. 1(2), 141-151.

Kannan K, Lakshminarashimhan C (1988) Survey of VAM of Maritime atrand plants of Point Calimere. In: Mycorrhizae for Green Asia (Ed) Mahadevan A, Raman N and Natarajan K, Proc 1st Asian Conf on Mycorrhizae, pp: 53-55.

Klopper JW, Gutierrez A, Mcinroy JA (2007) Photoperiod regulates elicitation of growth promotion but not induced resistance by plant growth promoting rhizobacteria. Can. J. Microbiol. 53, 159-167.

Kohler J, Hernandez JA, Caravaca F, Roldan A (2009) Induction of antioxidant enzymes is involved in the greater effectiveness of a PGPR versus AM fungi with respect to leuttce to severe salt stress. Environ. Expt. Biol. 65, 245-252.

Lait CG, Saelim S, Zwiazek JJ, Zheng Y (2001) Effect of basement pump effluent on the growth and physiology of urban black ash and green ash ornamental trees. J. Arboriculture. 27, 69-77.

Leon B (1975) Effect of salinity and sodicity on plant growth. Annu. Rev. Phytopathol. 13, 295-312.

Lopez-Berenguer C, Garcia-Vaguer C, Carvajal M (2006) Are root hydraulic conductivity responses to salinity controlled by aqua porins in broccoli plants? Plant Soil 279, 13-23.

Marx DH (1991) Practical significance of ectomycorrhizae in forest establishment. In: Ecophysiology of Ectomycorrhizae of Forest Trees, Proc. of Marcus Wallenberg Foundation Symposium, Sweden, pp 54-90.

Marx DH, Bryan WC, Cordell CE (1977) Survival and growth of *Pinus* seedlings with *Pisolithus* ectomycorrhizae after two years on reforestation sites in North Carolina and Florida. For. Sci., 16, 363-373.

Martinez-Ballesta MD, Martinez V, Carvajal M (2000) Regulation of water channel activity in whole roots and in protoplasts from roots of melon plants grown under saline conditions. Aust. J. Plant Physiol. 27, 685-691.

Mayak S, Tirosh T, Glick BR (2004) Plant Growth Promoting bacteria confer resistance in tomato plants to salt stress. Plant Physiol. Biochem. 42, 565-572.

Meagher RB (2000) Phytoremediation of toxic elemental and organic pollutants. Curr. Opin. Plant Biol., 3(2), 153–162.

Mirza MS, Mehnaz S, Normand P, Combaret CP, Loccoz YM, Balley R and Malik KA (2006) Molecular characterization and PCR detection of a nitrogen fixing *Pseudomonas* strain promoting rice growth. Biol. Fertil. Soil. 43, 163-170.

Mohan V (1995) Occurrence and importance of VAM association in sand dune and desert ecosystems in Rajathan. In: Proc. International Workshop on Forestry Research Methods, UNDP-ICFRE Project, Tropical Forest Research Institute, Jabalpur, M.P. (India), pp 49-51.

Mohan V (2002) Distribution of ectomycorrhizal fungi in association with economically important tree species in southern india. In: Frontiers of Fungal Diversity in India (Ed) Rao GP, Manoharachari C, Bhat DJ, Rajak RC and Lakhanpal TN, International Book Distributing Co., Lucknow, India, pp 863-872.

Mohan V (2005) Role of Ectomycorrhizal fungi as bio-fertilizers in tree nurseries" In: Current Trends in Mycological Research (Ed) Deshmukh SK, Quest Institute of Life Sciences, Nicholas Piramal India Limited, Mumbai, India, pp 111-133.

Mohan V (2008) Diversity of ectomycorrhizal fungal flora in the Nilgiri Biosphere Reserve (NBR) Area, Nilgiri Hills, Tamil Nadu. ENVIS Newsletter Microorg. Environ. Manage. 6(3), 5-9.

Mohan V (2013) Influence of beneficial microbes in conferring salt tolerance to *Casuarina* clones. Project Completion Report. ICFRE, Dehra Dun, pp. 1-160.

Mohan V, Natarajan K (1988) Studies on VAM association in sand dune plants in the Coromandel Coast. In: Mycorrhizae for Green Asia (Ed) Mahadevan A, Raman N and Natarajan K, Proc. 1ˢᵗ Asian Conf on Mycorrhiza, University of Madras, Madras, India, pp 73-76.

Mohan V, Senthilarasu G, Manokaran P (2011) Diversity of ectomycorrhizal fungal flora in association with important tree species in South India. In: Souvenir on International Day of Biological Diversity and Centenary of Fischer Herbarium, IFGTB, Coimbatore, pp 79-90.

Mohan V, Sangeetha Menon (2015) Diversity status of beneficial microflora in saline soils of Tamil Nadu and Pudhucherry in Southern India. J. Acad. Indus. Res. 3, 384-392.

Mohan V, Saranya devi K (2015) Selection of Potential Isolates of Plant Growth Promoting Rhizobacteria (PGPR) in Conferring Salt Tolerance under *in vitro*. J. Acad. Indus. Res. 4, 91-99.

Mohankumar V, Nirmala C.B, Ragupathy S, Mahadevan A (1988) Distribution of vesicular arbuscular mycorrhizae (VAM) in the sandy beach soils of Madras Coast. Curr. Sci. 57, 367-368.

Munns R (2005) Genes and salt tolerance: Bringing them together. New Phytol. 167, 645-663.

Naseri AA, Rycroft D (2002) Effect of swelling and overburden weight on hydraulic conductivity of restructured saline sodic clay, 17th World Congress of Soil Science, pp.14-21.

Natarajan K, Mohan V, Kaviyarasan V (1988) On some ectomycorrhizal fungi occurring in Southern India. Kavaka 16, 1-7.

Nautiyal CS (2000) An efficient microbiological growth medium for screening phosphate solubilising microorganisms. FEMM Microbiol. Lett. 182, 291-296.

Owusu-Bennoah E, Wild A (1979) Autoradiography of the depletion zone of phosphate around onion roots in the presence of vesicular-arbuscular mycorrhiza *Glomus mosseae*. New Phytol.. 82, 133-140.

Parida AK, Das AB (2005) Salt tolerance and salinity effect on plants: A review. Ecotoxicol. Environ. Saf. 60, 324-349.

Paul D, Sudha N (2008) Stress adaptations in a Plant Growth Promoting Rhizobacterium (PGPR) with increasing salinity in the coastal agricultural. J. Basic Microbiol. 48, 378-384.

Perrig D, Boiero ML, Masciarelli O, Penna C, Ruiz OA, Cassan F, Luna V (2007) Plant growth promoting compounds produced by two strains of *Azospirillum brasilense* and implications for inoculant formation. Appl. Microbiol. Biotechnol. 75, 1143-1150.

Perry DA, Molina R, Amaanthus MP (1987) Mycorrhizae, Mycorrhizae and reforestation: Current Knowledge and Research needs. Can. J. For. Res. 17, 929-940.

Powell CL (1982) Selection of efficient VA mycorrhizal fungi. Plant Soil. 68, 3-9.

Purdy PG, Macdonald SE, Lieffers VJ (2005) Naturally saline boreal communities as models for reclamation of saline oil sand tailings. Restor. Ecol. 13, 667-677.

Reinhold B, Thomas H, Ernst-Georg N, Istvan F (1986) Close association of *Azospirillum* and Diazotrophic rods with different root zones of Kallar grass. Appl. Environ. Microbiol. 52, 520.

Renault S, Croser C, Franklin JA, Zwiazek JJ (2001) Effects of NaCl and Na$_2$SO$_4$ on red-osier dogwood (*Cornus stolonifea* Michx) seedlings. Plant Soil. 233, 261-268.

Rhoades JD, Kandlab A, Mashal AM (1994) The use of saline water for crop production. Scientific Publishers, pp.5-23.

Reddell P (1986) The effects of sodium chloride on the growth and nitrogen fixation in *Casuarina obesa* Miq. New Phytol. 102, 397-408.

Saikia N, Brezbaruah B (1995) Iron-dependent plant pathogen inhibition through *Azotobacter* RRLJ 203 isolated from iron-rich acid soils. Indian J. Expt. Biol. 33, 571-575.

Selvaraj T, Subramanian G (1991) Survey of vesicular–arbuscular mycorrhizae in mangroves of Muthupet Estuary: Ecological implications. In: Proc. of the Second Asian Conference on Mycorrhiza (Ed) Soerianegara I and Supriyanto, pp 271.

Shannon MC (1977) Adaptation of plants to salinity. Adv. Agron. 60, 76-120.

Siddique S. (1997) Growth effects of *Triticum aestivum* seedling under NaCl stress after inoculating with plasmid free bacterial strains. Endevour. Biotechnol. 1(3), 97-112.

Smith GS (1988) The role of phosphorus nutrition in interations of vesicular arbuscular mycorrhizal fungi with soil borne nematodes and fungi. Phytopathology 78, 731-734.

Smith SE, John BJ, St Smith FA, Bromley JL (1986) Effect of mycorrhizal infection on plant growth, nitrogen and phosphorus nutrition of glasshouse-grown *Allium cepa* (L). New Phytol. 103, 359-373.

Sperberg JI (1958) The incidence of apatite-solubilizing organisms in the rhizosphere and soil. Aust. J. Agric. Res. 9, 778.

Subba Rao NS (1984) Phosphate solubilizing microorganisms. In: Bio-fertilizers in Agriculture, 2nd edition, Oxford and IBH Publishing Co. New Delhi. Bombay, Calcutta, India, pp: 126-132.

Tang M, Sheng M, Chen H, Zhang FF (2009) *In vitro* salinty resistance of three ectomycorrhizal fungi. Soil Biol. Biochem. 41, 948-953

Tian CY, Feng G, Li, XL, Zhang FS (2004) Different effects of arbuscular mycorrhizal fungal isolates from saline or non-saline soil on salinity tolerance of plants. Appl. Soil Ecol. 26, 143-148.

Tilak KVBR, Saxena AK (2005) *Azospirillum-* its impact on crop production. In: *Recent Advances in Biofertilizer Technology* (Ed) Yadav AK, Motsara MR and Ray Chaudri S, Society for Promotion and Utilization of Resources and Technology, New Delhi, pp 176-189.

Tilak KVBR, Ranganayaki N, Pal KK, De R, Saxena K, Nautiyal CS, Mittal S, Tripathi AK, Johri BN (2005) Diversity of plant growth and soil health supporting bacteria. Curr. Sci. 89, 136-150.

Trembeker DH, Gulhane SR, Somkuwar DO, Ingle KB, Kanchalwar SP, Upadhye MA, Bidwai UA (2009) Potential *Rhizobium* and Phosphate Solublisers as a bio-fertilizer from saline belt of Akola and Buldhana districts (India). Res. J. Agri. Biol. Sci. 5(4), 578-582.

Tripathi AK, Mishra BM, Tripathi K (1998) Salinity stress responses in plant growth promoting rhizobacteria. J. Biosci. 23, 463-471.

Upadhyay K, Singh DP, Salika R (2009) Genetic diversity of Plant Growth Promoting Rhizobacteria isolated from rhizospheric soil of wheat under saline condition. J. Curr. Microbiol. 59, 489-496.

Vijayakumar R, Prasada Reddy BV, Mohan V (1999) Distribution of Ectomycorrhizal fungi in forest tree species of Andhra Pradesh, Southern India – A new record. Indian For. 125(5), 496-502.

Warner A, Mosse B (1980) Independent spread of vesicular arbuscular mycorrhizal fungi in soil. Trans. Br. Mycol. Soc. 74, 407-410.

Widawati S, Suliasih, Latupapua HJD, Sugiharto A (2005) Bio-diversity of Soil Microbes from Rhizosphere at Wamena Biological Garden (WBiG), Jayawijaya, Papua. Biodiversitas. 6(1), 6-11.

Yang J, Joseph WK, Choong-Min R (2009) Rhizosphere bacteria help plants tolerate abiotic stress. Trends Plant Sci. 14(1), 1-4.

Yasmin A, Hasnian S (1997) Moderately halophilic bacteria associated with the roots of *Chenopdium album, Oxalis cornocula* and *Lyceum edgeworthii*. Pak. J. Zool. 29, 249-257.

Yosuke M, Fumio S, Kenichi N, Shin-ichiro I (2006) Effects of sodium chloride on growth of ectomycorrhizal fungal isolates in culture. Myco. Sci. 47(4), 212-217.

14

Mycorrhizal Fungi in Restoration of Coastal Forests

Sridhar, K.R.

Abstract

Sustainable landscape and forest management are of utmost concern in alleviating climate change. Fungal mutualism is one of the most significant positive biotic interactions responsible for forest structure and longevity. Coastal sand dunes imitate xeric habitats, while mangroves and river floodplains are often marshy. In spite of such extremities, mycorrhizal fungi are prominent component in coastal vegetation. Studies so far revealed that mycorrhizal association in herbaceous plant species is responsible for dune formation, stabilization and serve as potential inoculum in coastal habitats. Mycorrhizal association facilitates plant species to exploit larger volume of soil, supply nutrients (organic nitrogen, phosphorus and minerals), aggregates soil, provides tolerance to salinity stress, relieves from water stress and protects roots from pathogens. Human interference has severe impacts on plant-mycorrhizal mutualism in coastal habitats. Cultivation of exotic tree species and their uncontrolled exploitation further impoverish the habitats. Protection of native herbaceous flora, cultivation of native tree species and introduction of selected indigenous keystone species in disturbed habitats with mycorrhizal mutualism is an appropriate measure of rehabilitation. Such forestry practices have far reaching advantages in prevention of erosion, protection from storms, building ecosystem in favor of wildlife and other potential benefits (nutritional, medicinal and industrial). This review emphasizes importance of mycorrhizal and endophytic fungal association in coastal forests more specifically with experiences from the southwest coast of India.

Keywords: Coastal sand dunes, Ectomycorrhizae, Endomycorrhizae, Endophytic fungi, Mangroves, Rehabilitation, Wildlife protection.

1. Introduction

Over 30% of human population worldwide sheltered <100 km from the coastal region (Hoagland and Jin 2006), while two-third of the major cities are housed on the coast (Martínez *et al.* 2007). Approximately three-quarters constitute coastal dunes in world's shorelines (Bascom 1980). Obviously human interference adversely influences in degradation of coastal forests, which are the arteries of life sustaining system. It is inevitable to restore or re-vegetate the coastal region with sufficient forest cover to harness benefits. Several non-ecosystem based approaches are common in practice to prevent risks of floods and erosion (e.g. construction of groins, sea walls and dykes) (Temmerman *et al.* 2013). Most appropriate ecosystem based approaches are to manage sufficient cover of vegetation in beach dunes, tidal salt marshes, mangroves and sea grass beds to protect shores against several natural calamities. Besides protection, coastal forests provide several valuable ecological services such as supporting (soil formation, nutrient cycling and habitat protection), regulation (water purification, water storage and storm protection), provisioning (fresh water, nutrients, fibre, fuel and minerals) and heritage value (cultural, recreation, tourism and aesthetic values) (Silva *et al.* 2016).

Coastal forests could be generally classified in to three major divisions: i) coastal sand dune forests; ii) mangrove forests; iii) coastal river valley forests. This categorization is applicable for the virgin coastal landscapes without substantial alteration. However, these natural forest zones are considerably meddled due to encroachment, urbanization and industrialization. The major task ahead is how to maintain a balance and restore these forest ecosystems in spite of urbanization and industrialization in coastal region? One of the ecosystem-based approaches to be followed to develop and sustain coastal forests with appropriate vegetation is dependent on mycorrhizal fungi, which are the integral part of coastal forests. A variety of mycorrhizal mutualism serve as essential component of coastal forest ecosystem. Implications and role of forest management and silviculture in coastal region in British Columbia has been addressed by Outerbridge and Trofymow (2009). Table 1 lists some useful literature on development of coastal forest ecosystem with an ecosystem bias. In view to follow ecosystem-based approaches to sustain coastal forest ecosystem, the current short review tries to ascertain occurrence and role of mainly mycorrhizal association of fungi in coastal habitats.

Table 1: Selected publications dealt with different aspects of coastal sand dunes and mangroves.

Topic	Reference
Coastal sand dunes	
Biology	Maun (2009b)
Bioresources	Sridhar (2009)
Plant communities	Sridhar & Bhagya (2007); Maun (2009c)
Biotic interactions	Muhamed *et al.* (2013)
Restoration	Russell *et al.* (2014); Silva *et al.* (2016)
Ecology and conservation	Martinez & Psuty (2004)
Disturbance	Hesp & Martínez (2007)
Landscape changes	Malavasi *et al.* (2013)
Mycorrhizal fungi	Maun (2009a); Botnen *et al.* (2015)
Arbuscular mycorrhizal fungi	Beena *et al.* (2000, 2001); Sridhar & Beena (2001); Sengupta & Chaudhuri (2002); Ramos-Zapata *et al.* (2011); Aytok *et al.* (2013)
Ectomycorrhizal fungi	Montecchio *et al.* (2004); Tedersoo *et al.* (2007); Outerbridge & Trofymow (2009); Ashannejhad & Horton (2006); Obase *et al.* (2010, 2011); Auèina *et al.* (2011); Ma *et al.* (2012); Dutta *et al.* (2013); Sulzbacher *et al.* (2013); Pecoraro *et al.* (2014); Séne *et al.* (2015)
Endophytic fungi	Seena & Sridhar (2004); Min *et al.* (2014)
Mangroves	
Biology of mangroves	Kathiresan & Bingham (2001); Alongi (2016)
Arbuscular mycorrhizal fungi	Sridhar *et al.* (2011); D'Souza & Rodrigues (2013a, b)
Ectomycorrhizal fungi	Ghate & Sridhar (2015)
Endophytic fungi	Suryanarayanan *et al.* (1998); Kumaresan and Suryanarayanan (2001); Ananda & Sridhar (2002); Maria & Sridhar (2003); Karamchand *et al.* (2009)

2. Mycorrhizal Fungi

Although importance of mycorrhizal fungi is generally accepted as contributors in nutrient dynamics, carbon cycling, tree physiology and seedling growth, their importance in forest ecosystems on long-term basis is less understood. This section takes stock of the situation on occurrence and role of various mycorrhizal fungi in coastal regions. Most of the studies on mycorrhizal component in coastal vegetation are confined to enlist and diversity rather than implementation to harness their potential in raising coastal forests. However, mycorrhizal flora although exist as understory vegetation, they serve as potential source of inoculum for forest revegetation tasks. Besides, understory vegetation increased their positive effects under most severe environmental conditions in raising oak species in coastal dune habitats (Muhamed *et al.* 2013).

2.1 Arbuscular Mycorrhizal Fungi

Although arbuscular mycorrhizal (AM) fungi have mutualistic relationship with over 80% of terrestrial plant species (Smith and Read 2008), their assessment in coastal forest ecosystem is sporadic. Majority of available studies are confined to herbs and shrubs with a few tree species in the coastal sand dunes and mangroves (Table 2). Sridhar and Beena (2001) reviewed AM fungal species richness, diversity, seasonal periodicity and other ecological aspects in coastal sand dunes in different continents.

Table 2: List of representative tree species endowed with AM, EM and endophytic fungi in coastal forests of India.

Tree species	Location	Reference
Arbuscular mycorrhizal fungi		
Acacia planifrons	Tamil Nadu	Selvaraj & Kim (2004)
Aegiceras corniculatum	Goa	D'Souza & Rodrigues (2013a)
Avicennia officinalis	Goa	D'Souza & Rodrigues (2013a)
A. marina	Goa	D'Souza & Rodrigues (2013a)
Borassus flabellifer	Tamil Nadu	Selvaraj & Kim (2004)
Bruguiera cylindrica	Goa	D'Souza & Rodrigues (2013a)
B. gymnorrhiza	Goa	D'Souza & Rodrigues (2013a)
Casuarina equisetifolia	Tamil Nadu	Selvaraj & Kim (2004)
Ceriops tagal	Goa	D'Souza & Rodrigues (2013a)
Cocos nucifera	Tamil Nadu	Selvaraj & Kim (2004)
Excoecaria agallocha	Goa	D'Souza & Rodrigues (2013a, b)
Kandelia candel	Goa	D'Souza & Rodrigues (2013b)
Prosopis juliflora	Tamil Nadu	Selvaraj & Kim (2004)
Rhizophora apiculata	Goa	D'Souza & Rodrigues (2013a)
R. mucronata	Karnataka	Sridhar *et al.* (2011)
	Goa	D'Souza & Rodrigues (2013a, b)
Salvadora persica	Goa	D'Souza & Rodrigues (2013a)
Sonneratia alba	Karnataka	Sridhar *et al.* (2011)
	Goa	D'Souza & Rodrigues (2013a)
S. caseolaris	Goa	Rodrigues (2013a)
Ectomycorrhizal fungi		
Acacia auriculiformis	Karnataka	Ghate *et al.* (2014); Ghate (2016)
	West Bengal	Dutta *et al.* (2013)
Casuarina equisetifolia	Karnataka	Ghate *et al.* (2014); Ghate (2016)
Eucalyptus globulus	West Bengal	Dutta *et al.* (2013)
Rhizophora mucronata	Karnataka	Ghate & Sridhar (2015)
Endophytic fungi		
Aegiceras corniculatum	Tamil Nadu	Kumaresan & Suryanarayanan (2001)
Avicinnia marina	Tamil Nadu	Kumaresan & Suryanarayanan (2001)
A. officinalis	Karnataka	Ananda & Sridhar (2002)
	Tamil Nadu	Kumaresan & Suryanarayanan (2001)
Bruguiera cylindrica	Tamil Nadu	Kumaresan & Suryanarayanan (2001)

(Contd.)

Tree species	Location	Reference
Ceruios decabdra	Tamil Nadu	Kumaresan & Suryanarayanan (2001)
Excoecaria agallocha	Tamil Nadu	Kumaresan & Suryanarayanan (2001)
Lumnitzera racemosa	Tamil Nadu	Kumaresan & Suryanarayanan (2001)
Rhizophora apiculata	Tamil Nadu	Suryanarayanan *et al.* (1998); Kumaresan & Suryanarayanan (2002)
R. mucronata	Karnataka	Ananda & Sridhar (2002)
	Tamil Nadu	Suryanarayanan *et al.* (1998)
Sonneratia alba	Karnataka	Ananda & Sridhar (2002)

Figure 1 consists of spores of some representative AM fungi recorded in association with tree species in mangroves of southwest India. Spores of five AM fungi (*Gigaspora albida, G. margarita, Glomus dimorphicum, G. occultum* and *Scutellospora nigra*) were recovered from the rhizosphere of *Rhizophora mucronata*, while spores of three species (*Acaulospora taiwania, Glomus aggretatum* and *Scutellospora pellucida*) were found in rhizosphere of *R. mucronata* as well as *Sonneratia alba*. However, rhizosphere of *R. mucornata* as well as *S. alba* were composed of spores of 18 and 7 species of AM fungi, respectively in Nethravathi mangrove (Sridhar *et al.* 2011). In Nethravathi mangroves, the woody legume climber *Derris triflorum* composed of 34 species of AM fungi with highest species and spore richness. *Derris triflorum* is also a dominant legume in the coastal sand dunes of the

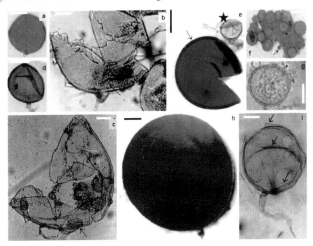

Fig. 1: Spores of arbuscular mycorrhizal fungi obtained from rhizosphere soils of mangrove tree species: a, *Acaulospora taiwania*; b, *Gigaspora albida*; c, *Gigaspora margarita* (suspensor arrowed); d, *Glomus aggretatum*; e, *Glomus dimorphicum* (big arrowed and small starred); f, Bunch of sporocarps of *Glomus dimorphicum* (small arrowed); g, *Glomus occultum*; h, *Scutellospora nigra*; i, *Scutellospora pellucida* (broken and separated wall layers arrowed) (Scale bar: a-f, h & i, 50 μm; g, 20 μm).

southwest India and being nitrogen fixing with rhizobia, mycorrhizal association confer stability and tolerance against adverse effects. Similarly, *Prosopis juliflora* another woody legume in coastal sand dunes of southwest India colonized by *Glomus macrocarpum* also showed high efficiency in saline-tolerance (Selvaraj and Kim 2004). Besides, *P. juliflora* being symbiotically associated with *Rhizobium* survive in disturbed sand dunes, which deserves to be considered for revegetation of coastal sand dunes.

Rizosphere of several mangrove tree species belonging to nine genera (*Aegiceras, Avicinnia, Bruguiera, Ceriops, Excoecaria, Kandelia, Rhizophora, Salvadora* and *Sonneratia*) in mangroves of Goa in the west coast yielded a variety of AM fungi (D'Souza and Rodrigues, 2013a 2013b). Soil and root samples of about 31 plant species occurring in mangroves of deltaic basin of River Ganges in Sundarban with exception of members of Cyperaceae consists of several AM fungal spores (Sengupta and Chaudhuri 2002).

Based on studies on a dominant sand-binding creeper, *Ipomoea pes-caprae* in the coastal sand dunes of southwest India, moderate levels of disturbance like erosion and accretion/burial facilitate dispersal of AM fungal propagules along with plant propagules to the new sites (See Beena *et al.* 2000). Moderate sand burial also confers many advantages to plant species as well as AM fungi to escape from predators, access to rich nutrients and access to sufficient moisture, provide additional room for growth and protect spores/seeds/seedling roots from elevated temperature/desiccation. Severe disturbance leads to complete elimination or elimination beyond regeneration of both host and AM fungi. The number of AM fungi in the coastal sand dunes was reduced to one-third due to lack of vegetation on the southwest coast of India (Kulkarni *et al.* 1997).

The AM fungi are known to form soil aggregates and three phases in this process which include: i) adherence of the soil particles by hyphae; ii) hyphal roots develop conditions favorable for formation of microaggregates; iii) roots and hyphae together entangle to bind the microaggregates into macroaggregates (Miller and Jastrow 1990). Polysaccharide produced by the AM fungi firmly bind microaggregates, for example, one gram stable macroaggregate possesses hyphae up to 50 m (Tisdall 1991). Weight of sand aggregates in non-mycorrhizal and mycorrhizal *Phaseolus vulgaris* with *Glomus* was between 10 and 54 g/kg (Sutton and Sheppard 1976). Such aggregate's weight and the size increases from mobile to fixed sand dunes as well as towards away from sea due to decreasing salinity and aggregates remain intact in dunes in spite of death of roots as well as mycorrhizal hyphae (Koske and Polson 1984). According to Lemoine *et al.* (1995), Acaulosporaceae and Glomaceae consist of extracellular polysaccharide (β-1-3 glucan) involving in soil aggregation. *Gigaspora gigantea* produces a stable glycoprotein (glomalin) resulting in soil aggregation and insoluble hydrophobic glomalin coating which has insulation ability to protect minerals, microbes and organic matter in aggregates (Wright and Upadhyaya, 1998).

2.2 Ectomycorrhizal Fungi

Rough estimate of ectomycorrhizal (EM) fungi worldwide comprise 7,000-10,000, however, majority of them are yet to be described (Taylor and Alexander 2005). These fungi contribute about one-third of the microbial biomass in association with roots of woody plants in wide geographic regions (Högberg and Högberg 2002). However, meagre information is available on occurrence of EM fungi in the coastal sand dunes and mangroves especially in the Indian Subcontinent (Table 2).

A total of seven EM fungi were associated with coastal sand dune *Acacia auriculiformis* and *Casuarina equisetifolia* in the southwest coast of India (Ghate *et al.* 2014, Ghate 2016). Figure 2 represents fruit bodies of four EM fungi (*Amanita* sp., *Macrolepiota dolichaula, Pisolithus albus* and *Scleroderma citrinum*) associated with *A. auriculiformis* and *C. equisetifolia* in coastal sand dunes, besides three more species (*Macrolepiota rhacodes, Macrolepiota* sp. and *Pisolithus* sp.) were also mycorrhizal. *Inocybe petchii and Thelephora palmata* were mycorrhizal in *Rhizophora mucronata* in mangroves of southwest India (Ghate and Sridhar 2015). Interestingly *T. palmata* is also a common EM fungus in coastal sand dunes and mangroves. In

Fig. 2: Fruit bodies of ectomycorrhizal fungi associated with tree species in coastal sand dunes and mangroves: a, *Amanita* sp.; b, *Inocybe petchii*; c, *Macrolepiota dolichaula*; d, *Pisolithus albus*; e, *Scleroderma citrinum*; f, *Thelephora palmata*.

mangroves of Sundarbans (India), *Pisolithus arhizus* was mycorrhizal in *Acacia auriculiformis* and *Eucalyptus globulus* (Dutta *et al.* 2013). Trees species in the high forests and low forests of coastal sand dunes in Brazil (e.g. *Eugenia ligustrina, Maytenus erythroxyla* and *Myrcia guianensis*) composed of several EM fungi belonging to six genera (*Amanita, Coltricia, Lactifluus, Russula, Scleroderma* and *Tylopilus*) (Sulzbacher *et al.* 2013).

Foster and Nicolson (1981) opined that the EM fungal hyphae extend up to 120 mm in soil, covered by extracellular polysaccharides and the hyphae have particles of soil enmeshed between them. Some mycorrhizal roots are also known to stimulate specific groups of bacteria, including those that produce extracellular polysaccharides (Meyer and Linderman 1986).

2.3 Other Mycorrhizal Fungi

Besides AM and EM fungi, does other mycorrhizal fungi (arbutoid, ectendo-, ericoid, monotropoid, orchid and septate endomycorrhizal) exist in coastal forests? Almost all of them are likely components at least in coastal river valley forests. However, arbutoid mycorrhizal fungi occurred in sand dunes associated with several host genera (e.g. *Arbutus, Arctostaphylos* and *Pyrola*) (Maun 1993). *Arctostaphylos uva-ursi*, the major component of dune vegetation was colonized by arbutoid mycorrhizal fungi in the foredunes as well as in transition zone between foredunes and forested areas in Great Lakes. Similarly, ericoid mycorrhizal fungi have been reported from ericaceous shrubs (e.g. *Vaccinium* sp.) occurring in foredunes and stable dunes (Maun 2009a). Orchid mycorrhizal fungi were also associated with orchid species in the dune complex of the Great Lakes (*Epipactis helleborine*), however *Cypripedium calceolus* was found in the transition zone between grassy and woody vegetation.

3. Endophytic Fungi

Several fungi hide within the live tissues of the plant species in coastal region without causing visible disease symptoms (e.g. foliar, bark, wood and root). Invariably they protect flora from water, salinity and herbivore stress. Besides many of them are producers of valuable metabolites of health and industrial interest. Similar to mycorrhizal symbiosis in coastal forests, endophytic studies are also confined to a few tree species in the coastal sand dunes. Some of the tree species screened for endophytic fungi in mangroves have been listed in Table 2.

Studies on endophytic fungi in the coastal sand dunes were confined to herbs, shrubs and creepers (e.g. *Ammophila, Elymus, Suaeda fruticosa* and salt marshes) (Fisher and Petrini 1987; Maciá-Vicente *et al.* 2008; Márquez *et al.*

2008). Two wild legumes (*Canavalia cathartica* and *C. maritima*) in the coastal sand dunes of southwest coast of India were colonized by 46 species of endophytic fungi (Seena and Sridhar, 2004). Among different age groups (seed, seedling and mature plant) and tissues classes (root, stem, leaf, seed coat and cotyledon), the highest endophytic fungi was seen in the seedlings as well as mature plants, while it was the lowest in seeds. *Chaetomium globosum* exhibited single species dominance in *Canavalia* spp. and it is known to produce several bioactive compounds (e.g. chaetomin, chaetoglobosins, chaetoquadrins, oxaspirodion, chaetospiron, orsellides, chaetocyclinones) (Sekita *et al.* 1976; Loesgen *et al.* 2007; Suryanarayanan *et al.* 2010). Besides, *C. globosum* also produces a nematicide the flavipin (Chitwood, 2002) and indicating its extensive colonization in *Canavalia* in coastal sand dunes have potential role in protection from nematodes.

Foliar endophytes in mangrove trees are dominated by single species: *Avicennia marina* (*Phoma* sp.), *Bruguiera cylindrica* (*Colletotrichum gloeosporioides*), *Rhizophora apiculata, R. mucronata* (*Sporormiella minima*) (see Sridhar 2012). But, multiple species dominance is shown by *Lumnitzera racemosa* (*Alternaria* sp., *Phomopsis* sp. and *Phyllosticta* sp.). *Avicennia officinalis, Rhizophora mucronata* and *Sonneratia caseolaris* tree species also showed multiple species dominance of endophytic fungi (Ananda and Sridhar 2002). Woody tissues and leaves of *Kandelia candel* were studied for endophytic fungi in Mai Po Nature Reserve of Hong Kong (Pang *et al.* 2008). It is interesting to note several entomopathogenic fungi exist as endophytes in the coastal sand dune and mangrove plant species (Sridhar 2012). Although endophytic assemblage was similar in bark and wood, it differed in leaf samples. The root endophytic fungal diversity of black pine (*Pinus thunbergii*: widely distributed in Korea, Indonesia and Japan) was studied in three coastal regions by Min *et al.* (2014). Eighteen species of endophytes showed salinity resistance, which consist of many of species of *Penicillium* and *Trichoderma*.

4. Indigenous Tree Species

The following information on native tree species of the coastal sand dunes and mangroves provide basis to protect, rehabilitate and manage forest cover in phase-wise with a blend of native and introduced tree species. In the interest of success of revegetation, invariably forest managers have to depend on mycorrhizal fungi to develop seedlings before transplanting them to required zone of landscape.

Rao and Meher-Homji (1985) have given a broad outline about coastal vegetation including scrubs (*Acrostichium aureum, Clerodendrum inerme, Dimorphocalyx glabellus, Halopyrum mucronatum, Myriostachya wightiana, Scaevola taccada, Syzygium ruscifolium* and *Tamarix articulata*) and tree species (*Acacia planifrons, Ardisia littoralis, Calophyllum inophyllum, Copparis cartilaginea, Euphorbia caducifolia, Hyphaene dichotoma, Messerschmidia argentea, Morinda citrifolia, Pandanus tectorius, Pemphis acidula, Premna serratifolia* and *Salvadora persica*) existing in about 7,000 km coastline of the Indian Subcontinent from the Rann of Kutch (Gujarat) to Sundarbans (West Bengal). Recently, some researchers surveyed the indigenous or introduced tree species growing in the Tamil Nadu coastal sand dunes of southern India and these tree species include: *Azadirachta indica, Borassus flabellifer, Casuarina equisetifolia, C. litorea, Cocos nucifera, Hibiscus tiliaceus, Mangifera indica, Moringa oleifera, Psidium guajava, Salvadora persica, Tectona grandis, Terminalia* spp. and *Thespesia populnea* (Muthukumar and Samuel 2011; Poyyamoli *et al.* 2011; Ramarajan and Murugesan 2014). A shrub *Prosopis juliflora* has wide distribution in the coastal sand dunes of Tamil Nadu. There are several tree species other than discussed in Section 4 are worth considering for the revegetation of coastal sand dunes especially in the Indian context. For instance, *Careya arborea, Caryota urens, Erythrina variegata, Ficus arnottiana, Mimusops elengi, Pandanus mangalorensis, P. tectorius* and *Pongamia pinnata*. Besides, plantation and wild tree species like *Anacardium occidentale, Arotcarpus heterophyllus* and *A. hirsutus* adapted well to the coastal sand dunes. It is worth considering some of the Dipterocarpaceae members as their population is currently dwindling in the Western Ghats and other tropical regions. In fact, *Hopea ponga, H. parviflora* and *Macaranga peltata* have adapted to the lateritic scrub jungles of southwest India.

More than 100 mangrove tree species are listed by Chapman (1976). About 50-60 among 80 true mangrove tree and shrub species (in 30 genera and 20 families) significantly contribute to the structure of mangrove forests (Tomlinson 1986; Field 1995). Accordingly, the mangrove plant species have been classified into three categories: i) true mangroves (~80 tree and shrub species restricted to high tides and neap spring tides; e.g. *Avicennia, Kandelia, Rhizophora* and *Sonneratia*); ii) minor vegetation (rarely forming pure stands; e.g. *Aegiceras* and *Excoecaria*); iii) mangrove associates (salt-tolerant plant species restricted to landwards and seawards; e.g. *Acanthus, Dalbergia* and *Derris*). For rehabilitation of mangroves, the above plant species are suitable and in addition trees of *Kandelia candel* need to be encouraged as their population is severely dwindling. Other tree species of value in revegetating mangrove habitats include *Borassus flabellifer, Caryota urens, Ficus callosa, Pandanus mangalorensis* and *P. tectorius*.

In the interest of revegetation of degraded river valleys, several tree species can be encouraged. Some of the potential tree species include *Artocarpus heterophyllus, A. hirsutus, Bamboosa* spp., *Borassus flabellifer, Careya arborea, Caryota urens, Garcinia indica, Holigarna* spp., *Macaranga peltata, Mangifera indica, Morinda citrifolia* and *Pandanus* spp. (Shetty and Kaveriappa 2001). It is likely several dipterocarp tree species are also suitable for river valley forestry revegetation (Appanah and Turnbull 1998).

During 1992-94, an arboretum has been established in three hectares of lateritic region of a temporary streamlet basin in Mangalore University with 57 tree species, two species of bamboo, 23 species of shrubs (woody climbers) and 16 endemic species of herbs/under shrubs (Shetty and Kaveriappa, 2001). The rare and threatened species in the arboretum include: six tree species (*Holigarna grahamii, Hopea canarensis, Myristica fatua, M. malabarica, Pterospermum reticulatum* and *Vepris bilocularis*), three scandent shrubs (*Aspidopteris canarensis, Grewia heterotricha* and *Rubus fockei*) and two herbs (*Crotalaria lutescens* and *Paracautleya bhatii*). As the arboretum is over two decades old, the extent of survival of planted endemic and endangered tree species provide practical example how to vegetate tropical coastal region with these plant species. It is interesting to note that five dipterocarp tree species that grew in the arboretum were *Dipterocarpus indicus, Hopea canarensis, H. parviflora, H. ponga* and *Vateria indica. Hopea ponga* has well adapted to grow in lateritic scrub jungles of south west coast and ectomycorrhizal with *Astraeus odoratus* (Greeshma *et al.* 2015; Pavithra *et al.* 2015). *Vateria indica* has also well adapted to the coastal lateritic scrub jungles in southwest India (Shetty and Kaveriappa 2001). Similarly, another highly ectomycorrhizal dipterocarp tree species *Shorea robusta* is well adapted to the lateritic regions and mangroves of Sundarbans in the West Bengal (Pradhan *et al.* 2012; Dutta *et al.* 2013). *Hopea* spp. are found suitable as second storey crop along with teak (*Tectona grandis*) plantations (Weinland 1998).

Lee (1998) has given a review on the ectomycorrhizal dipterocarp species in different forests more specifically in southeast Asia. According to him, 14 genera of dipterocarps are ectomycorrhizal: *Anisoptera* (6 spp.), *Cotylelobium* (2 spp.), *Dipterocarpus* (23 spp.), *Dryobalanops* (5 spp.), *Hopea* (14 spp.), *Marquesia* (2 spp.), *Monotes* (2 spp.), *Neobalanocarpus* (1 sp.), *Parashorea* (3 spp.), *Pentacme* (2 spp.), *Shorea* (46 spp.), *Vatica* (6 spp.), *Vateria* (1 sp.) and *Vateriopsis* (1 sp.). The most dominant EM fungal genera on dipterocarp tree species in southeast Asia include *Amanita, Boletus* and *Russula*. Appanah and Turnbull (1998) in their edited book provided an excellent review of dipterocarp taxonomy, biogeography, ecology and silviculture potential. Dipterocarps being dominant in the Southeast Asia, they have adaptation to a variety of habitats (e.g. coast, riverine, freshwater swamps and dry land) (Symington 1943; Wyatt-Smith 1963).

5. Threats and benefits

Human interference of coastal forests has far reaching consequences in the future. Felling, clear cutting, constructions and industrial activities devastate coastal forests, which are the green wall against storms and floods. Possible impacts on ecosystem-based and non-ecosystem based approaches in coastal sand dune dynamics have been interpreted by Sridhar (2009). He has addressed mainly impact of disturbance on organic matter input, biological nitrogen fixation, phosphate solubilization and interaction of microbes and plant species on the coastal sand dune habitats. In the larger context, non-ecosystem based approaches of protection of coastal sand dunes impairs the major biogeochemical cycles.

Beena et al. (2000, 2001) dealt with various human interference on the coastal sand dunes of the southwest coast of India. A large number of timber industries established in mangrove locations are misusing backwaters to cure timbers and wood products. Sand mining is a major threat that goes on in the coastal sand dunes, mangroves and river valleys imparts major impact on the entire ecosystem interaction and stability. The ecosystem services of ecosystem-based approaches of protection of coastal forests are much more than we normally assume. The major economic gains of green wall along the coast are protection from storms, tides and erosion. Appropriate vegetation build the ecosystem in favour of existence and interactions of flora, fauna and microbes. Besides, coastal forests provide potential food, fodder and bioactive compounds of pharmaceutical value (Sridhar and Bhagya 2007). Additional benefits are dealt under the section on endophytic fungi. At present many macrofungi on the coastal sand dunes or mangroves are edible and utilized by the native dwellers (*Amanita* sp., *Lentinus*

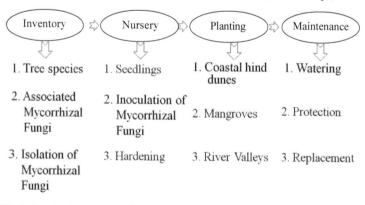

Fig. 3: Possible basic steps to be undertaken for reforestation of coastal regions.

squarrosulus, Lecoperdon decipiens, L. utriforme, Macrolepiota rachodes, Pleurotus flabellatus and *Termitomyces* spp.) (e.g. Ghate *et al.* 2014; Ghate and Sridhar 2015; Ghate 2016).

Conclusions and Vision

There are several important tasks to support the coastal forests: i) maintenance of existing vegetation stand without further degradation; ii) re-vegetation of maximum available areas with appropriate tree vegetation; iii) educate the coastal dwellers to understand the value and benefits of coastal forests; iv) to open up for tourists on sufficient forest cover with indigenous/endangered tree species to generate revenue for future developments. There is an urgent need to take stock of the situation of existing coastal forests and appropriate space available for re-vegetation using remote sensing and other modern approaches. Some broad basic steps necessary to be undertaken are given in Fig. 3.

Practicing large-scale tourism in coastal region without developing sustainable forest cover further aggravates denudation, pollution and accumulation of solid wastes. On sufficient forest cover sustaining the ecosystem, may support tourism and help establishing marsh/mangrove/dune parks. Such approaches help tourists to understand how important the coastal forests and wild life in human affairs. Industries already established on the zones of coastal forests should be demanded to conserve and re-vegetate for sustainability.

The extent of human interference on coastal forests worldwide denotes that it is the last chance to move towards sustainable development in management, restoration and development. Supporting endangered and endemic plant species for conservation would be more tangible to derive more benefits and encourage perpetuation of wildlife. Quick action could be achieved through community participation, institutional involvement and public participation in the drive of reforestation of coastal regions. Such movements should not be mere action, it should move forward towards gaining what we have lost for several years.

Acknowledgement

Author is grateful to the Mangalore University for providing facilities to carry out research on mycorrhizal fungi and the University Grants Commission, New Delhi for the award of UGC-BSR Faculty Fellowship. I appreciate help rendered by Keshavachandra, Karun Chinnappa and Sudeep Ghate.

References

Alongi DM (2016) Mangroves. In: Encyclopedia of Estuaries (Ed) Kennish MJ, Springer, Netherlands, pp 393-404.

Ananda, K, Sridhar KR (2002) Diversity of endophytic fungi in the roots of mangrove species on the west coast of India. Can. J. Microbiol. 48, 871-878.

Appanah S, Turnbull JM (1998) A Review of Dipterocarps Taxonomy, Ecology and Silviculture. Centre for International Forestry Research, Bogor, Indonesia.

Ashkannejhad S, Horton TR (2006) Ectomycorrhizal ecology under primary succession on coastal sand dunes: interactions involving *Pinus contorta,* suilloid fungi and deer. New Phytol. 169, 345-354.

*Auèina*A, Rudawska M, Leski T, Ryliškis D, Pietras M, Reipšas E (2011) Ectomycorrhizal fungal communities on seedlings and conspecific trees of *Pinus mugao* grown on the coastal dunes of the Curonian Spit in Lithuania. Mycorrhiza 21, 237-245.

Aytok Ö, Yilmaz T, *Orta°*I, Çakan H (2013) Changes in mycorrhizal spore and root colonization of coastal dune vegetation of the Seyhan Delta in the postcultivation phase. Turk. J. Agric. For. 37, 52-61.

Bascom W (1980) Waves and beaches. Anchor Press/Doubleday Publishing, New York.

Beena KR, Arun AB, Raviraja NS, Sridhar KR (2001) Association of arbuscular mycorrhizal fungi with plants of coastal sand dunes of west coast of India. Trop. Ecol. 42, 213-222.

Beena KR, Raviraja NS, Arun AB, Sridhar KR (2000) Diversity of arbuscular mycorrhizal fungi on the coastal sand dunes of the west coast of India. Curr. Sci. 79, 1459-1466.

Botnen S, Kauserud H, Carlsen T, Blaalid R, Høiland K (2015) Mycorrhizal fungal communities in coastal sand dunes and heaths investigated by pyrosequencing analyses. Mycorrhiza 25, 447-456.

Chapman VJ (1976) Mangrove Vegetation. Cramer Vadyz, Liechtenstein, Germany.

Chitwood DJ (2002) Phytopathological based strategies for nematode control. Ann. Rev. Phytopathol. 40, 221-249.

D'Souza J, Rodrigues BF (2013a) Biodiversity of Arbuscular Mycorrhizal (AM) fungi in mangroves of Goa in West India. J. For. Res. 24, 515-523.

D'Souza J, Rodrigues BF (2013b) Seasonal Diversity of Arbuscular Mycorrhizal Fungi in Mangroves of Goa, India. Int. J. Biodivers. Article # 19657, DOI: http://dx.doi.org/10.1155/2013/196527.

Dutta AK, Pradhan A, Basu SK, Acharya K (2013) Macrofungal diversity and ecology of the mangrove ecosystem in the Indian part of Sundarbans. Biodivers. 14, 196-206.

Field C (1995) Journey amongst mangroves. International Society for Mangrove Ecosystems, Okinawa, Japan.

Fisher PJ, Petrini O (1987) Location of fungal endophytes in tissues of *Suaeda fruticosa*: a preliminary study. Trans. Br. Mycol. Soc. 89, 246-249.

Forster SM, Nicolson TH (1981) Microbial aggregation of sand in a maritime dune succession. Soil Biol. Biochem. 13, 205-208.

Ghate SD (2016) Studies on Macrofungi and Aquatic fungi of Selected Wetlands of the Southwest India. PhD Thesis, Mangalore University, Karnataka, India.

Ghate SD, Sridhar KR (2015) Contribution to the knowledge on macrofungi in mangroves of the Southwest India. Pl. Biosys. DOI: 10.1080/11263504.2014.994578.

Ghate SD, Sridhar KR, Karun NC (2014) Macrofungi on the coastal sand dunes of south-western India. Mycosphere 5, 144-151.

Greeshma AA, Sridhar KR, Pavithra M (2015) Macrofungi in the lateritic scrub jungles of southwestern India. J. Threat. Taxa 7, 7812-7820.

Hesp PA, Martínez ML (2007) Disturbance processes and dynamics in coastal dunes. In: Plant Disturbance Ecology: The Process and Response (Ed) Johnson EA and Miyanishi K, Academic Press, pp 215-247.

Hoagland P, Jin D (2006) Accounting for economic activities in large marine ecosystems and regional. United Nations Environment Programme.

Högberg MN, Högberg P (2002) Extramatrical ectomycorrhizal mycelium contributes one-third of microbial biomass and produces, together with associated roots, half the dissolved organic carbon in a forest soil. New Phytol. 154, 791-795.

Karamchand KS, Sridhar KR, Bhat R (2009) Diversity of fungi associated with estuarine sedge *Cyperus malaccensis* Lam. J. Agric. Technol. 5, 111-127.

Kathiresan K, Bingham BL (2001) Biology of mangrove ecosystem. Adv. Mar. Biol. 40, 81-251.

Koske RE, Polson WR (1984) Are VA mycorrhizae required for sand dune stabilization? BioScience 34, 420-424.

Kulkarni SS, Raviraja NS, Sridhar KR (1997) Arbuscular mycorrhizal fungi of tropical sand dunes of west coast of India. J. Coast. Res. 13, 931-936.

Kumaresan V, Suryanarayanan TS (2001) Occurrence and distribution of endophytic fungi in a mangrove community. Mycol. Res. 105, 1388-1391.

Kumaresan V, Suryanarayanan TS (2002) Endophyte assemblage in young, mature and senescent leaves of *Rhizophora apiculata*: Evidence for the role of endophytes in mangrove litter degradation. Fungal Divers. 9, 81-91.

Lee SS (1998) Root symbiosis and nutrition. In: A Review of Dipterocarps Taxonomy, Ecology and Silviculture (Ed) Appanah S and Turnbull JM, Centre for International Forestry Research, Bogor, Indonesia, pp 99-114.

Lemoine MC, Gollotte A, Gianinazzi-Pearson V (1995) Localization of β (1-3) Glucan in walls of the endomycorrhizal fungi *Glomus mosseae* (Nicol. & Gerd.) Gerd. & Trappe and *Acaulospora laevis* Gerd. & Trappe during colonization of host roots. New Phytol. 129, 97-105.

Loesgen S, Schloerke O, Meindl K, Herbst-Irmer R, Zeeck A (2007) Structure and biosynthesis of chatocyclinones, new polyketides produced by and endosymbiotic fungus. Eur. J. Org. Chem. 132, 2191–2196.

Ma D, Zang S, Wan L, Zhang D (2012) Ectomycorrhizal community structure in chronosequenes of *Pinus densiflora* in eastern China. Afr. J. Microbiol. Res. 6, 6204-6209.

Maciá-Vicente JG, Jansson H-B, Abdullah SK, Descals E, Salinas J, Lopez-Llorca LV (2008) Fungal root endophytes from natural vegetation in Mediterranean environments with special reference to *Fusarium* spp. FEMS Microbiol. Ecol. 64, 90-105.

Malavasi M, Santoro R, Cutini M, Acosta ATR, Carranza ML (2013) What has happened to coastal dunes in the last half century? A multitemporal coastal landscape analysis in Central Italy. Landsc. Urb. Plann. 119, 54-63.

Maria GL, Sridhar KR (2003) Endophytic fungal assemblage of two halophytes from west coast mangrove habitats, India. Czech Mycol. 55, 241-251.

Márquez SS, Bills GF, Zabalgogeazcoa I (2008) Diversity and structure of the fungal endophytic assemblages from two sympatric coastal grasses. Fungal Divers. 33, 87-100.

Martínez ML, Intralawan A, Vázquez G, Pérez-Maqueo O, Sutton P, Landgrave R (2007) The coasts of our world: ecological, economic and social importance. Ecol. Econ. 63, 254-272.

Martínez ML, Psuty (2004) Coastal Dunes, Ecology and Conservation. Springer-Verlag Berlin Heidelberg.

Maun MA (1993) Dry coastal ecosystems along the Great Lakes. In: Ecosystems of the World Dry coastal ecosystems: Africa, America, Asia and Oceania (Ed) van der Maarel E, Elsevier, Amsterdam, pp 299–316.

Maun MA (2009a) Mycorrhizal fungi. In: The Biology of Coastal Sand Dunes (Ed) Maun MA, Oxford University Press, Oxford, pp 134-152.

Maun MA (2009b) The Biology of Coastal Sand Dunes. Oxford University Press, Oxford.

Maun MA (2009c) Plant communities. In: The Biology of Coastal Sand Dunes (Ed) Maun MA, Oxford University Press, Oxford, pp 164-180.

Meyer JR, Linderman RG (1986) Selective influence on populations of rhizosphere or rhizoplane bacteria and actinomycetes by mycorrhizas formed by Glomus fasciculatum. Soil Biol. Biochem. 18, 191-196.

Miller RM, Jastrow JD (1990) Hierarchy of root and mycorrhizal fungal interactions with soil aggregation. Soil Biol. Biochem. 22, 579-584.

Min YJ, Park MS, Fong JJ, Quan Y, Jung S, Lim YW (2014) Diversity and saline resistance of endophytic fungi associated with Pinus thunbergii in coastal shelterbelts of Korea. J. Microbial. Biotechnol. 24, 324-333.

Montecchio L, Causin R, Rossi S, Accordi SM (2004) Changes in ectomycorrhizal diversity in a declining Quercus ilex coastal forest. Phytopathol. Mediterr. 43, 26-34.

Muhamed H, Touzard B, Bagousse-Pinguet YL, Michalet R (2013) The role of biotic interactions for the early establishment of oak seedlings in coastal dune forest communities. For. Ecol. Manage. 297, 67-74.

Muthukumar K, Samuel AS (2011) Coastal sand dune flora in the Thoothukudi District, Tamil Nadu, southern India. J. Threat. Taxa 3, 2211-2216.

Obase K, Cha JY, Lee JK, Lee SY, Lee JH, Chun KW (2010) Investigation of ectomycorrhizal fungal colonization in Pinus thunbergii seedlings at a plantation area in Gangneung, using mophotyping and sequencing and rDNA internal transcribed spacer region. J. Korean For. Soc. 99, 172-178.

Obase K, Lee JK, Lee SY, Chun KW (2011) Diversity and community structure of ectomycorrhizal fungi in Pinus thunbergii coastal forests in the eastern region of Korea. Mycoscience 52, 283-391.

Outerbridge RA, Trofymow JA (2009) Forest management and maintenance of ectomycorrhizae: A case study of green tree retention in south-coastal British Columbia. BC J. Ecosys. Manage. 10, 59-80.

Pang K-L, Vrijmoed LLP, Goh TK, Plaingam N, Jones EBG (2008) Fungal endophytes associated with Kandelia candel (Rhizophoraceae) in Mai Po Nature Reserve, Hong Konga. Bot. Mar. 51, 171-178.

Pavithra M, Greeshma AA, Karun NC, Sridhar KR (2015) Observations on the Astraeus spp. of Southwestern India. Mycosphere 6, 421-432.

Pecoraro L, Angelini P, Arcangeli A, Bistocchi G, Gargano ML, La Rosa A, Lunghini D, Polemis E, Rubini A, Saitta S, Venanzoni R, Zervakis GI (2014) Macrofungi in Mediterranean maquis along seashore and altitudinal transects. Plant Biosys. 148, 367-376.

Poyyamoli G, Padmavathy K, Balachandran N (2011) Coastal sand dunes-vegetation structure, diversity and disturbance in Nallavadu village, Puducherry, India. Asian J. Water Environ. Pollut. 8, 115-122.

Pradhan P, Dutta AK, Roy A, Basu SK, Acharya K (2012) Inventory and Spatial Ecology of Macrofungi in the Shorea Robusta Forest Ecosystem of Lateritic Region of West Bengal. Biodiversity 13, 88-99.

Ramanujan S, Murugesan AG (2014) Plant diversity on coastal sand dune flora, Tirunelveli District, Tamil Nadu. Ind. J. Plant Sci. 3, 42-48.

Ramos-Zapata J, Zapata-Trujillo R, Ortíz-Díaz, Guadarrama P (2011) Arbuscular mycorrhizas in a tropical coastal dune system in Yucatan, Mexico. Fungal Ecol. 4, 256-261.

Rao TA, Meher-Homji VM (1985) Strand plant communities of the Indian sub-continent. Proc. Indian Acad. Sci. (Plant Sci.) 94, 505-523.

Russell W, Sinclair J, Michels KH (2014) Restoration of coast redwood (*Sequoia sempervirens*) forests through natural recovery. Open J. Forest. 4, 106-111.

Seena S, Sridhar KR (2004) Endophytic fungal diversity of 2 sand dune wild legumes from the southwest coast of India. Can. J. Microbiol. 50, 1015-1021.

Sekita S, Yoshihira K, Natori S (1976) Structures of chaetoglobisins C, D, D, and F, cytotoxic indole-3-yl-(13) cytochalasans from *Chaetomium globosum*. Tetrahed. Lett. 17, 1351-1354.

Selvaraj T, Kim H (2004) Ecology of vesicular-arbuscular mycorrhizal (VAM) fungi in coastal areas of India. Agric. Chem. Biotechnol. 47, 71-76.

Séne S, Avril R, Chaintreuil C, Geoffroy A, Ndiaye C, Diédhiou AG, Sadio O, Courtecuisse R, Sylla SN, Selosse MA, Bâ A (2015) Ectomycorrhizal fungal communities of *Coccoloba uvifera* (L.) L. mature trees and seedlings in the neotropical coastal forests of Guadeloupe (Lesser Antilles). Mycorrhiza 25, 547-559.

Sengupta A, Chaudhuri S (2002) Arbuscular mycorrhizal relations of mangrove plant community at the Ganges river estuary in India. Mycorrhiza 12, 169-174.

Shetty BV, Kaveriappa KM (2001) An arboretum of endemic plants of Western Ghats at Mangalore University Campus, Karnataka, India. Zoos' Print Journal 16, 431-438.

Silva R, Martínez ML, Odériz I, Mendoza E, Feagin RA (2016) Response of vegetated dune-beach systems to storm conditions. Coast. Eng. 109, 53-62.

Smith SE, Read DJ (2008) Mycorrhizal Symbiosis. 3rd Edition, Academic Press, San Diego.

Sridhar KR (2009) Bioresources of Coastal Sand Dunes - Are they Neglected? In: Coastal Environments: Problems and Perspectives (Ed) Jayappa KS and Narayana AC. I.K. International Publishing House Pvt. Ltd., New Delhi, India, pp 53-76.

Sridhar KR (2012) Aspect and prospect of endophytic fungi. In: Microbes: Diversity and Biotechnology (Ed) Sati SC and Belwal M, Daya Publishing House, New Delhi, India, pp 43-62.

Sridhar KR, Beena KR (2001). Arbuscular mycorrhizal research in coastal sand dunes –A review. Proc. Nat. Acad. Sci. 71, 179-205.

Sridhar KR, Bhagya B (2007) Coastal sand dune vegetation: a potential source of food, fodder and pharmaceuticals. Livestock Res. Rural Develop. 19, Article # 84: http://www.cipav.org.co/lrrd/lrrd19/6/srid19084.htm

Sridhar KR, Roy S, Sudheep NM (2011) Assemblage and diversity of arbuscular mycorrhizal fungi in mangrove plant species of the southwest coast of India. In: Mangroves: Ecology, Biology and Taxonomy (Ed) Metras JN, Nova Science Publishers Inc., New York, pp 257-274.

Sulzbacher MA, Giachini AJ, Grebenc T, Silva BDB, Gurgel FE, Loiola MIB, Neves MA, Baseia IG (2013) A survey of an ectotrophic sand dune forest in the northeast Brazil. Mycosphere 4, 1106-1116.

Suryanarayanan TS, Kumaresan V, Johnson JA (1998) Foliar fungal endophytes from two species of the mangrove *Rhizophora*. Can. J. Microbiol. 44, 1003-1006.

Suryanarayanan TS, Thirunavukkarasu N, Govindarajulu MB, Sasse F, Jansen R, Murali TS (2010) Fungal endophytes and bioprospecting. Fungal Biol. Rev. 23, 9-18.

Sutton JC, Sheppard BR (1976) Aggregation of sand dune soil by endomycorrhizal fungi. Can. J. Bot. 54, 326-333.

Symington CF (1943) Foresters Manual of Dipterocarps. Forest Department, Kuala Lumpur.

Taylor AFS, Alexander IJ (2005) The ectomycorrhizal symbiosis: life in the real world. Mycologist 19, 102-112.

Tedersoo L, Suvi T, Beaver K, Kõljalg U (2007) Ectomycorrhizal fungi of the Seychelles: diversity patterns and host shifts from the native *Vateriopsis seychellarum* (Dipterocarpaceae) and *Intsia bijuga* (Caesalpiniaceae) to the introduced *Eucalyptus robusta* (Myrtaceae), but not *Pinus caribea* (Pinaceae). New Phytol. 175, 321-333.

Temmerman S, Meire P, Bouma TJ, Herman PM, Ysebaert T, De Vriend HJ (2013) Ecosystem-based coastal defence in the face of global change. Nature 504, 79-83.

Tisdall JM (1991) Fungal hyphae and structural stability of soil. Aust. J. Soil Res. 29, 729-743.

Tomlinson PB (1986) The botany of mangroves. Cambridge University Press, Cambridge.

Weinland G (1998) Plantations. In: A Review of Dipterocarps Taxonomy, Ecology and Silviculture (Ed) Appanah S and Turnbull JM, Centre for International Forestry Research, Bogor, Indonesia, pp 151-185.

Wright SF, Upadhyaya A (1998) A survey of soils for aggregate stability and glomalin, a glycoprotein produced by hyphae of arbuscular mycorrhizal fungi. Plant Soil 198, 97-107.

Wyatt-Smith J (1963) Manual of Malayasian silviculture for in land forests, Volume 1, Malayan Forest Record # 23. Forest Research Institute, Kepong, Malaysia.

15

Biological Degradation of Crude Oil Contaminants and its Application in Indonesia

Asep Hidayat and Maman Turjaman

Abstract

Since the first oil well drilled successfully in 1859, crude oil has been a serious problem around world, including Indonesia. As a source of all petroleum products, crude oil is very complex component and divided into aliphatic, aromatic, resin and asphaltene. Petroleum products have been used for daily life and created large amounts of pollutants in environment and ecosystems around the globe. The pollutants should be treated by certain method; among them, bioremediation has several advantages to resolve the problem. In 2003, Ministry of Environmental (MoE) – Republic Indonesia was endorsed with a new regulation about biological treatment or bioremediation for crude oil contaminated soil (MoE regulation No. 128/2003). The most important to be noted in bioremediation process is the presence of microbes with certain metabolite capabilities. The degradation rate also can be improved and sustained by ensuring the adequate required factors. This article describes the removal of crude oil by utilization of microbial potentials, factors affecting biodegradation, and how they are applied in Indonesia.

Keywords: Bioremediation, Crude oil, Microbes, Regulation

1. Introduction

Environment in which we live is a gift of almighty God which is priceless. On a normal environment, the plant is able to transform inorganic materials into organic through photosynthesis, while the microbes act otherwise; converting back organic to inorganic forms. For example, when a leaf falls from a tree, the decomposition by microbes will occur. Soil will be rich in organic materials that plants utilize to grow. In 1860, Pasteur found that there is a close relationship between conversions of the chemical structures and microbial activity. The term is better known as biodegradation. In 1952, Gayle studied the role of microbes in degrading of hazardous organic compound with the microbial infallibility hypothesis. He concluded that when environment is contaminated with any conceivable organic pollutants, there exists a microorganism that can degrade it under the right conditions (Alvarez and Illman 2006). If the degradation process is a failure, it is because the researchers failed. Nowadays the environment is being threatened; land, sea, water, and air have been decreasing drastically in their function and quality. The water that we use for daily life has became scarce and polluted (Dale *et al.* 1965; Tanabe *et al.* 1990; Aislabie *et al.* 1997; Sudaryanto *et al.* 2002). In Indonesia, the oil exploration has started since 1924 and since then environmental problems generated. Crude oil spills in Indonesia were recorded in the province of East Kalimantan, Riau and West Java from 1999 to 2014. These facts showed the potential of crude oil spill in Indonesia which was significant, and their removal considered to be important because they are carcinogenic, mutagenic, and toxic (Shi-Zhong *et al.* 2009).

The utilization of metabolic potential of microbes to degrade crude oil has been shown to be a viable, effective method, and widely used around the world (Vidali 2001; Erdogan and Karaca 2011). The term itself is known as bioremediation. Biological degradation by microbes can be distinguished by the presence of the end products such as CO_2, H_2O and other metabolic products. Although, bioremediation process shows several advantages, but it has limitations leading to failure in the degradation process. This article discusses biological process of crude oil degradation, as well as the regulations and its application in Indonesia.

2. Crude Oil

Crude oil is known as petroleum hydrocarbon. The term of petroleum is derived from the word *petra* (rock) and *oleum* (oil), and is naturally occurring and commonly processed by refinery into various products. Crude oil usually consists of hydrocarbons (compounds containing carbon and hydrogen atoms), heteroatom compounds (containing carbon and hydrogen atoms with heteroatoms such as sulfur, nitrogen, or oxygen), and relatively small concentrations of metallic constituents (Weisman 1998; Adekunle and Adebambo 2007). General categories of petroleum are aliphatic, aromatics, asphaltenes, and resins. The aliphatic fraction consists of straight-chain alkanes (normal alkanes), branched alkanes (isoalkanes), and cycloalkanes (naphthenes). The aromatic fraction includes volatile monoaromatic hydrocarbons such as benzene, toluene, and xylenes; polycyclic aromatic hydrocarbons (PAHs); naphthenoaromatics; and aromatic sulfur compounds, such as thiophenes and dibenzothiophenes. The asphaltene (phenols, fatty acids, ketones, esters, and porphyrins) and resin (pyridines, quinolines, carbazoles, sulfoxides, and amides) fractions consist of polar molecules containing N, S, and O_2. Asphaltenes are large molecules dispersed in oil in a colloidal manner, whereas resins are amorphous solids truly dissolved in oil.

The relative distribution of these fractions depends on many factors, i.e. the source, age, geological history, migration, and alteration of crude oil (Singh 2006). Regardless of the complexity, petroleum compounds can be generally classified into two major component categories: hydrocarbons (measured as THP) and non-hydrocarbons. Figure 1 summarizes the different categories and subclasses. Hydrocarbon coincided in crude oil is so complex, more than 10 thousand hydrocarbon compounds. Although crude oil has been separated but it is usually still very complex mixture of molecules, i.e jet fuels contains more than 300 different hydrocarbon compounds (Alvarez and Illman 2006).

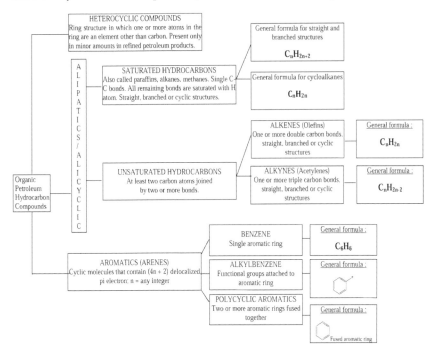

Fig. 1: Summarization of different categories and subclasses of total petroleum hydrocarbons (Adapted from Hidayat 2013).

Crude oil spills are the most serious type of contaminants that polluted the environment such as soil, water and ocean. The common source contamination include leaking and routine washings underground storage tanks, offshore platform, drilling rigs, well accident, transportation accident, ruptured pipelines and natural oil seepage (Harayama *et al.* 1999; Kvenvolden and Cooper 2003; Chaineau *et al.* 2005; Minai-Tehrani *et al.* 2006; Elshafie *et al.* 2007; Hasanuzzaman *et al.* 2007).

As the maritime nation, the Indonesia territory is known for its busy trade route, traffic tanker carrying millions tons barrels of crude oil. These conditions cause Indonesia as a region very susceptible to be contamination by oil spill. If the crude oil is spilled in ocean, oil will float on the surface. The oil will spread everywhere including beaches, mangroves and brackish water. The marine ecosystem will be disturbed, and become poisonous, inhibiting the growth, reproduction and other physiological processes on coral reefs, mangrove and some water biota. Crude oil spill in Indonesia were noted in the province of East Kalimantan, Riau and West Java. In the year 1999, there were 4,000 barrels of crude oil spilled from the tanker (MT King Fisher) to Sea Cilacap - Central Java. In 2009 the oil spill occurred in Manyar, Coarse sand - East Java. In 2014, oil spills occurred in an area Indarmayu and 7 km of coastline have been polluted. In the same year, 200 barrels of crude oil from wells accident in Bakau Area reached into the swamp river in Siak district, Riau province. Our prediction is the oil spill in Indonesia will occur periodically in future in many places. All those accidents have potential for carcinogenic, mutagenic, or toxic effects and changes in composition can lead to variations in overall physical properties and chemical toxicity (Shi-Zhong *et al.* 2009).

3. Bioremediation

Bioremediation, described as a "biological response to environmental abuse" (Hamer 1993), is the use of living organisms, primarily microorganisms, to degrade environmental pollutants into less toxic forms or non-toxic compounds. It is also the use of naturally occurring bacteria, fungi, or plants to purify hazardous materials (Boopathy 2000; Vidali 2001). Bioremediation utilizes the metabolic potential of microorganisms by enzymatic system reaction (Watanabe 2001).

Remediation offers the possibility to reduce or destroy hazardous contaminant before they are released into the environment at relatively low cost and is a low-tech approach for soil contaminated with petroleum-base compound, and can often be carried out on site (Bogan and Lamar 1996; Kang and Charles 1996; Vidali 2001). Costs of some methods to treat soil contaminated hazardous wastes are incineration (975 US $/m^3), landfilling (350 US $/m^3), thermal desorption (125 US $/m^3), soil wash (237 US $/m^3) and bioremediation (95 US

$/m^3$), (Bioremediation report, 1993 in Alvarez and Illman 2006). Although, bioremediation costs cheaper than other methods it has several limitations. Bioremediation can only be used where environmental conditions permit microbial growth and activity. In the other word, it needs specific microbes in the specific place with proper environmental factor for degradation to occur. Application of bioremediation often involves some manipulation of the environment to enhance microbial growth and improve its rates for the target substrates.

Different techniques of bioremediation could be chosen depending on the degree of saturation and aeration in impacted area. Generally, bioremediation can be classified as *ex situ* and *in situ*. Ex situ techniques are those that involve the physical removal of the contaminated material for treatment. In situ techniques are those applied to contaminated sites with minimal disturbance. Table 1 summarizes the advantages and disadvantages of each technique.

4. Role of Microbes on Degradation of Crude Oil

Many varieties of microorganisms, including bacteria, fungi, and algae have been shown to detoxify pollutants. Fungi are believed to have survived for about 5300 years (Haselwandter and Ebner 1994). In any ecosystem, fungi are among the major decomposers of plant polymers such as cellulose, hemicellulose and lignin. Fungi also have the ability to mineralize, release and store various elements and ions and accumulate toxic materials. Fungi have achieved the removal, degradation, and mineralization of phenols and chlorinated phenolic compounds, petroleum hydrocarbons, polycyclic aromatic hydrocarbon, polychlorinated bisphenyls, chlorinated insecticides and pesticides, dyes, biopolymer, and other substances in various matrices (Singh 2006).

More than 100 genera of fungi and 90 genera of bacteria have the ability to decompose petroleum hydrocarbon (Prince and Walters 2007). Fungi and bacteria usually have a wide range of hydrocarbon decomposition, but some of them are also found to degrade specific hydrocarbons. The majority of degradation of petroleum hydrocarbons generally used carbon as energy alternative. Some studies reported that fungus *Fusarium* sp. F092 (Hidayat andTachibana 2012; Hidayat and Tachibana 2013), *Pestalotiopsis* sp. NG007 (Yanto and Tachibana 2013; Yanto and Tachibana 2014a,b), and *Polyporus* sp. S133 (Ayu *et al.* 2011) had ability to degrade crude oil on liquid and soil medium. *Fusarium* sp. F092 showed capacity to decompose three types of crude oil, type-3 (98%), type-2 (72%) and type-1 (49%) after 60 days with the initial concentration 1000 ppm (Hidayat and Tachibana 2012). Futhermore, F092 also showed its capacity to degrade crude oil type-2 about 60% and 49% on soil and sea sand (Hidayat and Tachibana 2013). The asphalts degradation (initial concentration 1000 ppm) by *Pestalotiopsis* sp. NG007 on liquid medium was about 50% and the degradation

Table 1: Summary of bioremediation strategies

Technology	Parameter	Remarks
In situ (no removal)	Advantage	The most cost efficient, non-invasive, relatively passive, natural attenuation processes.
	Limitations	Environmental constrains, extended treatment time, difficulties in monitoring.
	Factors to consider	Biodegradative abilities of indigenous microorganisms, presence of metals and other inorganic compounds, environmental parameters, biodegradability of pollutants, chemical solubility, geological factor, distribution of pollutants.
	Examples	*Bioventing* : Supplying air and nutrient through wells to stimulate indigenous microorganisms.
		Biosparging : The injection of air under pressure to increase groundwater oxygen concentrations, enhance biodegradation rates and increase the contact between soil and ground water.
		Bioaugmentation : The addition of pollutants degrader microorganisms.
Ex situ	Advantages	Cost efficient, low cost, can not be done on site.
	Limitations	Space requirements, extended treatment time, need to control abiotic loss, mass transfer problem, bioavailability limitation.
	Factors to consider	Similar to those of *in situ*.
	Examples	*Landfarming* : solid-phase treatment system for contaminated soil.
		Composting : Thermophilic treatment process in which contaminated material is mixed with a bulking agent.
		Biopiles : A hybrid of landfarming and composting.

Adapted from Hidayat (2013)

increased on medium with salinity 35 ppt after 30 days (Yanto and Tachibana 2013). Consortium culture between *Pestalotiopsis* sp. NG007 and *Polyporus* sp. S.133 (1:1) was also showed better result for the degradation of asphalt (initial concentration 30000 ppm) on soil medium (Yanto and Tachibana 2014a). The degradation of crude oil by *Polyporus* sp. S.133 in different concentration 1000, 5000, 7500, 10000, and 15000 ppm was about 93%, 46%, 37%, 25%, and 19% on liquid medium (Ayu *et al.* 2011). *Fusarium* sp. and *Pestalotiopsis* sp. are categorized as a non-ligninolytic fungi, while *Polyporus* sp. is a ligninolytic fungus. The degradation processes for both fungi are different because enzymatic system secreted was also different.

The intensive studies have been conducted with bacteria as the biological agent on degradation of crude oil. *Bacillus* sp. degraded crude oil (80-89%) in the initial concentration 5000 ppm after 5 days (Margesin and Schinner 2001). Consortium culture, *Peudomonas* sp. PSP01, PSP05 and *Bacillus* sp. PSP03 (1:1:1) with addition of nutrition, aeration and bulking agent (10%), resulted in 91% degradation after 45 days (initial concentration 100000 ppm, Munawar and Zaidan 2013). *Streptomyces* sp. ERI-CPDA-1 also showed its capacity to decompose fuel diesel (initial concentration 1000 ppm) to 98% after 7 days (Balachandran *et al.* 2012). *Dietzia cinnamea, Hoyosella altamirensis, Vibrio alginolyticus* isolated from Kuwait coast degraded crude oil (initial concentration 1000 ppm) to about 39% after 14 days (Al-Awadhi *et al.* 2012). Table 2 shows the ability of various fungi and bacteria to degrade crude oil.

Crude oil contains very complex compounds. Based on their solubility, crude oil is divided into several fractions: aliphatic, aromatic, resin, and asphaltene (Mishra *et al.* 2001). Based on their physicochemical properties and the compatibility model of petroleum hydrocarbons (Yanto and Tachibana 2014a), placed oil aliphatic fraction in the outer layer of crude oil, while the other factions are placed afterwards, aromatic, resin and asphalt, respectively. The higher the ratio of asphaltene on crude oil, the higher the effect for sorption in soil, leading to lower bioavailability and biodegradability by microorganisms.

4.1 Aliphatic Degradation

Aliphatic, especially alkanes, are a major group in crude oil and their removal from oil contaminated fields have become an environmental priority and been considered useful for enhancing recovery (Binazadeh 2009). The degradation of alkanes by microbes could be done in several pathways, among those are terminal oxidation, subterminal oxidation and via alkyl hydroperoxide. Enzymatic and/or non-enzymatic reactions might also occur in that process. Harayama *et al.* (1999), as shown on Figure 2, describes the process as follows: 1) the methyl group of alkanes transformed via mono- and di-terminal oxidation giving

Tabel 2: Crude oil degrading microbes

Microbes	Substrates	Culture	Degradation (%)	References
Fungi				
Candida lipolytica *	Oil	Soil	-	Kulichevskaya et al. 1995
Candida tropicalis PFS-95*	Crude oil	Soil	69 (16 days)	Ijah 1998
Phanerochaete chrysosporium*	TPHs	Soil	78 (365 days)	Yateem et al. 1998
Pleurotus ostreatus*			53	
Coriolus versicolor*			69	
Eupenicillium javanicum*	TPH	Liquid	30 (30 days)	Oudot et al. 1993
Graphium sp., Tetracoccosporium*	Fuel oil No. 2 & No. 4	Liquid	Assimilated (15 days)	Snellman et al. 1998
Fusarium sp. F092	Crude oil type 1, 2, 3 (1000 ppm)	Liquid	98%, 72%, 46% (60 days)	Hidayat & Tachibana 2012, 2013
	Crude oil type-2 (1000 ppm)	Soil	49% (60 days)	
		Sea sand	60% (60 days)	
Pestalotiopsis sp. NG007	Aspalt, crude oil type A & C (1000 ppm)	Liquid	50%, 92% & 77% (30 days)	Yanto & Tachibana 2013
Polyporus sp. S.133	Crude oil	Liquid	93% (60 days)	Ayu et al. 2011
Aspergillus ochraceus NCIM-1146	Kerosene (10%)	Liquid	77-83% (20 days)	Saratale et al. 2007
Pseudomonas aeruginosa DQ8	Crude oil (10%)Diesel Fuel (2%)	Liquid	60% (10 day)83%	Zhang et al. 2011
Bacteria				
Bacillus sp.	Crude oil (5000 ppm)	Liquid	89-90% (5 days)	Margesin & Schinner 2001
Dietzia cinnamea, Hoyosella altamirensis dan Vibrio alginolyticus	Crude oil (1000 ppm)	Liquid	39% (14 days)	Al-Awadhi et al. 2012

(Contd.)

Microbes	Substrates	Culture	Degradation (%)	References
Peudomonas sp. PSP01, PSP05 dan *Bacillus* sp. PSP03 (1:1:1) [Consortium]**	Crude oil (100000 ppm)	Soil	91% (45 days)	Munawar & Zaidan 2013
Etreptomyces sp. ERI-CPDA-1	Diesel fuel (1000 ppm)	Liquid	98% (7 days)	Balachandran et al., 201
Breviobacterium, Rhodococcus ***	-	-	-	Murygina et al. 2005; Ouyang et al. 2005
Actinomycetes ***	-	-	-	Rahman et al. 2002
Acetobacterium ***	-	-	-	Watanabe et al. 2002
Clostridium ***	-	-	-	Gaylarde et al. 1999
Petrotoga miotherma ***	-	-	-	Bonch-Osmolovskaya et al. 2003

*) Adapted from Singh 2006

**) Additional treatments by aeration, nutrient and bulking agent

***) Adapted from Yemashova et al. 2007

TPHs : Total petroleum hydrocarbons

rise to alcohols, alkanals and fatty acids (mono and di-carboxylic acid); 2) alkanes converted to secondary alcohols, ketones and to fatty acids; 3) alkanes transformed to alkyl peroxides, and these molecules would be further metabolized to the corresponding aldehyde. The degradation of aliphatics has generally been ranked in the following order of decreasing susceptibility: *n*-alkanes > branched alkanes > cyclic alkanes (Leahy and Colwell 1990; Wengler *et al.* 2002; Chaineau *et al.* 2005; Prince and Walters 2007). In microbial degradation processes, enzymes such as di- and/or monooxygenase are involved in catalyzing hydrocarbons biodegradation, and non-enzymes such as free radicals compounds might occur.

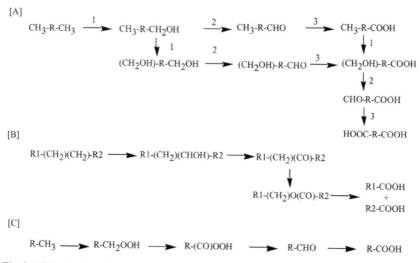

Fig. 2: Aliphatic degradation pathways. [A] terminal oxidation, transformed via mono- and di-terminal by monooxygenase (1) to form alcohol, dehydrogenase or oxidase to form alkanals (2) and dehydrogenase to form fatty acid (3); [B] subterminal oxidation; [C] via alkyl hydroperoxides (Adapted from Harayama *et al.* 1999).

Several microbes including fungi and bacteria have been widely studied for aliphatic degradation. The degradation pathways of aliphatic compounds by fungi *Fusarium* sp. F092, *Pestalotiopsis* sp. NG007, and the bacterium *Pseudomonas aeruginosa* DQ8 have been occurred either via terminal oxidation, subterminal oxidation or via alkyl hydroperoxide (Hidayat and Tachibana 2013; Yanto and Tachibana 2014b; Zhang *et al.* 2011). Other bacteria such as *Rhodococcus* sp., *Brevibacterium* sp., *Corynebacterium* sp.*Actinobacteria*, *Dietzia* sp. HOB, *Pseudomonas balearica* Strain BerOc6, *Desulfatibacillum aliphaticivorans CV2083T, and Marinobacter* sp. BP42 also showed their capability to degrade aliphatic compounds (Pirnik *et al.* 1974; Woods and Murrell 1989; Bonin *et al.* 2004; Cravo-Laureau *et al.* 2004; 2005; Grossi *et al.* 2008; Alonso-Gutiérrez *et al.* 2011).

4.2 Aromatic Degradation

Aromatics have one or more aromatic rings, and among them benzene is the simplest one. Aromatics present in crude oil with or without an alkyl substituent and their fraction are considered as the second major group after aliphatic fraction in crude oil. Aromatics with two or more fused benzene rings are called as polycyclic aromatic hydrocarbons (PAHs); which include a large group of xenobiotic pollutants those are common, persistent, and recalcitrant contaminants. They are also potentially hazardous since several of them are highly mutagenic or carcinogenic (Cerniglia *et al.* 1985; Dipple *et al.* 1990; Bispo *et al.* 1999; Li *et al.* 2008).

Fungi and bacteria degrade some of PAHs, and result in either complete degradation to form carbon dioxide or some metabolite products without carbon dioxide formation. Determination of degradation pathways of PAHs is important to know that the result of metabolite products is no longer toxic. The degradation processes are depended on enzymes secreted by microbes as shown in Fig. 3. Dioxygenases and monooxygenases in bacteria catalyze PAHs to form *cis*-dihydrodiols and *trans*-dihydrodiols (Cerniglia and Sutherland 2001). Furthermore, Cerniglia and Sutherland (2001) described that degradation PAHs by fungi is divided into ligninolytic and non-ligninolytic. Cytochrome P450

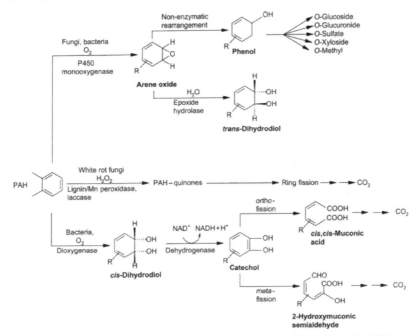

Fig. 3: Aromatic degradation pathways by microbes (Adapted from Cerniglia 1993).

monooxygenase and epoxide hydrolase in non-ligninolytic fungi catalyze PAHS to form *trans*-dihydrodiols, while ligninolytic fungi degrade PAHs by non-specific radical oxidation, catalyze by extracellular ligninolytic enzymes (laccases), that leads primarily to PAH quinones.

Fluorene, a PAH with 2 benzene rings, was degraded by *Pseudomonas aeruginosa* DQ8 (Zhang *et al.* 2011), & *Cunninghamella elegans* (Pothuluri *et al.* 1993). The degradation of fluorene (initial concentration 45 ppm) by *Pseudomonas aeruginosa* DQ8 was 41%, and then converted to form: 9-fluorenol, 9-fluorenone, 3-(1- oxo-2,3-dihidro-1H-inden-2-yl) propanoic acid, 2-(1-oxo- 2,3-dihydro-1H-inden-2-yl) acetic acid, and 1-indanone by the role of oxygenase (mono- and di-) enzymes to transform fluorene at C-9, C-3, and C-4 (Zhang *et al.* 2011).

Previous studies reported that phenanthrene, PAH with 3 benzene ring, was degraded by *Trametes* sp. AS03 (Hidayat and Tachibana 2014), *Trichoderma* sp. (Hadibarata *et al.* 2007), *Ganoderma lucidum* (Ting *et al.* 2011), *Polyporus* sp. S133 (Hadibarata *et al.* 2011) and *Anthracophyllum discolor* (Acevedo *et al.* 2011). The degradation of phenanthrene by AS03 resulted in five intermediate compounds: phenanthrene 9,10-dihydrodiol, 2,2'-diphenic acid, benzoic acid, 4-hydroxybenzoic acid, and phthalic acid (Hidayat and Tachibana 2014). The initial oxidation of phenanthrene by *Trametes* sp AS03 occurred through cytochrome P450, and resulted in the formation of hydroxylated products, and further ligninolytic enzymes would act as catalyst to form other metabolite products.

Chrysene, a PAHs with 4 benzene rings, is categorized as a high-molecular-weight PAHs including fluanthene, pyrene, benz(a)anthracene, benzo(a)pyrene and dibenz(a,h)anthrane. The studies on degradation of chrysene have been conducted by several fungi such as *Fusarium* sp. F092 (Hidayat *et al.* 2012), *Polyporus* sp. S133 (Hadibarata *et al.* 2009), *C. elegans* (Pothuluri *et al.* 1995), *Syncephalastrum racemosum* (Kiehlmann *et al.* 1996), *Penicillium* spp. (Pinto and Moore 2000). In culture medium of *Fusarium* sp. F092, the occurrence of chrysene 1,2-oxide, chrysene *trans*-1,2-dihydrodiol, 1-hydroxy 2-naphtoic acid and catechol (Hidayat *et al.* 2012) were identified. Chrysene would be attacked by F092 to form chrysene 1,2-oxide and chrysene trans-1,2-dihydrodiol *via* the activities of dioxygenases. F092 converted them into chrysene trans-1,2-dihydrodiol and 1-hydroxy 2- napthoic acid. The resultant dihydrodiol was then attacked by a dehydrogenase and a 1,2-dioxygenase. These reactions are considered to yield 1-hydroxy 2-napthoic acid (Kiyohara *et al.* 1994; Pinyakong *et al.* 2000). The analyses of metabolites revealed that F092 could also convert 1-hydroxy 2- napthoic acid to 1,2 dihydroxynaphthalene, which was further metabolized via salicylic acid (Evans *et al.* 1965). Salicylic acid could also be converted via the formation of catechol, after the aldose and hydroxylase reactions, to tran-o-hydroxybenzyliden epyruvate (tHBPA) which was further mineralized to produce carbon dioxide and water. This in turn underwent ring fusion to form tricarboxylic acid (TCA) cycle intermediates (Gibson and Subramanian 1984; Houghton and Shanley 1994).

4.3 Resin and Alphaltene Degradation

In contrast to the aliphatic and aromatic fractions, both resin and asphaltene fractions contain non-hydrocarbon polar compounds. Resin and asphaltene consist primarily of carbon and hydrogen in addition with trace amounts of nitrogen, sulfur and/or oxygen. Asphaltenes consist of high-molecular-weight compounds, which are not soluble in a solvent such as n-heptane, while resins are n-heptane soluble polar molecules (Harayama *et al.*, 1999). Resin and asphaltenes are very complex structures, which are difficult to analyze, and enriched by biodegradation (Venosa and Zhu 2003; Prince and Walters 2007). They were degraded approximately up to 15% to 20% at low concentrations (Venosa and Zhu 2003; Chaineau *et al.* 2005). The molecular weight of Asphaltenes is in the range of 600 to 2,000,000 DA (Strausz *et al.* 1999). Asphaltene structure is combination of aromatic and aliphatic structures with alkyl substitutes and sulfur, nitrogen and oxygen covalently linked (Pineda-Flores and Mesta-Howard (2001). Structure model of aspaltene was developed by Speight and Moschopedis (1981) and Murgich *et al.* (1999) as described in Pineda-Flores and Mesta-Howard (2001).

Bacteria like *Pseudomonas* spp. TMU2-5, *Bacillus licheniformis* TMU1-1, *Bacillus lentus* TMU5-2, *Bacillus cereus* TMU8-2 and *Bacillus firmus* TMU6-2 when applied, individually or as consortiums, could degrade asphaltene to more than 40% (Tavassoli *et al.* 2012). *Bacillus, Brevibacillus, Staphylococcus,* and *Corynebacterium* have capability to mineralize and utilize asphaltene as carbon and energy source alternative (Pineda-Flores *et al.* 2004). The degradation pathways of asphaltene initially occurred by photo-oxidation, and resultant products could be degraded through biochemical reactions, aliphatic, hetero polyaromatic and aromatic by oxidations ranging from several days to approximately 33 months (Pineda-Flores and Mesta-Howard 2001).

5. Factors Affecting Bioremediation

Microbes as bioremediation agents have been studied intensively, both indigenous (local) and exogenous (introduction). In many cases, they are only effective when the degradation process is conducted in a laboratory. It is because some environmental conditions could be controlled. Therefore, most pollutant contaminants do support microbes to grow and to carry out degradation in the fields. Bioremediation process should be started in understanding some critical factors, which affect the degradation process. The most important is the presence of microbes having the capacity to produce enzymes that are responsible in catalyzing the target pollutants. If all requirement factors are fulfilled, the degradation of hazardous material will quickly occur and degradation process will be done naturally.

Some of pollutants including crude oil may serve as an energy source. When degradation occur by oxidation, an electron acceptor is needed, such as oxygen. The moisture, pH, temperatures, nutrients, concentrations of pollutants, metabolites product, and protozoa are among factors that affect the sustainability of bioremediation. These factors are described on the pyramidal fashion (Alvarez and Illman 2006) as displayed in Figure 4. Other studies brought out that several factors will affect the improvement of degradation rate of target substrate such as addition of glucose (Hadibarata *et al.* 2008; Ayu *et al.* 2011), nitrogen (Hadibarata *et al.* 2008; Ayu *et al.* 2011; Hidayat *et al.* 2012), Mn^{2+} (Hidayat and Tachibana 2013; Yanto and Tachibana 2014b), agitation (Hidayat *et al.* 2012), surfactants (Hidayat *et al.* 2012; Yanto and Tachibana 2014b) and aeration (Ayu *et al.* 2011).

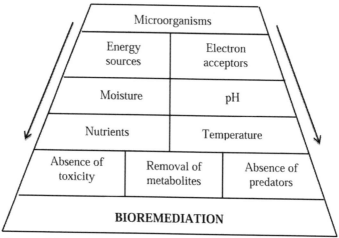

Fig. 4: Requested factors for bioremediation (Adapted from Cookson 1995 in Alvarez and Illman 2006).

6. Application of Bioremediation in Indonesia

In 1970 the bioremediation was applied for the first time to degrade oil spill in Ambler, Pennsylvania. Four years after that, Richard Raymond got a patent on bioremediation. In Indonesia bioremediation has been endorsed by the Ministry of Environment Regulation, MoE No. 128 in 2003. The bioremediation was actually initiated by PT Chevron Pacific Indonesia (CPI) in collaboration with Ministry of Environment in 1994. This company worked on degrading crude oil contaminated soil. In 2011 PT. CPI claimed that their bioremediations effort have successfully treated more than 500,000 m³ soil; and the treated soil was used in reforestation of 60 ha land in Riau Province, Sumatra (Anonymous 2014).

The MoE decree No. 128/2003 describes that bioremediation should be done especially for crude oil waste and crude oil contaminated soil. According to that regulation, bioremediation technologies for oil spill could be approached based on principle of bioaugmentation, in which known oil-degrading microbes are added to enrich the existing microbial population, and the growth of indigenous oil degraders is stimulated by the addition of nutrients or other growth-limiting co-substrates, including land farming, bio-pile and composting.

PT. CPI chose biostimulation by land farming as their bioremediation techniques. During the process, monitoring on degradation rate was conducted every 2 weeks for 8 months. The company also conducted monitoring for each treated soil (treated soil having total petroleum hydrocarbon (TPH) < 1%) every 6 months for 2 years to ensure that treated soil has no more toxic substrates (Anonymous 2012). In the end of bioremediation process, PT. CPI will receive a Letter of Contaminated Soil Completion Status (SSPLT) as a certificate of verification from Ministry of Environment.

So far, the regulation of bioremediation in Indonesia targeted for only crude oil. Other pollutants including PAH, chlorinated dyes and some persistent organic pollutants (POP) are excluded. Furthermore, with the dispute of bioremediation case at PT. CPI as it was suspected to fictitious, the bioremediation activities in Indonesia more and less were affected. Strict and tight monitoring should be conducted to ensure that bioremediation activities are done by correct method, non-fictitious and low cost. Each strategy of bioremediation process has certain specific advantages and disadvantages, which need to be considered for each situation (Megharaj et al. 2011). The Indonesian Ministry of Environment decree No. 128/2003 should be revised based on detailed reexamination and review by experts in the area of bioremediation. to minimize the negative impact. The experts from different fields with adequate knowledge and experience in the areas of biological degradation, strong integrity, high competency, honesty, and responsibility should be invited to do this revision as bioremediation is multidisciplinary subject. The experts should consist of microbiologist, environmentalist, analytical chemist, geologist and also an architect.

Conclusion

Crude oil contaminating environment is a huge problem in Indonesia. Recovering such environment is not simple. Indigenous microbes often show their success to degrade pollutants but they have several limitations. That is why understanding the role of enzyme systems involved in degradation becomes crucial. Biological degradation could be used only when environmental conditions permit microbial growth and activity. Proper microbes present in suitable site and environment are responsible for the success of the bioremediation activities. Microbes as a biological agent in degradation have to be considered as a main key in bioremediation strategy for crude oil recovery. The success of bioremediation program in the field are particularly dependent on the presence of suitable microbes, bioavailability of the pollutant, and external and internal factors such as temperature, aeration, nutrients, presence of toxic metabolite substances, pH and electron acceptor. Choosing proper bioremediation technique to be applied in certain site and condition require pre-studies at that particular site.

Acknowledgement

The authors would like to thank their colleagues for support and help during the preparation of the manuscript.

References

Acevedo F, Pizzul L, Castillo MP, Cuevas R, Diez MC (2011) Degradation of polycyclic aromatic hydrocarbons by the Chilean white-rot fungus *Anthracophyllum discolor*. J. Hazard. Mater. 185, 212-219.

Adekunle AA, Adebambo OA (2007) Petroleum hydrocarbon utilization by fungi isolated from *Detarium Senegalense* (J. F Gmelin) seeds. J. Am. Sci. 3, 69-76.

Aislabie JM, Richards NK, Bould HL (1997) Microbial degradation of DDT and its residues: a review. NZ J. Agric. Res. 40, 269-282.

Al-Awadhi H, Dashti N, Kansour M, Sorkhoh, N, Radwan S (2012) Hydrocarbon-utilizing bacteria associated with biofouling materials from offshore waters of the Arabian Gulf. Int. Biodeterior. Biodegrad. 69, 10-16.

Alonso-Gutiérrez J, Teramoto M, Yamazoe A, Harayama S., Figueras A, Novoa B (2011). Alkane-degrading properties of *Dietzia* sp. H0B, a key player in the Prestige oil spill biodegradation (NW Spain). J. Appl. Microbiol. 111, 800-810.

Alvarez PJJ, Illman WA (2006) Bioremediation and Natural Attenuation: Process Fundamentals and Mathematical Models. John Wiley & Sons, Inc., Hoboken, New Jersey.

Anonymous (2012) Chevron Pacific Indonesia Bioremediation Program. Fact Sheet. http://www.chevron.com/documents/pdf/BioremediationFactSheetEnglish.pdf. Accesed on Agust 12, 2015.

Anonymous (2014) Bioremediation program. http://infobioremediasi.com/en/case-analysis/chronology/. Accesed on September 10, 2015.

Ayu KR, Hadibarata T, Toyama T, Tanaka Y, Mori K (2011) Bioremediation of Crude Oil by White Rot Fungi *Polyporus* sp. S133. J Microbiol. Biotechnol. 21, 995-1000.

Balachandran C, Duraipandiyan V, Balakrishna K, Ignacimuthu S (2012) Petroleum and polycyclic aromatic hydrocarbons (PAHs) degradation and naphthalene metabolism in *Streptomyces* sp. (ERI-CPDA-1) isolated from oil contaminated soil. Bioresour. Technol, 112, 83-90.

Binazadeh M, Karimi IA, Li Z (2009) Fast biodegradation of long chain n-alkanes and crude oil at high concentrations with *Rhodococcus* sp. Moj-3449. Enzyme Microbial Technol. 45, 195-202.

Bispo A, Jourdain MJ, Jauzein M, (1999) Toxicity and genotoxicity of industrial soils polluted by polycyclic aromatic hydrocarbons (PAHs). Org. Geochem. 30, 947-952.

Bogan BW, Lamar RT (1996) Polycyclic aromatic hydrocarbon-degrading capabilities of *Phanerochaete laevis* HHB-1625 and its extracellular ligninolytic enzymes. Appl. Environ. Microbiol. 62, 1597-1603.

Bonin P, Cravo-Laureau C, Michotey V, Hirschler-Réa A (2004) The anaerobic hydrocarbon biodegrading bacteria: An overview. Ophelia 58, 243-254.

Boopathy R (2000) Factors limiting bioremediation technologies. Bioresour. Technol. 74, 63-67.

Cerniglia CE (1993) Biodegradation of polycyclic aromatic hydrocarbons. Curr. Opin. Biotechnol. 4, 331-338.

Cerniglia CE, Sutherland JB (2001) Bioremediation of polycyclic aromatic hydrocarbons by ligninolytic and non-ligninolytic fungi. In: Fungi in Bioremediation (Ed) Gadd GM, Cambridge University Press, New York, pp 136-187.

Cerniglia CE, White GL, Heflich RH (1985) Fungal metabolism and detoxification of polycyclic aromatic hydrocarbons. Arch. Microbiol. 143, 105-110.

Chaineau CH, RougeuxG. Yepremian C, Oudot J (2005) Effect of nutrient concentration on the biodegradation of crude oil and associated microbial populations in the soil. Soil Biol. Biochem. 37, 1490-1497.

Cravo-Laureau C, Grossi V, Raphel D, Matheron R, Hirschler-Réa A (2005) Anaerobic n-alkane metabolism by a sulfate-reducing bacterium, *Desulfatibacillum aliphaticivorans* strain CV2803T. Appl. Environ. Microbiol. 71, 3458-3467.

Cravo-Laureau C, Matheron R, Cayol JL, Joulian C, Hirschler-Réa A (2004) *Desulfatibacillum aliphaticivorans* gen. nov., sp. nov., an n-alkane- and n-alkene-degrading, sulfate-reducing bacterium. Int. J. Syst. Evol. Microbiol. 54, 77-83.

Dale WE, Copeland MF, Hayes WJ (1965) Chlorinated insecticides in the body fat of people in India. Bull World Health Org. 33, 471-477.

Dipple A, Cheng CC, Bigger CAH (1990) Polycyclic aromatic hydrocarbon carcinogens. In: Mutagens and Carcinogens in the Diet (Ed) Pariza MW, Aeschbacher HU, Felton JS, Sato S and Wiley-Liss, New York, pp 109-127.

Elshafie A, AlKindi AY, Al-Busaidi S, Bakheit C, Albahry SN (2007) Biodegradation of crude oil and n-alkane by fungi isolated from Oman. Mar Pollut Bull. 54, 1692-1696.

Erdogan EE, Karaca A (2011) Bioremediation of crude oil polluted soils. Asian J. Biotechnol. 3, 206-213.

Evans WC, Fernley HN, Griffiths E (1965) Oxidative metabolism of phenanthrene and anthracene by soil *Pseudomonads*. The ring-fission mechanism. Biochem J. 95, 819-831.

Gibson DT, Subramanian V (1984) Microbial degradation of aromatic hydrocarbons. In: Microbial Degradation of Organic Compounds (Ed) Gibson DT, Dekker, New York, pp. 181-252.

Grossi V, Cravo-Laureau C, Guyoneaud R, Ranchou-Peyruse A, Hirschler-Réa A. (2008) Metabolism of n-alkanes and n-alkenes by anaerobic bacteria: A summary. Org Geochem. 39, 1197-1203.

Hadibarata T, Tachibana S, Askari M (2011) Identification of Metabolites from Phenanthrene Oxidation by Phenoloxidases and Dioxygenases of *Polyporus* sp. S133. J. Microbiol. Biotechnol. 21, 299-304.

Hadibarata T, Tachibana S, Itoh K (2007) Biodegradation of phenanthrene by fungi screened from nature. Pak. J. Biol. Sci. 10, 2535-2543.

Hadibarata T, Tachibana S, Itoh K (2009) Biodegradation of chrysene, an aromatic hydrocarbon by *Polyporus* sp. S.133 in liquid medium. J Hazar Mater. 164, 911-917.

Hamer G (1993) Bioremediation: a response to gross environmental abuse. Trends Biotechnol. 11, 317-319.

Harayama S, Kishira H, Kasai Y, Shutsubo K (1999) Petroleum biodegradation in marine environments. J. Mol. Microbiol. Biotechnol. 1, 63-70.

Hasanuzzaman M, Ueno A, Ito H, Ito Y, Yamamoto Y, Yumoto I, Okuyama H (2007) Degradation of long-chain n-alkanes (C36 and C40) by *Pseudomonas aeruginosa* strain WatG. Int. Biodeterior. Biodegrad. 59, 40-43.

Haselwandter K, Ebner MR (1994) Microorganisms surviving for 5300 years. FEMS Microbiol. Lett. 166, 189-194.

Hidayat A (2013) Biodegradation of Polycyclic Aromatic Hydrocarbons (PAHs), Polychlorinated Aromatic Compounds (PACs), Polylactic Acid (PLA)/Kenaf Composite, and Crude Oil by Fungi Screened from Nature. Dissertation, The United Graduate School of Agricultural Science – Ehime University, Japan.

Hidayat A, Tachibana S (2012) Biodegradation of aliphatic hydrocarbon in three types of crude oil by *Fusarium* sp. F092 under stress with artificial sea water. J Environ Sci Technol. 5, 64-73.

Hidayat A, Tachibana S (2013) Crude oil and *n*-octadecane degradation under saline conditions by *Fusarium* sp. F092. J. Environ. Sci. Technol. 6, 29-40.

Hidayat A, Tachibana S (2014) Decolorization of azo dyes and mineralization of phenanthrene by *Trametes* sp. AS03 Isolated from Indonesian mangrove forest. Int. J. Fores. Res. 1, 67-75.

Hidayat A, Tachibana S, Itoh K (2012) Determination of chrysene degradation under saline conditions by *Fusarium* sp. F092, a fungus screened from nature. Fungal Biol. 116, 706-714.

Houghton JE, Shanley MS (1994) Catabolic potential of Pseudomo- nas: a regulatory perspective. In: Biological Degradation and Bioremediation of Toxic Chemicals (Ed) Caudhry GR, Chapman & Hall, London, pp 11-32.

Kang SH, Charles SC (1996) Evaporation of petroleum products from contaminated soils. J. Environ. Eng. 122, 384-387.

Kiehlmann E, Pinto L, Moore M (1996) The transformation of chrysene to *trans*-1,2-dihydrochrysene by filamentous fungi. Can. J. Microbiol. 42, 604-608.

Kiyohara H, Torigoe S, Kaida, N, Asaki T, Iida T, Hayashi H, Takizawa N (1994) Cloning and characterization of a chromosomal gene cluster, pah, that encodes the upper pathway for phenanthrene and naphthalene utilization by *Pseudomonas putida* OUS82. J. Bacteriol. 176, 2439-2443.

Kvenvolden KA, Cooper CK (2003) Natural seepage of crude oil into the marine environment. Geo-Marine Lett. 23, 140-146.

Leahy JG, Colwell RR (1990) Microbial degradation of hydrocarbon in environment. Microbiol. Rev. 54, 305-315.

Li X, Xi P, Lin X, Zhang C, Li Q, Gong Z (2008) Biodegradation of aged polycyclic aromatic hydrocarbons (PAHs) by microbial consortia in soil and slurry phases. J. Hazard. Mater. 150, 21-26.

Margesin R, Schinner F (2001) Biodegradation and bioremediation of hydrocarbons in extreme environments. Appl. Microbiol. Biotechnol. 56, 650-663.

Megharaj M, Ramakrishnan B, Venkateswarlu K, Sethunathan N, Naidu T (2011) Bioremediation approaches for organic pollutants: A critical perspective. Environ. Int. 37, 1362-1375.

Minai-Tehrani D, Minooi S, Azari-Dehkordi F, Herfatmanesh A (2006) The effect of triton X-100 on biodegradation of aliphatic and aromatic fractions of crude oil in soil. J. Applied Sci. 6, 1756-1761.

Mishra S, Jyot J, Kuhad RC, Lal B (2001) In situ bioremediation potential of an oily sludge-degrading bacterial consortium. Curr. Microbiol. 43, 328-335.

Munawar Z (2013) Bioremediasi limbah minyak bumi dengan teknik biopile di lapang Klamono Papua. J Sains Matematika 1, 41-46.

Pineda-Flores G, Boll-Argüello G, Lira-Galeana C, Mesta-Howard AM (2004) A microbial consortium isolated from a crude oil sample that uses asphaltenes as a carbon and energy source. Biodegradation 15, 145-151.

Pineda-Flores G, Mesta-Howard AM (2001) Petroleum asphaltenes: Generated problematic and possible biodegradation mechanisms. Rev. Latinoam. Microbiol. 43, 143-150.

Pinto LJ, Moore MM (2000) Release of polycyclic aromatic hydrocarbons from contaminated soils by surfactant and remediation of this effluent by Penicillium spp. Environ. Toxicol. Chem. 19, 1741-1748.

Pinyakong O, Habe H, Supaka N, Pinpanichkarn P, Juntongjin K, Yoshida T, Furihata K, Nojiri H, Yamane H, Omori T (2000) Identification of novel metabolites in the degradation of phenanthrene by Sphingomonas sp. Strain P2. FEMS Microbiol. Lett. 191, 115-121.

Pirnik MP, Atlas RM, Bartha R (1974) Hydrocarbon metabolism by Brevibacterium erythrogenes: normal and branched alkanes. J. Bacteriol. 119, 868-878.

Pothuluri, JV, Selby A, Evans FE, Freeman JP, Cerniglia CE (1995) Transformation of chrysene and other polycyclic aromatic hydrocarbon mixture by the fungus Cunninghamella elegans. Can. J. Bot. 73, 1025-1033.

Prince RC, Walters CC (2007) Biodegradation of Oil Hydrocarbons and its Implications for Source Identification. In: Oil Spill Environmental Forensics (Ed) Wang Z and Stout SA, Academic Press Publication, California, USA, pp 349-403.

Saratale G, Kalme S, Bhosale S, Govindwar S (2007) Biodegradation of kerosene by Aspergillus ochraceus NCIM-1146. J. Basic Microbiol. 47, 400-405.

Si-Zhong Y, Hui-Jun J, Zhi W, Rui-Xia H, Yan-Jun J, Xiu-Mei L, Shoa-Peng Y (2009) Bioremediation of oil spill in cold environments: a review. Pedosphere 19, 371-381.

Singh H (2006) Mycoremediation: Fungal Bioremediation. John Wiley & Sons, Inc. United State of America.

Strausz OP, Mojelsky TW, Faraji F, Lown EM (1999) Additional structural details on Athabasca asphaltene and their ramifications. Energy Fuels 13, 207-227.

Sudaryanto A, Takahashi S, Monirith I, Ismail A, Muchtar M, Zheng J, Richardson BJ, Subramanian A, Prudente M, Hue ND, Tanabe S (2002) Asia-Pacific mussel watch: monitoring of butyltin contamination in coastal waters of Asian developing countries. Environ. Toxicol. Chem. 2, 2119-2130.

Tanabe S, Gondaira F, Subramanian A, Ramesh A, Mohan D, Kumaran P, Venugopalan VK, Tatsukawa R (1990) Specific pattern of persistent organo chlorine residues in human breast milk from South India. J. Agric. Food. Chem. 18, 899-903.

Tavassoli T, Mousavi SM, Shojaosadati SA, Salehizadeh H (2012). Asphaltene biodegradation using microorganisms isolated from oil samples. Fuel 93, 142-148.

Ting WTE, Yuan SY, Wu SD, Chang BV (2011) Biodegradation of phenanthrene and pyrene by Ganoderma lucidum. Int. Biodeterior. Biodegrad. 65, 238-242.

Venosa AD. Zhu X (2003) Biodegradation of crude oil contaminating marine shorelines and freshwater wetlands: Review. Spill Sci. Technol. Bull. 8, 163-167.

Vidali M (2001) Bioremediation. An overview. Pure Appl Chem. 73, 1163-1172.

Watanabe K (2001) Microorganisms relevant to bioremediation. Curr. Opin. Biotechnol. 12, 237-241.

Weisman W (1998) Analysis of petroleum hydrocarbons in environmental media. Amherst Scientific Publishers, United States of America.

Wenger LM, Davis CL, Isaken GH (2002) Multi control on petroleum biodegradation and impact on oil quality. SPE Reserv. Eval. Eng. 5, 375-383.

Woods NR, Murrell JC (1989) The metabolism of propane in *Rhodococcus rhodochrous* PNKB1. J. Gen. Mircobiol. 135: 2335-2344.

Yanto DHY, Tachibana S (2013) Biodegradation of petroleum hydrocarbons by a newly isolated *Pestalotiopsis* sp. NG007. Int. Biodeterior. Biodegrad. 85, 438-450.

Yanto DHY, Tachibana S. (2014a) Potential of fungal co-culturing for accelerated biodegradation of petroleum hydrocarbons in soil. J. Hazard. Mater. 278, 454-463.

Yanto DHY, Tachibana S (2014b) Enhanced biodegradation of asphalt in the presence of Tween surfactants, Mn^{2+} and H_2O_2 by *Pestalotiopsis* sp. in liquid medium and soil. Chemosphere 103, 105-113.

Yemashova NA, Murygina VP, Zhukov DV, Zakharyantz AA, Gladchenko MA, Appanna V, Kalyuzhnyi SV (2007) Biodeterioration of crude oil and oil derived products: A review. Rev Environ. Sci. Biotechnol. 6, 315-337.

Zhang Z, Hou Z, Yang C, Ma C, Tao F, Xu F (2011) Degradation of n-alkanes and polycyclic aromatic hydrocarbons in petroleum by a newly isolated *Pseudomonas aeruginosa* DQ8. Bioresour. Technol. 102, 4111-4116.

Printed and bound by CPI Group (UK) Ltd, Croydon, CR0 4YY

17/10/2024

01775681-0003